Arie S. Issar
Mattanyah Zohar

**Climate Change – Environment and History
of the Near East**

Arie S. Issar
Mattanyah Zohar

Climate Change –

Environment and History
of the Near East

2nd Edition

With 34 Figures

 Springer

Prof. Arie S. Issar
Ben Gurion Univ. of the Negev
13 Hameshoreret Rachel St.
96348 Jerusalem
Israel

Dr. Mattanyah Zohar
PO Box 82548
90805 Mevaseret
Israel

Library of Congress Control Number: 2006939920

ISBN 978-3-540-69851-7 Springer Berlin Heidelberg New York
ISBN 978-3-540-21086-5 (1st edition) Springer Berlin Heidelberg New York

Springer is a part of Springer Science+Business Media
springer.com

© Springer-Verlag Berlin Heidelberg 2007

Cover design: deblik, Berlin
Production: Almas Schimmel
Typesetting: LE-TEX Jelonek, Schmidt & Vöckler GbR, Leipzig

Printed on acid-free paper   30/3141/as   5 4 3 2 1 0

*To Margalit and to Aviva
the same and even more!*

Contents

List of Illustrations

Figures

Plates

List of Tables

List of Abbreviations

ABD   The Anchor Bible Dictionary, ed. D.N. Friedman, 6 vols. New York, London, Toronto, Sydney, Auckland, Doubleday, 1992

AGU   American Geophysical Union

ANET   Ancient Near Eastern Texts Relating to the Old Testament, edited by James B. Pritchard, 3rd edition with supplement, (Princeton, N.J., Princeton University Press, 1969) abridged edition The Ancient Near East, An Anthology of Texts and Pictures (Princeton, N.J., Princeton University Press, 1958)

BAR   Biblical Archaeology Review

BASOR   Bulletin of the American School of Oriental Research

CAD   The Assyrian Dictionary of the Oriental Institute of the University of Chicago, 24 volumes, (incomplete), Chicago & Glueckstadt, starting 1964

COWA   Chronologies in Old World Archaeology, ed. Robert W. Ehrich, 3rd edition, 2 volumes, Chicago-London, The University of Chicago Press, 1992

GSA   Geological Society of America

GSI   Geological Survey of Israel

HUJI   Hebrew University Jerusalem Israel

IEJ   Israel Exploration Journal

IJES   Israel Journal of Earth Sciences

NEAEHL   The New Encyclopedia of Archaeological Excavations in the Holy Land, ed. E. Stern, 4 vols., Jerusalem, The Israel Exploration Society-Carta, 1993

OCA   The Oxford Companion to Archaeology, ed. Brian M. Fagan, New York-Oxford, 1996

Palaeo 3   Palaeogeography, Palaeoclimatology, Palaeoecology

PEQ   Palestine Exploration Quaterly

Copyright and Authors Acknowledgements

Figures 3, 3a, & 3b. M. Bar-Matthews, A. Ayalon and A. Kaufman, "Middle to Late Holocene (6500-Year Period) – Paleoclimate in the Eastern Mediterranean Region from Stable Isotopic Composition of Speleothems from Soreq Cave, Israel," in *Water, Environment and Society in Times of Climate Change*, eds. A. Issar and N. Brown (Dordrecht, Kluwer Academic Publishers, 1998).

A. Frumkin, M. Magaritz, I. Carmi and I. Zak, "The Holocene Climatic Record of the Salt Caves of Mount Sedom, Israel", *The Holocene* 1, no. 3 (1991) Journals-Department-Arnold.

A. Raban and E. Galili, "Recent Maritime Archaeological Research in Israel – A Preliminary Report", *The International Journal of Nautical Archaeology and Underwater Exploration* 14, no. 4 (1985) Academic Press.

G. Lemcke and M. Sturm, "$\delta^{18}O$ and Trace Element as Proxy for the Reconstruction of Climate Changes at Lake Van (Turkey): Preliminary Results" *Third Millennium B.C. Climatic Change and Old World Collapse*. G. Dalfes, G. Kukla and H. Weiss, eds., (NATO ASI Series I: Global Environmental Change, Vol. 49. Berlin, Springer-Verlag, 1997)

R. Bookman (Ken-Tor), Y. Enzel, A. Agnon and M. Stein, "Late Holocene lake levels of the Dead Sea", Geological Society of America Bulletin v. 116 no. 5/6, (2004) p. 561

Figure 12. D.C. Kypris, 1996, "Cyclic Climatic Changes in Cyprus as Evidenced from Historic Documents and One Century's Rainfall Data," In A.N. Angelakis and A. Issar, eds. *Diachronic Climatic Impacts on Water Resources*, (NATO ASI Series I, Vol. 36, Berlin, Springer-Verlag. 1996).

Introduction

When the first edition of this book was published in 2004, the following year 2005 has happened to have been the warmest year since 1880, when the first reliable worldwide instrumental records came into existance. Claiming no linkage between the publication of our book and the temperature record, yet this record demonstrates the trend of increase in the global surface temperatures during the past 20 years, reinforced by evidence of rise of atmosphere's and oceans' temperatures, and increased melting of ice and snow in the arctic and antarctic regions as well as on mountain tops. All these observations are paralleled by the increase in the quantity of heat trapping gases in the atmosphere, causing most probably, the global greenhouse effect.

In order to try and predict, what might be the impact of this effect on the on the natural and human environments of the Near East, (Figs. 1–1d) the authors adopted the saying that the past is the key for the future. The practical conclusion of this principle says that the acquiring knowledge of the impact of past climate changes on the nature and human societies, may allow conclusions with regard to future possible impact of climate changes. By correlating proxy data of all types, paleo-sea and lake levels, paleo-hydrology, pollen profiles, environmental isotopes as well as archaeological and historical documents, the authors tried to collect as much as possible of this knowledge. The region investigated spans the Fertile Crescent in the wider sense which arcs from northeastern Africa and Egypt to Syria-Palestine and Mesopotamia, skirting the Anatolian, Iranian and Caucasian highlands. Since the so-called "Holy Land" has attracted western scholarly attention for nearly two centuries, and resulted in extensive and intensive historical and archaeological research, most data mentioned in the book were derived from this area. It is, therefore, natural that the historical documentation of the book reflects the uneven distribution of western research over the last two centuries with centers of gravity in the Levant, Egypt, and Mesopotamia.

The presentation of the archaeological and historical material follows a broad timetable beginning with the origin of mankind in Africa and its spread across the rest of the world, all seen as resulting from ever-changing climates and environments. Despite the fact that the archaeological evidence includes most of the major excavation sites of the Fertile Crescent, old and new, and often goes into detail, particularly in the formative years of the ancient civilizations in these areas, the principal aim was to convey an overall picture of cultural development of the entire region and clarify the importance of climate change during the process. It goes without saying that it was not every climate change automatically entailed a cultural and historical change. However, there can be little doubt that extreme climate changes influenced the welfare of pre-industrial cultures and civilizations whose subsistence entirely dependents on agriculture and husbandry, especially in a semi-arid region.

The understanding of the role of climate change on major transitions in human history gained increasing recognition by the general public during the late 1980's and throughout the 1990's. Studies of the African droughts have shown that human activity was clearly of secondary importance to the desertification of the Sahel and the magnitude of the subsequent human suffering. Albeit there is little doubt that the impact of agriculture and pastoralism on the natural environment was, and still is, catastrophic, yet it is the negative climate change, which triggers the non-sustainable processes, such as the invasion of the sand dunes to the coastal plain of Palestine during the warm phase starting in the 7th century A.D.[1]

Concern about the possible impact of the Global Change on the world-wide hydrological cycle brought about this research in the framework of the International Hydrologic Program (IHP) UNESCO and WMO. The results of this work, which was on a global scale were reported by Issar in a previous book.[2] Its basic conclusions were that major climate changes occurred during the Holocene and that these changes had influenced the hydrological cycle in the different parts of the globe in different ways. In regions having a Mediterranean type climate, warm periods spelled dryness while cold periods were humid. The contrary was the rule in regions with "monsoon" type climates. In regions along the margins of climate belts these changes had a decisive impact on the history of the inhabitants, as the shift of the belts spelled either dryness thus desertification or abundant rains, which spelled lushness. This phenomenon was especially recognizable in the history of the Eastern Mediterranean region.

In parallel to the studies of Issar, Zohar was studying the ethnography of the transhumant and semi-nomadic pastoralists of the Old World and their interaction and effect on the agrarian and urban cultures of the Fertile Crescent. He found that these effects were most apparent in the so-called 'intermediate' or 'transitional' periods, sometimes called "Dark Ages", intervals between the times of flourishing of urban civilizations in the ancient Near East in a seemingly periodic pattern. Excavated sites with archaeological layers dated to this periods often show signs of increased violence, such as destruction layers. They are roughly contemporary in all parts of the region and beyond but with distinct characteristics and variable durations.[3]

The conformity between the conclusions of Issar and Zohar's investigations brought them to compile the first edition of the present book and in which

---

[1] A.S. Issar, *Geology of the Subterranean Water Horizons of the Shephela and of the Sharon Regions* Ph.D. thesis. Hebrew University, Jerusalem (1961, Hebrew with summary in English).

[2] A.S. Issar, *Climate Changes during the Holocene and their Impact on Hydrological Systems*, Cambridge University Press, Cambridge UK (2003).

[3] M. Zohar, *Early Transition Periods in the Archaeology of Syria-Palestine.* unpublished Ph.D. thesis. The Hebrew University Jerusalem (1993).

the Neo-Deterministic Paradigm is presented. This paradigm argues that the principal cause for major developments and several decisive events of the Middle Eastern history were often accompanied by climate changes, while human intervention played a secondary role, attenuating or intensifying the effects of the natural impact. This conclusion seemingly rejuvenated the Deterministic Paradigm prominent at the beginning of the 20th century, which argued that climate change by itself could explain the birth, the flourishing and the demise of the ancient civilizations.[4] The modification of this paradigm by Issar and Zohar concern the role of the human ingenuity to invent devices and tools helping them to mitigate the impact of climate changes in the positive cases, while in the negative cases over-exploitation of natural resources in a non-sustainable way amplify the impact of the natural hazards leading to decline and collapse.

As could have been foreseen, the suggestion to swing back the pendulum of paradigms from that of blaming peoples for their misfortunes and putting the blame on climate changes was not acceped by most historians and archaeologists. Since the beginning of historical writings in antiquity, the humanities have considered the rise and decline of human societies as the outcome of acts of gods, God or men. The Enlightenment and the industrial revolution prepared the ground for a more realistic world view based on the natural sciences which engendered the Deterministic Paradigm of the 19th and early 20th century. The majority of archaeologists and historians then discarded the Deterministic Paradigm since the thirties of the 20th century and returned to the Anthropogenic Paradigm, according to which all blame was put on human society for its failures. One of the major faults was the human interference with natural processes thus causing environmental, economic, and political calamities.

This book will discuss the major climate changes that affected the Near East over the last ten thousand years, as determined by time series of proxy-data. The response of the societies to these changes will be investigated through an examination of their cultural and socioeconomic structures as well as the characteristics of the supporting natural system. We will not exonerate the human race entirely from its grave sins against its environment and the natural world. We do not claim that a few years of crop failures toppled any society. However, we shall demonstrate that major changes in civilizations did, indeed, coincide with major changes in the global climate.

In its very general aspects, the swing of the pendulum of paradigms from its deterministic peak in the first decades of the twentieth century to the opposite peak of anthropogenic disposition in the mid-twentieth century, and the beginning of a neo-deterministic trend corresponds with the process suggested by Kuhn.[5] In a nutshell, this process says that scientists think and build their theories within the general framework of the prevailing "truths" in their society. Also, the reluctance of most contemporary archaeologists, historians and geographers to accept conclusions based on new data with regard to the involvement of climate fits well with Kuhn's model. Yet, in this special case,

---

[4] E. Huntington, *Palestine and its Transformation*. Houghton Mifflin Company, New York (1911).

[5] T. Kuhn, *The Structure of Scientific Revolution* Chicago University Press, Chicago (1970).

there is an additional aspect, which has to do with Snow's conclusion about the schism between "the two cultures," i.e. the physical-natural sciences versus the humanistic sciences.[6] This schism was illustrated by the divorce between the scientist familiar with the global importance of the second law of thermodynamics and the Shakespearean scholar versed in interpreting Hamlet. Regarding archaeological research, classical archaeology, as a branch of the science of history and linguistics, belongs to the humanities. On the other hand, most of the paleo time-series proxy data is based on investigations in the physical (mainly environmental isotopes), geological (mainly sedimentological) and biosciences (mainly pollen and dendro-chronology). The evolution of the world of sciences does not promise closure of the breach between the two intellectual cultures, as the general trend is towards further reductionism and increasing expertise in narrow fields of specialization. Thus, a future divergence within and between the two cultures seems inevitable.

Investigation of the reasons for the swing of the pendulum reveals that in our case the instrument enforcing the "Kuhn's model" transformation from one paradigm to the other was the development of specialization in the sciences. On the one hand, this brought further specialization, but on the other hand, specialists who looked beyond the walls of their expertise could see other fields in which their special methods could be applied. It was up to these experts to open their minds to test new methods, and, if successful, apply them. Thus, the field of gravity enforcing the swing of the paradigm pendulum was the evolution of science, while the force of friction hindering this motion was the reluctance of scientists to introduce new methods not part of their expertise.

One example of this is the absence in most scholarly works about the archaeology, history and geography of the Near East, of a correlation between the archaeological findings and the results of investigations of recent paleo-climates conducted by isotope experts (except for the use of ^{14}C dating), geologists and botanists. The recent change to the new paradigm, only slowly gaining ground in the last three decades of the twentieth century, can be observed in the increasing appearance of interdisciplinary literature. This new approach began during the 1970's, with the increasing involvement of paleo-botanists in environmental interpretations[7] (although in various reports the trend to put the

---

[6] C.P. Snow, *The Two Cultures and a Second Look*. New American Library, New York (1963).

[7] S. Bottema, "Late Glacial in Eastern Mediterranean and the Near East" in *The Environmental History of the Near and Middle East Since the Last Ice Age* W.C. Brice (ed.) Academic Press, London, pp. 15–28 (1978).

A. Horowitz, "Palynology-climate and Distribution of Settlements in Israel" *Qadmoniot* 13/3–4:51–52 (1980, Hebrew).

A. Leroi-Gourhan, "Diagrammes polliniques de sites archéologiques au Moyen-Orient" *Beihefte zum Tübinger Atlas des Vorderen Orients* W. Frey, H.P. Uerpmann, and A. Reihe, (eds.) *Beiträge zur Umweltgeschichte des vorderen Orients*, Tübingen pp. 121–133 (1981).

A. Leroi-Gourhan and F. Darmon, "Analyses Palynologiques de Sites Archéologiques du Pléistocène Final dans la Vallée du Jourdain" *IJES* 36:65–72 (1987).

W. Van Zeist and S. Bottema, "Vegetational History of the Eastern Mediterranean and the Near East During the Last 20,000 Years" in *Palaeoclimates, Palaeoenvironments and Human Communities in Eastern Mediterranean Region in Later Prehistory* British Archaeological Reports, International Series 133:277–321 (1982).

blame on man rather than on climate still prevailed)[8]. The interdisciplinary approach is exemplified also by involvement of other humanistic sciences, such as anthropology, sociology, economics, etc. The widening of the interdisciplinary movement involves a closer interaction with the physical sciences, as exemplified by the symposium on the decline of the Early Bronze Civilization of northern Mesopotamia mentioned earlier as well as books comprising of a number of inter-discplinary studies.[9]

In our case archaeologists, not to speak about historians educated in the faculties of humanities, the evidence of climate changes based on proxy data can be compared to discussing 'Schroedinger's cat paradox" in a seminar of theologians. An illustration to the Kuhn's and Snow's theories is the difference between the reviews on the 1st edition of this book. On the one hand by a senior archaeologist, who works for many years in the region and sticks to the old paradigm[10] and on the other a young pedologist who investigated soil profiles of the deserted city of Abila, one of the Decapolis sites in Jordan.[11]

Another example could be seen during an international conference, sponsored by the Center for Old World Archaeology and Art at Brown University, was held in May 1990.[12] In his opening presentation, the historian W.W. Hallo from Yale University concluded:

"I thus reject all field theories that threaten to obscure the boundaries between natural history and human history ... The traditional hypotheses for explaining the crisis of the 12th century B.C.E. are mostly concerned with natural disasters such as earthquakes, famine, or climatic change. But all these rest on the chance recording of what are basically perennial factors. The transition from Bronze Age to Iron Age should be seen rather in terms of human role."[13]

---

[8] U. Baruch, "The Late Holocene Vegetational History of Lake Kinneret [Sea of Galilee], Israel" *Paléorient* 12/2:37–48 (1986).

N. Liphschitz and Y. Waisel, "The Effects of Human Activity on Composition of the Natural Vegetation During Historic Periods" *Le-Yaaran* 24:9–15 (Hebrew), 27–30 (English abstract) (1974).

R. Rubin , "The Debate Over Climatic Changes in the Negev, Fourth – Seventh Centuries CE". *Palestine Exploration Quarterly* 121:71–78 (1989).

S.A. Rosen, "The Decline of Desert Agriculture: A View from the Classical Period Negev", *Symposium: Agriculture in Arid Environments: Archaeological Perspectives World Archaeological Congress 4* University of Cape Town (1999).

[9] G. Dalfes, G. Kukla and H. Weiss, (eds.) *Third Millennium B.C. Climatic Change and Old World Collapse.* NATO ASI Series, Sub series I Global Environmental Change, (1997).

H. Fischer, T. Kumke, G. Lohmann, G. Floser, H. Miller, H. von Storch and J.F.W. Negendank (eds.) *The climate in historical times: towards a synthesis of Holocene proxy data and climate models.* Springer, Berlin (2004).

[10] O. Bar-Yosef, "Issar & Zohar Book review" – *The Holocene* 15/6:933–934 (2005).

[11] B. Lucke, *Abila's Abandonment* M. S, Thesis BTU, Cottbus, Germany, Yarmuk University, Irbid (2002).

[12] W.A. Ward and M.S. Joukowsky (eds.) *The Crisis Years: The 12th Century B.C. From Beyond the Danube to the Tigris.* Kendall/Hunt Publishing Co.,Dubuque, Iowa, p. 208 (1992).

[13] W.W. Hallo, "From Bronze Age to Iron Age in Western Asia: Defining the Problem" in *The Crisis Years,* 1–9.

In contrast, the present authors argue that the agricultural evolution was generated in principle by the warming and aridization of the Near East, with human societies reacting to survive these changes. Similarly, the urban revolution and flourish of the Early Bronze, the renewal of relative prosperity during the Middle Bronze and of the Iron Age were due primarily to the abundant precipitation that enabled the accumulation of resources by all levels societies. Decline came when these conditions worsened.

A similar case will be made here to draw the background of the natural environment – in particular, the role of climate change on the historical events discussed in the above mentioned conference. In agreement with the conference's keynote address, which aimed to "synthesize" and asked the participants *"to venture beyond the boundaries of their own specializations"*, the present authors recommend to trespass the boundary of the natural habitat in order to encompass the broadest spectrum of all potential causes, natural as well as anthropogenic.

To cross the boundaries and get a rather detailed picture about the natural habitat the time series of proxy data related to this period should be examined in detail. Further progress, however, in this direction is still needed, and is constantly coming forth

The present work attempts to take an interdisciplinary approach in which the data from the fields of research of its authors (hydro-geology and archaeology) are interwoven to construct the environmental-cultural picture of the past. Simultaneous with this construction, they conducted a dialogue explaining their respective techniques, which helped each to arrive at certain similar or distinct conclusions. This approach will be followed in the ensuing chapters of this book, particularly in Chap. 2, where it will enable readers from the two different banks of the chasm separating between the "two cultures" to understand the different methodologies of the fields.

The authors are convinced that the conservative negative attitude is slowly changing, mostly due to the ever increasing quantity and quality of scientific research of the earth's past, mainly by geologists, geochemists, botanists, climatologists etc. The data obtained by these investigations cannot be ignored and will force the traditionally opposing academic worldviews to accept the obvious: All human cultures and civilizations were, and still are, a product of their environment. In the temperate and the arid zone it was, above all, the availability of water, which had the most profound effect on the development of human societies.

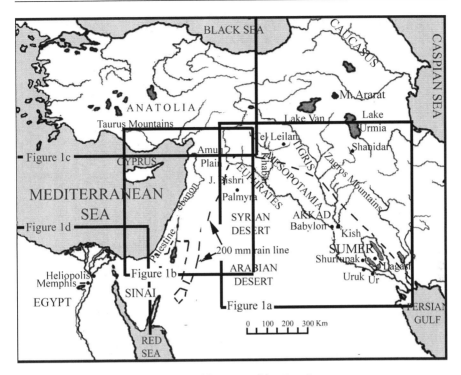

Fig. 1. General key map of the Near East

Fig. 1a. Mesopotamia

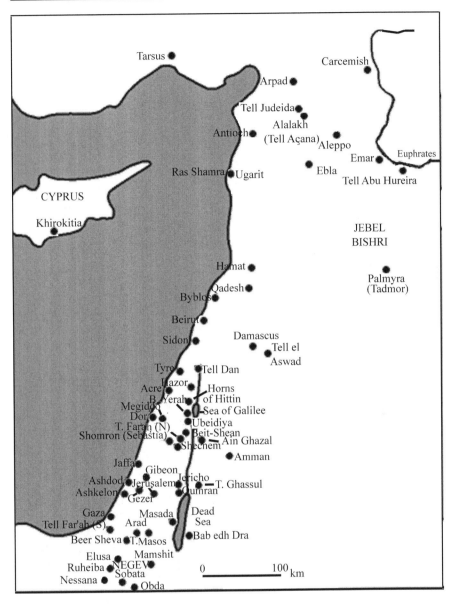

Fig. 1b. Levant

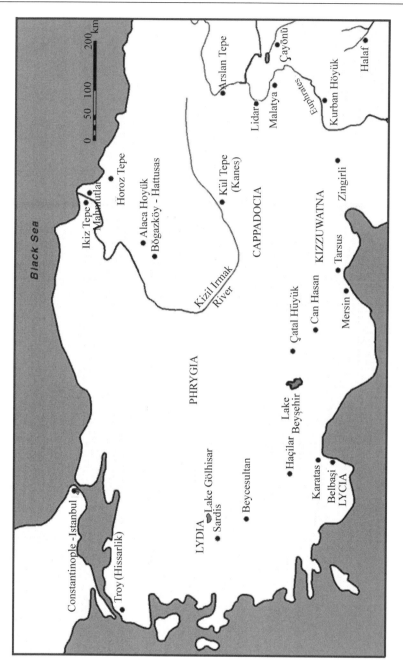

Fig. 1c. Anatolia

Fig. 1d. Egypt

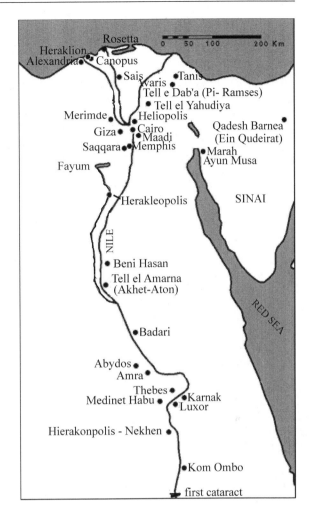

Acknowledgements

We wish to express our appreciation for their commentaries and support to Dr. Katharina Galor, Visiting Assistant Professor, Center for Old World Archaeology and Art, Brown University, Providence, USA, Professor Izhar Hirshfeld, blessed be his memory, Institute of Archaeology, The Hebrew University, Jerusalem, Israel, and Prof. Marinus J.A. Werger, Dept. of Plant Ecology, Utrecht University, Utrecht, The Netherlands.

CHAPTER 1

The Pendulum of Paradigms

"But where shall wisdom be found? And where is the place of understanding?"

(Job 28:12)

In the year 1909, Yale University geologist and geographer Ellsworth Huntington was granted a leave of absence and funds to tour Palestine and its neighboring countries. Huntington was not a stranger to Asia; within the previous ten years he had investigated the flood patterns of the upper Euphrates in eastern Turkey, traveled with a research team to regions of present-day Uzbekistan and Turkmenistan in Central Asia, and toured the Sinkiang province of Western China.

At that time, the Near East was part of the by then tottering Ottoman Empire. Turkish governors still ruled the main urban centers, but the rural areas were controlled by local *sheiks* frequently engaged in petty warfare among them. The nomadic bedouin eked out a meager living by their traditional practices of herding sheep, goats and camels, exacting tribute from passing caravans and, from time to time, raiding their neighbors, both nomadic and sedentary. Travelers to the more remote areas of certain regions thus had to secure protection from the bedouins, usually at the cost of a hefty *baksheesh*.

Upon arriving in Palestine, Huntington embarked upon a risky journey into the Negev Desert far to the south and also visited the eastern regions bordering the desert of Transjordan and southern Syria. The ruins of Petra, Ruhaiba, 'Auja el-Hafir (Nessana-Nezzana-Nizanna), Palmyra and Jerash greatly impressed him, and he concluded that only a profound change in climate could account for the large-scale desertion of these once-flourishing cities. He concluded:

Extensive travels in Asia Minor, Persia, India and central Asia led the author to adopt certain theories as to changes of climate and their relation to history. Descriptions of Palestine suggested that the same changes of climate have taken place there. Hence it seemed that in no other country could the theories be so well tested; for not only is Palestine so situated that climatic variations would there produce notable variations in habitability, but also its known history extends back to remote antiquity.[1]

---

[1] Huntington op. cit.

Huntington compiled his observations from Central Asia, including changes in caravan routes, levels of the Caspian Sea and of the Nile and other rivers and so forth, and correlated them with the rise and fall of ancient Near East civilizations. He added to his own observations those of scholars and educated travelers to many different countries and concluded that all the historical transformations of the time were consequences of global climate changes. He sought to explain these changes in the cyclical pulses of solar activity and claimed that changes in the physical environment affected the quality of life and nature of man and, therefore, human history.

Huntington used these conclusions to support his "deterministic" paradigm, which held that the physical geographic conditions of the earth's regions mold the spiritual and physical character of the peoples in those regions. In *Mainsprings of Civilization*, Huntington states:

> *"Our first conclusion is that we live in the midst of an intricate series of cycles, some of which are closely associated with atmospheric differences. How far atmospheric electricity and ozone are causes or merely concomitants of the cycles in business and in the reproduction of animals we do not know, but clearly the field for further study is wide and alluring.*
>
> *Long cycles as well as short cycles have engaged our attention for quite a while. During the present century the evidence of cycles with a length of hundreds of years has gradually become clearer. One of their chief characteristics is variation in the number and intensity of ordinary cyclonic storms. This opens the way to a study of specific periods such as the Golden Age of Greece, the Dark Ages in Ireland, and Revival of Learning in Western Europe. These give an idea of the way in which climatic cycles appear to have influenced the activity of the human mind as well as the vigor of the body, the production of food, and the capacity of a region to support people."*[2]

Huntington's Deterministic School, as it became known, was widely accepted from the beginning of the twentieth century until the start of World War II. The Belgian mathematician, statistician, and astronomer Lambert Adolphe Jacques Quételet (1796–1874) laid some of the earliest foundations of this school of mind by introducing the application of statistical methods in biology, anthropology, and social studies. Quételet claimed that, using statistical methods, one could distinguish the "average human being" (*homme moyen*) in different societies living in environments with similar characteristics and concluded that the physical environment shaped the average "profile" of its inhabitants, including their moral code.

Quételet's theory was adopted by the British history-philosopher Henry Thomas Buckle (1821–1862), who suggested that the general laws of history include physical laws, such as climate, soil type, etc., which decide the character and moral code of human society. The German geographer Friedrich Ratzel further developed these ideas in a series of books and articles he

---

[2] E. Huntington, *Mainsprings of Civilization*. Arno Press, New York (1972).

published between 1885 and 1904 on the influence of the natural environment on man and society. Unfortunately, one of Ratzel's terms *"Der Lebensraum"*, which he used as the title for a book published in 1904, was perverted by German Nazi ideologists to justify their conquest of neighboring lands. Ratzel himself never justified conquest and expansion based on racist ideology.

In time, the geographic deterministic school joined forces with that of the Darwinian evolutionists who maintained that the natural environment influences the character of societies through the process of mutation and selection, favoring the best adapted, and therefore, fittest mutant. Applied to human society, this theory maintained that over the millennia, through selection and survival of the fittest, environments with harsh living conditions produce people better able to cope with hostile circumstances.

In the late 1930s, for various reasons, the deterministic paradigm lost credibility among geographers. For one, misuse of the school's scientific conclusions and terminology (racial characteristics, *Lebensraum*, etc.) by Nazi ideologists prior to World War II fostered contempt of this worldview and its terminology.

Second, several world-renowned archaeologists began to assign much greater importance to human factors and less to forces of nature as the major determinant of the fate of societies and countries. Among them was Sir Flinders Petrie, professor of Egyptology at the University of London and one of the fathers of modern archaeology, who carried out many excavations in Egypt and southern Palestine in the first half of the twentieth century.

Sir Leonard Woolley, the famous explorer and archaeologist who discovered the city of Ur in Sumer, also favored anthropogenic causes of change. In the years before World War I, Woolley and T.E. Lawrence (known to many as "Lawrence of Arabia") collaborated in surveying the deserts of the Sinai Peninsula and the Negev. Summarizing their findings in *The Wilderness of Zin*,[3] the two dispute Huntington's conclusions about desertion of the cities in this region.

The eminent American archaeologist W.F. Albright, who conducted excavations at numerous sites in the Levant, also denounced Huntington's theories and wrote in *The Archaeology of Palestine*:

> *"In his famous book, Palestine and Its Transformation (1912), the late Elsworth Huntington explained most of the historical vicissitudes of Palestine in accordance with his hypothesis of cyclic oscillations of climate and rainfall. By an uncritical combination of data from literary sources with a superficial study of archaeological remains, then inadequately understood even by professional archaeologists, he concluded that there had been a series of drastic shifts in the water supply of the land since the second millennium B.C. Systematic archaeological research has proved that all his deductions were wrong. . . On many such erroneous inferences Huntington built up an elaborate superstructure of historical interpretation."*[4]

---

[3] C.L. Woolley and T.E. Lawrence, *The Wilderness of Zin*, Jonathan Cape, London (1936).

[4] W.F. Albright, *The Archaeology of Palestine*. Penguin Books, Hardsworth (1949).

The American archaeologist, N. Glueck, who conducted an extensive survey of Transjordan and the Negev, agrees with Albright. In his book *Rivers in the Desert* he writes:

> *"The conclusion seems inescapable, wherever it has been possible to check, that the major factors affecting the course of human history certainly in the Near East, and probably elsewhere, during the last ten thousand years, are those over which in general there is a large measure of human control."*[5]

Another American, soil scientist and agricultural engineer W.C. Lowdermilk, also helped undermine the deterministic paradigm and establish the axiom "blame the human". Working in China after World War I to help the Chinese fight drought and famine, Lowdermilk concluded that man is responsible for catastrophic soil erosion and economic disasters. He preached his "eleventh commandment" against the sin of causing land wastage from erosion due to improper methods of soil tilling and recommended soil conservation to counter the severe erosion in many areas in the United States that resulted from improper tilling and irrigation. He attributed the decline of the agricultural societies of the Near East to invasion and conquest by Arabs – desert people who lacked knowledge of soil and water management.[6]

The anthropogenic argument was strengthened by the environmental and socio-economical catastrophe taking place in the Great Plains of the United States extending over parts of Colorado, Kansas, Texas, Oklahoma and New Mexico. The region was labeled "Dust Bowl" in the 1930s, when strong winds carried off the topsoil in heavy dust storms that blocked the sun and occasionally swept across the entire country to the east coast. At the time, agronomists and soil scientists blamed the devastation on the farmers' agricultural practices, giving little weight to the severe drought that has triggered the process, beginning in 1931 and lasting for seven years. The drought and the soil erosion caused by the windstorms destroyed the agricultural economy of the region, forcing thousands of bankrupt families to abandon their farms.[7]

The Israeli botanist M. Evenari and his collaborators, water engineer L. Shenan and plant ecologist N. Tadmor, having for many years studied the ecology of the Negev and the irrigation methods of its ancient inhabitants, blamed the invading desert people for the desertion of the region's cities and agriculture.[8]

Palestine's Jewish population strongly backed anthropogenic causes of desertification, as it seemingly supported Zionist ideology. Archaeological evidence indicated that the area was densely populated and, therefore, enjoyed prosperity and a flourishing agriculture during the second half of the first millennium B.C.E. when Judaea was first an independent Jewish state and later under Roman occupation. If desertification was the result of Arab conquest and subsequent centuries of neglect, then human enterprise could correct the

---

[5] N. Glueck, *Rivers in the Desert*, Norton & Co., New York (1968).

[6] W.C. Lowdermilk, *Palestine, Land of Promise* Golancz, London (1946).

[7] D.E. Worster, *Dust Bowl: The Southern Plains in the 1930s*, Oxford University Press, NY (1982).

[8] M. Evenari, L. Shannan, and N. Tadmor, *The Negev: the Challenge of a Desert*. Harvard University Press, Cambride USA (1971).

damages to the environment and make the land habitable for millions of Jewish refugees and immigrants.[9] Indeed, the entry on Huntington in the 1960 edition of the *Encyclopaedia Hebraica* states that

"Contrary to Huntington's suggestions, the present accepted opinion is that there is no proof of the occurrence of remarkable climate changes during the period of history. The level of the lakes in Palestine and Syria did not go down while years of severe droughts are known from ancient periods prior to the drying up of the Levant."[10]

New data on climate change and its impact on European history began to be published in the early 1970s, challenging the consensus that a stable climate prevailed in past historical periods and that human activity was the cause of environmental deterioration. The British archaeologist R. Carpenter argued that the extreme historical transformations in Greece were due to climate changes.[11] In France, the historical-geographers such as F. Braudel, Le Roy Laudurie and others in the school associated with *Annales* developed similar ideas.[12] In Britain, the climatologist H.H. Lamb led the school that began to investigate the relationship between climate change and history. He demonstrated the effects of climate change on Europe's environment and history and presented ample evidence to support the occurrence of the "Little Ice Age" between the mid-16th and mid-19th centuries of the Common Era (C.E.). Other investigations supported his conclusions.[13] In the Near East, the geographer D. Amiran interpreted the archaeological findings at Tell Arad and suggested that this city flourished prior to the 27th century B.C.E. i.e. during the Early Bronze Age, due to a more humid climate, which changed later. Also the archaeologist H. Ritter-Kaplan suggested in 1974 that a severe climate change in the 3rd millennium B.C.E. negatively impacted settlements in the region. A few years later H. Weiss presented his view that the decline of the Late Bronze Age civilization in the Near East was the result of climatic change. Subsequently, the climatologist J. Neumann as well as the archaeologist R. Amiran presented observations to correlate climate changes and historical events.[14]

---

[9] I. Troen, "Calculating the "Economic Absorbtive Capacity" of Palestine: A Study of the Political Uses of Scientific Research" *Contemporary Jewry* 10/2:19–38 (1989).

[10] *"Encyclopaedia Hebraica"*, Encyclopaedia Pub. Co., Jerusalem Vol. 14:816. (1960, Hebrew).

[11] R. Carpenter, *Discontinuity in Greek Civilization.* Cambridge University Press, Cambridge (1966).

[12] N. Brown, "Approaching the Medieval Optimum, 212 to 1000 A.D." in *Water, Environment and Society in Times of Climate Change*, A. Issar and N. Brown (eds.), Kluwer Academic Publishers, Dordrecht. pp. 69–97 (1998).
E. Le Roy Laudurie, *Times of Feast Times of Famine. A History of Climate Since the Year 1000*, Doubleday, New York (1971).

[13] H.H. Lamb, *Climate History and the Modern World*, Methuen, London (1982).
H.H. Lamb *Climate: Present, Past and Future.* Princeton University Press, (1985).
M.L. Parry, *Climatic Change, Agriculture and Settlement*, Archon Books, Folkstone, Dawson (1978).
S. Leroy, and D. Stewart conference on "Environmental Catastrophes and Recoveries in the Holocene", *Abstracts: Department of Geography & Earth Sciences*, Brunel University Uxbridge, UK (2002).

[14] D.H.K. Amiran, "The Climate of the Ancient Near East: The Early Third Millennium B.C. in the Northern Negev of Israel" *Erdkunde* 45/3:153–167 (1991).

Once the idea of impact of climate on the history of human society became acceptable and in the same time more data from the Mediterranean Region, archaeological as well as paleo-environmental, became available, more publications relating climate changes to history were published.[15] Although the present authors do not agree with all the conclusions arrived at in these publications, yet the fact that climate change is considered as a factor, which shoud not be overlooked, spells a "Kuhnian type" revolution.

In the same time, on the global scale, paleo-environmental research prompted by studies of the greenhouse effect and by the development of objective research tools, such as environmental isotopes (oxygen 18, deuterium, carbon 13 etc., see Appendix II), helped to draw attention to the decisive impact of climate change on the natural as well as the human environment. Analyses of the chemical and isotopic composition of ice cores from both Greenland and Antarctica provided the first objective data on climate changes in the last millennia, which could be correlated with historical events. Although most of the data available concerns climate changes during the Pleistocene, some of the investigations concentrated on the Holocene and enabled a better understanding of the impact of climate changes during this period on human history. An important step was taken when the WMO and UNEP established the Intergovernmental Panel on Climate Change (IPCC). The task of this panel is to assess scientific, technical and socio-economic information relevant for the understanding of climate change, its potential impacts and options for adaptation and mitigation.[16]

H. Ritter-Kaplan "The Crisis of the Dryness in the 3rd Millennium B.C.E. and Its Applications According to Excavations in Tel-Aviv Exhibition Garden" in *The Land of Israel, Braver Book*, 17: 333–338 (1974, Hebrew).

H. Weiss, "The Decline of Late Bronze Age Civilization as a Possible Response to Climatic Change" *Climatic Change* 4:173–198 (1982).

H. Weiss, M.A. Coutry, W. Wetterstrom, F. Guichaard, L. Senior, R. Meadow and A. Curnow, "The Genesis and Collapse of Third Millennium North Mesopotamian Civilization" *Science* 261:995–1004 (1993).

J. Neumann and S. Parpola, "Climatic Change and the Eleventh-Tenth Century Eclipse of Assyria and Babylonia" *Journal of Near Eastern Studies* 46:161–182 (1987).

J. Neumann and R.M. Sigrit, "Harvest Dates in Ancient Mesopotamia as Possible Indicators of Climatic Variations" *Climatic Change* 1:239–252 (1978).

R. Amiran, "The Fall of the Early Bronze Age II City of Arad," *IEJ* 36:74–76 (1986).

[15] N. Roberts, T. Stevenson B. Davis, R. Cheddadi, S. Brewster and A. Rosen, "Holocene climate, environment and cultural change in the circum-Mediterranean region". In: *Past Climate Variability through Europe and Africa*. R.W. Battarbee, F. Gasse and C.E. Stickley (eds.), Springer, Dordrecht, pp. 343–362 (2004).

A. Rosen, "Climate change, landscape, and shifting agricultural potential during the occupation of Tel Megiddo". In: *Megiddo IV*. I. Finkelstein, D. Ussishkin, and B. Halpern (eds.) Eisenbrauns. pp. 441–449 (2006).

A. Miller Rosen, *Civilizing Climate: The Social Impact of Climate Change in the Ancient Near East*. AltaMira Press (2006).

[16] W. Dansgaard, S.J. Johnsen, H.B. Clausen, and C.C. Langway, "Climatic Record Revealed by the Camp Century Ice Cores" in *The Late Cenozoic Glacial Ages*, K.K. Turekian (ed.), Yale University Press, New Haven, pp. 37–56 (1971).

W. Dansgaard, S.J. Johnsen, N. Reeh, N. Gundstrep, H.B. Clausen, and C.U. Hammer,"Climatic changes, Norsemen and Modern Man" *Nature* 255:24–28 (1975).

Yet, as Thomas Kuhn has forecasted, albeit all new observations the majority of historians and archaeologists refuse, or in the best case hesitate, to accept a causal relationship between climate change and history. The reluctance to adopt a more deterministic paradigm is partly due to their unfamiliarity with the research tools that have been developed in the physical, chemical and biological sciences. These include the use of environmental stable isotopes (such as oxygen 18, hydrogen 2 i.e. deuterium and carbon 13) also in lake and cave sediments, pollen and tree ring sections, geo-morphological reconstruction of ancient levels of seashores, lakes and rivers, and so forth.

A more fundamental reason involves the contrast of philosophy intrinsic in the humanities, including history and archaeology, with that of the physical sciences. Whereas physicists and many other natural scientists adopt the principle of Occam's razor – that is, to seek the shortest and simplest answer (mathematical if possible) to a complex assemblage of phenomena, most humanistic scholars investigating the human environment are happy to describe the picture in all its complexity and avoid the vexing question of 'why' or 'what caused it'. In this work we try to synthesize these two schools by using the tools of the physical and biosciences to investigate past climates and to correlate them with the complex story of human society.

The endeavor undertaken in this book has a practical edge, as it enables conclusions to be drawn about the future impact of the global climate change on the Near East. Over the last few decades the discussion about the global climate change and its possible impact on the global environment has broken from the narrow field of scientific investigation to become a topic of general concern, particularly the threat of global warming due to the greenhouse effect prompting a series of international conferences organized by WMO (World Meteorological Organization) and UNEP (United Nations Environmental Program) in the framework of IPCC (Intergovernmental Panel on Climate Change).

Focusing on the Mediterranean region, these organizations had by 1990 established a task force to prepare a regional overview on the possible impacts of the predicted climate change on the ecological systems as well as on the socioeconomic systems of the region. The resulting publication[17] contains much important contemporary climatic, oceanographic, ecological and hydrological data. Yet it concentrates on the more western part of the region and almost entirely lacks the historical dimension required for use as a tool to understand past developments and predict future ones. Due to the geographical limited scale of the Near East to the scale applied by the Global Meteoric Models (GCM) as well as its complex climatic nature, the prediction of the future impact of the green house effect on this region is problematic. Also a special report of

W. Dansgaard, S.J. Johnsen, D. Clausen, N. Dahl-Jensen, C.U. Gundstrep, H. Hammer, and H. Oeschger, "North Atlantic Oscillations Revealed by Deep Greenland Ice Cores" in *Climate Processes and Climate Sensitivity*, J. E., Hansen and T. Takahashi (eds.), A.G.U. pp. 288–298 (1984). IPCC Reports, IPCC Secretariat Geneva (1992–2001).

[17] L. Jeftic, J.D. Milliman, and G. Sestini, (eds.), *Climatic Change and the Mediterranean* Edward Arnold, London (1992).

the Intergovernmental Panel on Climate Change convened by the U.N. failed to give a concrete prediction.[18]

In Syria's Tell Leilan region, T.J. Wilkinson of the University of Chicago's Oriental Institute studied the vulnerability to climate change of an agricultural system along the margin of a desert. He based his study on the relationship of present-day land-use to the environment. Projecting the output of his contemporary modeling on the past he concluded that:

> *"Bronze Age settlement and land-use systems of upper Mesopotamia were brittle systems and were therefore vulnerable to collapse. Under conditions of maximum production it is likely that even a short-term dry period may have resulted in collapse but certainly a significant run of dry years would have resulted in considerable production deficits which in turn could precipitate collapse."*[19]

The significance of the "resiliency" and "brittleness" of the agricultural systems on which ancient society almost entirely depended is apparent from this study. It proved to be particularly important in the volatile belt between the sown land and the desert where agricultural settlements spread during wetter periods and declined during drier ones. Wilkinson points out the importance of negative feedback effects on driving collapse in the peripheral regions – for example, continuous cultivation downgrades the soil's fertility and structure, which in turn fosters wind and water erosion. He notes that although such processes could have resulted from the natural increase in population, a drastic change of climate would heighten the vulnerability of human societies as well as of the natural system.

The Tell Leilan excavations provided additional evidence for the role of climate change in the collapse of the Mesopotamian civilization towards the end of the third millennium B.C.E. and resulted in a special NATO workshop on this issue.[20] Other meetings which dealt with the problem of the impact of climate changes during the Holocene on the history of human societies andrepresentating various disciplines provided evidence from other regions, all showing a trend towards a more climate impact oriented paradigm.[21]

---

[18] IPCC (WMO–UNEP) Special Report on The Regional Impacts of Climate Change, An Assessment of Vulnerability – Chapter 7: Middle East and Arid Asia. (2001).

[19] J.T. Wilkinson, "Environmental Fluctuations, Agricultural Production and Collapse: A View from Bronze Age Upper Mesopotamia" in *Third Millennium B.C. Climatic Change and Old World Collapse.* G. Dalfes, G. Kukla and H. Weiss, (eds). NATO ASI Series, Sub series I Global Environmental Change, pp. 67–106 (1997).

[20] G. Dalfes, et al. op. cit. (1997).

[21] A.N. Angelakis and A. Issar, (eds.), *Diachronic Climatic Impacts on Water Resources*, NATO ASI Series, Vol. 36, Springer Verlag (1996).
A.S. Issar and N. Brown, (eds.), Water, Environment and Society in *Times of Climate Change*, Kluwer Academic Publishers, Dordrecht (1998).
M. Yoshino, M. Domros, A. Dougueedroit, J. Paszynski and L.C. Nkemdirim, (eds.), *Climates and Societies – A Climatological Perspective*, Dordrecht, Kluwer Academic Publishers (1997).
J.J. Duran, B. Andreo and F. Carrasco, (eds.) "Karst, climate chane and groundwater" *Proceedings of the Internatonal Congress on groundwater in the Mediterranean Countries* at Malaga in 2006, Serie; Hidrogeologia y aguas subterraneas, No. 18, Madrid (2006).

Zohar gathered evidence from a different angle. For more than twenty years he studied the ethnography of the transhumants and pastoral nomads of the Old World and their interaction and effect on the agrarian and urban cultures of the Fertile Crescent.[22] Their influence was most apparent in the so-called 'intermediate' or 'transitional' periods, often called "Dark Ages", intervals between the times of flourishing of urban civilizations in the ancient Near East. The very fact that civilizations rose and fell in a rhythmic pattern around every thousand years or slightly more suggests that factors far beyond the scope of human influence were at work.

Similar developments in contemporaneous cultures in neighboring, as well as more distant, parts of the globe could be explained only by climate change. The impact of shepherd tribes on their sedentary neighbors – considered so important by the earlier archaeologists – had to be reevaluated. Pastoral nomadic way of life was certainly one of the most important and basic socioeconomic components in the culture and history of ancient Near Eastern societies, but not as it was traditionally perceived.

The introduction of large herds of domesticated livestock by the nomads – particularly the black goats which are crucial to nomads for clothing, tents, rugs, and, in fact, their very survival – was, indeed, very detrimental to the vegetation, and thus, to the farmers. Furthermore, by burning the native bush and tree vegetation in order to promote the growth of fresh fodder, the nomads destroyed the natural plant cover in the most arid and sensitive landscapes (and continue to do so). Therefore, to Wilkinson's point that continuous cultivation exacerbated soil erosion, Zohar would add the deleterious effect of herds grazing on the stubble of fields – normally seen as a vision of pastoral tranquility that can be seen all over the area of the Fertile Crescent towards the end of a rainless summer. Although the herd's droppings fertilize the fields a bit, their trampling turns the dry soil to dust, carried away by the strong winds of autumn and winter (any bargain between the shepherd and the farmer apparently has always been at the expense of the latter). Still, all the damages inflicted by the animals and their owners are only aggravating factors in a chain of events whose ultimate cause is climate change itself.

The emerging "neo-deterministic" paradigm incorporates a variety of factors that may be considered in measuring the impact of a historical climate change on the natural and human systems. This framework views man and nature as interacting components of one system at a certain locus on space-time-information coordinates. The space coordinate defines the geographical zone; the time coordinate refers to the historical period; and the information coordinate describes the general anthropological character of the society – that is, the cultural and genetic level a society has reached as a result of co-evolutionary processes.

W.H. Durham describes these processes in detail. In his view of the evolution of human society, two information systems – genetic and cultural – characterize and differentiate societies. The genetic characteristics of human society are determined through Darwinian selection and adaptation to the environment;

---

[22] Zohar (1993) op. cit.

the cultural characteristics are derived from traditions passed from generation to generation through behavioral and verbal codes. It is important to stress that the two processes of evolution are complementary.[23]

The present authors also believe that the survival of human societies in problematic regions and during periods of stress requires special emphasis on the faculty of creativity and invention, which in itself probably derives from co-evolutionary processes. Creativity can also be defined as the level of intelligence achieved by the majority of population. Intelligence, as distinguished from instinct, is the ability to develop a novel way of reacting when an entirely new situation comes up.[24] It can be variously manifested – for example, by a technical innovation, such as a new method to assure a steady supply of water in time of drought or to adopt a new crop in times of famine (See Appendix I) The resiliency of the natural system in an arid zone largely depends on its physical properties for storing and delivering water. These can be present in surface water, subsurface water (groundwater), or a combination of the two. The storage capacity of rivers and lakes often depends on the storage of local groundwater, which overflows as springs to become rivers and lakes. In some cases, the storage capacity of a surface water system depends on the amount of precipitation in remote parts of the drainage basin.

Thus, for example, the extensive drainage basin of the Nile, fed by rains falling in the tropical and sub-tropical zones has a high storage capacity which is further augmented by the enormous swamps of the Sudd in the southern Sudan. In contrast, the flow of the Euphrates and Tigris is regulated by the storage capacity of the limestone aquifers of the Taurus and Zagros Mountains. This storage capacity of aquifers feeding springs that give rise to rivers or lakes depends on the permeability and storage coefficients of the water-bearing rocks as well as on the geological structure – i.e. the topographical and geographical distribution of the water-bearing strata. For instance, the ancient settlement of Jericho was able to withstand many climate changes because its supply of water for irrigation derives from a spring fed by a limestone aquifer drawing from a large area of the eastern Judaean hills. Localities near such springs often maintain their ancient names for millennia, attesting to the permanency of the habitation despite many population changes.

In conclusion, the ability of a society to withstand the impact of climate change and its consequences depends on the totality of the resiliencies of its sub-systems, both societal and natural. Yet, the magnitude of the impact, its duration, and the accumulation of secondary damages also play decisive roles.

---

[23] W.H. Durham, *Coevolution, Genes, Culture and Human Societies*. Stanford University Press, Stanford (1992).

[24] M.E. Bitterman, "Phyletic differences in learning" *American. Psychologist* 20:396–410 (1965).
M.E. Bitterman, "The Evolution of Intelligence" *Scientific American* 92–94 (January 1965).

CHAPTER 2

Constructing the Jigsaw Puzzle of Palaeo-Climates

"Who is the wise man that may understand this? And who is he to whom the mouth of the Lord hath spoken, that he may declare it, for what the land perished and is burned up like a wilderness, that none passeth through."

(Jeremiah, 9:11)

2.1
The Climatological and Historical Background (Table 1)

This chapter will relate the development of the neo-deterministic paradigm and paint a canvas of the climatological, hydrological, and historical setting of the Near East. The resulting picture should facilitate an understanding of the interaction between the physical geographical character of this region and its human history – an interaction owed to the region's situation in a transition zone between two global climatic belts.

The cyclonic rainstorms' belt of the westerlies and the anti-cyclonic Sahara Desert belt shift annually and multi-annually south and northward, deciding to a great extent the character of the physical environment. This change, as will be demonstrated in the chapters to follow, had a crucial influence on the prosperity and fate of the people living in this region, for better and for worse.

Despite all the advances in technology in the last years, the ultimate reasons for many contemporaneous climate phenomena still elude us. For many scientists, the eleven years rhythm of sunspot activity ranks high on the list of primary causes. Others prefer a narrow spacing of volcanic eruptions due to increased tectonic activity. It is perfectly possible that the two phenomena, and possibly others, are somehow interlinked.

One thing has become clear in the last years: Global weather conditions are expressions of the relationship between the atmosphere and the world's great oceans, with their currents and counter-currents. Two related, but unequal, great systems control our weather. The bigger one, known as the "Southern Oscillation," is a phenomenon of the tropics with its home province in the world's largest oceans – the Pacific, the Indian and the southern Atlantic Oceans. As the monsoon and the maverick "El Niño" fall into its "domain of responsibility", this system, called El Niño-Southern Oscillation, or ENSO, is critically important for most human populations in the tropics.

Table 1. A general historical-archaeological timetable of the Near East for the last 10,000 years

	Egypt	Syria-Palestine	Mesopotamia	Anatolia
2000	Mamluk/Ottoman	Mamluk/Ottoman	Seljuk/Ottoman	Seljuk/Ottoman
1000	···· Early Arab Period ········· Early Arab Period ············ Early Arab Period ······························			
C.E.	Roman/Byzantine P.	Roman/Byzantine P.	Parthian/Sassanian	Roman/Byzantine P.
B.C.E.	··· Ptolemaic Period ········ Persian/Hellenistic P.········· Persian/Hellenistic ········ Persian/Hellenistic··			
	Late Period	Iron Age II	Asyyrian/Neo-Babylon	Iron Age
1000	······································ Iron Age I ································			
	New Kingdom	Middle/Late Bronze Age	Old/Middle Babylonian	Middle/Late Bronze
2000	··· Middle Kingdom ...... Intermediate Bronze Age .... Akkad/Ur III/Isin ···················			
	Old Kingdom	Early Bronze Age II/II	Early Dynastic I–III	Early Bronze Age
3000	·································· Early Bronze Age I ············· Jemdet Nasr ····················			
	Archaic Period	Mature Chalcolithic	Gawra (N)/Uruk (S)	Late Chalcolithic
4000	··Pre Dynastic Period ···			
		Early Chalcolithic	Ubaid (N & S)	Middle Chalcolithic
5000	·· Early Chalcolithic··			
	Neolithic Period	Pottery Neolithic A/B	Halaf (N)/Ubaid (S)	Ceramic Neolithic
6000	··Early Ceramic ···· Neolithic			
7000	Various ····· Epi-Paleolithic ····· Cultures	Pre-Pottery Neolithic B	Hassuna/Samarra (north only)	
		Pre-Pottery Neolithic A	Pre-Pottery Neolithic (north only) ···	Aceramic Neolithic
	·· Natufian ···································· Epi-Paleolithic.... Period			
		Epi-Paleolithic:Kebaran		

The other system is the "North Atlantic Oscillation" (NAO). Although smaller, this system has as great an effect and is even more complex than ENSO. NAO is directly responsible for the weather in the Mediterranean region, as on the border between it and ENSO, the "Inter-tropical Convergence Zone" (ITCZ) has formed. The constantly fluctuating and even volatile interaction between these two global systems is largely responsible for the formation and subsequent increase or decrease of the desert belt ranging from northern Africa via the Near East to central Asia and Mongolia.[1]

Because of the importance of the North Atlantic Oscillation as the origin of the westerly winds bringing rain to the Near East, let us briefly remind the reader of its *modus operandi*. Sir Gilbert Walker coined the term in the 1920's based on his observations of interrelations between the low pressure near Iceland and the high pressure near the Azores.[2] It is supported by the Gulf

[1] M. Glantz, *Currents of Change: El Niño's Impact on Climate and Society*, Cambridge University Press, Cambridge, UK (1996).
W.S. Broecker and G.H. Denton, "What Drives Climatic Cycles?" *Scientific American*, 262/1:49–56 (1990).
R. Kandel, *Water from Heaven*, Columbia University Press, NY pp. 101–123 (2003).
[2] G. Walker, "Correlations in Seasonal Variations of Weather IX" *India Meteorological Department Memoirs* 24/9 (1924).

Stream carrying warm and salty water from the Caribbean Sea northeast along the eastern shores of northern America and then across the northern Atlantic. Before reaching the coasts of Europe, the stream splits into two. A southern current passes the coasts of Portugal, the Azores and northern Africa, and then rejoins the northern Brazil Current and heads back to the Caribbean Sea.

The northern branch continues in a northeasterly direction as the North Atlantic Current until Spitzbergen, the Barents Sea, and the Arctic Basin. South of Iceland, some of the warm waters turn west, drawn by the cold East Greenland Current, which runs along the eastern shores of Greenland from the northwest to the southeast and into the Labrador Sea. There, they join another cold flow from the north, the Labrador Current, famous for its many floating icebergs due to the increase of warmer water.

The Gulf Stream-North Atlantic Current warms western and northern Europe. The bulk of this salty, and therefore heavier water, eventually sinks deep into the ocean where it joins the deep "Great Ocean Conveyor Belt" on its way south. It is this sinking and down-welling of the salty waters that keeps the circulation of the Gulf Stream in motion. Any disturbance, such as abnormal dilution of the salty waters by melting waters from the northern ice, or other influence on the mechanism that might slow down or even stop the sinking would have enormous consequences on the oscillation between the Atlantic 'highs' and 'lows' and on the fluctuation of the jet stream bringing rain to the western Mediterranean area. If this proves to be correct, the melting of the Arctic ice cap observed in 1999/2000 might slow down the sinking of the Gulf Stream and in its wake bring colder weather to Europe and the Near East over the next few years. On the other hand, Issar's investigations have shown that global warming causes warming and drying of the Mediterranean region. Thus, a severe downward fluctuation of temperatures may be just the first stage of the general trend of warming and drying if at all, as stronger climatic forces may nullify it.Other elements in the general picture are the jet streams, fast moving air currents at high altitudes. The polar jet influences the climate over the Mediterranean, flowing from west to east and changing its position from north to south according to the seasons.

While it is clear that the interplay among these factors determines contemporary changes of climate within the seasonal range and a range of a couple of years, the question remains as to what could have caused the changes of climate in the range of historical periods, from a few decades to a few centuries and even a few millennia. The driving force may have been similar to that responsible for the other changes during the Quaternary, including the glacial and interglacial periods. It is beyond the scope of this work to detail the cause and effects of the global glaciations and de-glaciations that occurred during the Quaternary. The reader who is interested in this subject is advised to consult the list of references.[3]

---

[3] J. and K.P. Imbrie, *Ice Ages – Solving the Mystery*. Macmillan, London (1979).

A. Berger, R.E. Dickinson, J.W. Kidson, "Understanding Climate Change", *Geophysical Monographs* 52, IUGG Volume 7, A.G.U. Washington, DC (1989).

R.S. Bradley, "Palaeoclimatology. Reconstructing Climates of the Quaternary", *International Geophysics series*, 2nd ed., vol. 64, Academic Press, San Diego, USA, pp. 11–46 (1999).

In brief, however, the glacial and interglacial phenomena of the Pleistocene portion of the Quaternary are due primarily to the obliquity of the Earth's axis and variations in its orbit of the sun. In 1941, the Serbian astronomer Milankovitch calculated the minimum values of radiation due to minimum obliquity of Earth's bipolar axis and high eccentricity of its orbit around the sun and proposed that this was the reason for glaciations phenomena during the Quaternary.[4] The time spans for these occurrences were calculated at 185,000 years, 115,000 and 70,000 years. Later investigations, based on interpretation of the isotopic composition of marine microfossils taken from sea bottom cores, confirmed Milankovitch's general calculations, but specified the periodicity around 100,000, 41,000 and 19,000 to 23,000 years. Thus Milankovitch's theory does explain climate cyclic and long-term fluctuations in the order of magnitude of hundreds of thousands to a few tens of thousands of years. However, in the case of forces responsible for short-term climate changes within a few millennia, centuries or decades, different factors seem to be at work. These may be sunspots, which affect levels of solar radiation, or volcanic explosions, which have a negative effect on the transparency of the atmosphere. High temperatures, on the other hand, are sometimes associated with years of major El-Niño events, which are minimized when volcanic explosions occur during these years.[5] The possibility that volcanic explosions may have played an important role in the climate changes of the Holocene was brought up by Issar as a working hypothesis.[6]

2.2
Time Series of Proxy-Data to Decipher Climates of the Past

While the investigation of contemporaneous climate changes are based on direct data, such as air and sea temperatures, atmospheric pressures, wind directions and velocities, investigation of climatic changes of the past are based on proxy data derived from geology (the nature and distribution of sediments), geochemistry (chemical and isotopic composition of strata), biology (botanical and faunal assemblages, tree rings), other sources, such as archaeology (material remains of past cultures) as well as historical records, are of equal importance.

In the following chapters we also dare to discuss some of the myths an intrinsic part of the cultures of the ancient inhabitants of the Old World. We

---

R. Kandel, op. cit. pp. 66–82 (2003).

W.S. Broecker, *The Ocean's Role in Climate Yesterday, Today and Tomorrow.* Eldigio Press, NY, (2005).

[4] M.M. Milankovitch, "Canon of Insolation and the Ice-age Problem" (*Kanon der Erdbestrahlung und seine Anwendung auf das Eiszeitproblem*) Koenigliche Serbische Akademie, Edition Specializat 133 (1941) (English translation published by Israel Program for Scientific Translations, Jerusalem; and U.S. Department of Commerce, Washington, D.C., 1969).

[5] Glantz, (1996), op. cit.

[6] A.S. Issar, "The driving force behind the cold climate spells during the Holocene" in *Holocene and Late Vistulian Paleogeography and Paleohydrology* A. Kotarba (ed.) Polska Akademia Nauk, Prace Geograficzne Nr. 189. pp. 291–297 (2003).

refer to these myths with due caution, as "proxy-proxy" data, exemplified and discussed in detail later on. What should be emphasized at the outset is the need for an interdisciplinary approach that involves comparison of simultaneous events and an attempt to correlate them. An example, which will be discussed in some detail in the following chapters, is evaluation of proxy data indicating a humid period that caused floods in Mesopotamia during the 3rd millennium B.C.E. and which probably remained in the memory of the people of this region as the Biblical Flood.

The various sources of proxy data and how to interpret them will be discussed at length in this and other chapters. The authors will try to avoid excessively technical explanations by introducing the data while recounting their role in authors' investigations and conclusions. This exposition will also serve to shed light on the natural and cultural environments of the Near East and the changes they underwent during periods of climate change.

Issar first became involved with the question of dating and explaining desertification processes, during the late 1950s, when he investigated the Quaternary geology of Israel's Coastal Plain (Plate 1). The research was part of a general investigation of the groundwater resources of this region. In his conclusions he adhered to the conventional paradigm when he stated that the recent coverage of this region by sand dunes was a result of the invasion by the Arab tribes. As will be told later on, Issar had to admit that this conclusion was wrong.

Using sedimentological and paleontological methods, the investigation revealed that the subsurface of the Coastal Plain is built of alternate layers of sandstones, sands, loam and clays. This was deciphered into a series of transgressions and regressions of the sea. Sands and sandstones were deposited during the transgression periods, while during the regressions, clays were deposited and loam was formed under terrestrial conditions by weathering of the sandstones. This sequence was often datable by fossils and archaeological remains. The sands were brought from the Nile Delta by the counter-clockwise sea currents, deposited along the shore and driven inland by the winds.[7]

The transgressions were caused by the melting of the glaciers and, therefore, coincided with warm interglacial periods while the regressions occurred in glacial periods when water froze and was captured by the glaciers. The general conclusion is that the invasion of sand dunes, which is a result of a sea transgression, occurred during warm interglacial periods. The same investigation revealed that the recent sand dunes, started to penetrate around 700 C.E., i.e. the post-Byzantine/Early Arab Period, and by now have covered a vast area, extending to about 10 km from the shore and reaching, in some cases, a height of about 30 meters. The very recent age of the dunes was deduced from

---

[7] A. Issar, (1961) op. cit.

A. Issar, "Geology of the Central Coastal Plain of Israel" *IJES* 17:16–29 (1968).

Z. Reiss and A. Issar, "Subsurface Quaternary Correlations in the Tel Aviv Region" *GSI Bulletin* 32 (1960).

A. Issar, "Stratigraphy and Palaeoclimate of the Pleistocene of Central and Northern Israel" *Palaeo 3* 29:261–280 (1979).

A. Issar and U. Kafri, "Neogene and Pleistocene Geology of the Western Galilee Coastal Plain" *GSI Bulletin* 53:1–13 (1972).

Plate 1. Astronaut's photograph of the Levant (Courtesy NASA)

Byzantine and Early Arab pottery shreds and coins found in the loam below the dunes.

As mentioned already, Issar erroneously suggested that the cause for the invasion of the sands was a result of the gradual replacement of the local farmers by the invading Arab nomads who overtook the land and turned it into pastures for their goats, sheep and camels. This resulted in destruction of the natural and agricultural vegetation, which until that time prevented inland movement of the sands. This explanation was in accordance with the general anthropogenic paradigm, which was dominant during the period in which Issar carried out research for his Ph.D. dissertation. Support came in the early 1970s from satellite images that clearly distinguished Israel's Negev Desert, which appeared as a dark area, from Egypt's Sinai Desert, which was bright due to the lack of vegetal cover (Plate 1). The reason for the difference simply was the restrictions on wandering placed on the bedouin tribes living on the Israeli side of the border.

When the anthropogenic model seemed to have been well established and supported by field and satellite evidence, the new tool of analysis of the composition of stable environmental isotopes, especially ^{18}O and ^{2}H (deuterium) in water, helped to undermine the old conceptual model (see Appendix II).

During the late 1960's a team of hydrogeologists and hydrochemists headed by Issar discovered that the Nubian sandstones of the Lower Cretaceous age in the Negev and Sinai Peninsula contain vast amounts of groundwater, dated to the Last Glacial Period. The isotopic "fingerprints" i.e. the oxygen 18 to hydrogen 2 ratios of this water were different than those of contemporary water. While the latter accords with the Mediterranean Meteoric Line, the newly discovered water had the characteristic "fingerprints" of the Global Meteoric Line (Fig. 2), which could have meant that during the glacial period the storms arrived from the Atlantic Ocean rather than the Mediterranean. There was something wrong, however: The water was isotopically too heavy to have come from the far away Atlantic and too loaded with calcium and magnesium sulfate as well as sodium chloride salts.

In order to understand this anomaly, special samplers for rainwater and floodwater were developed and spread throughout the Negev Desert and the isotopic composition of the water of each storm was correlated with the trajectory of the storm

As expected, the isotopic composition of the contemporary rainwater brought by most storms fell on the Mediterranean Meteoric Line. However,

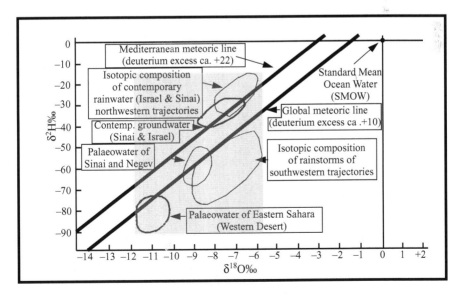

Fig. 2. Distribution of the environmental isotopes ^{18}O and ^{2}H in rainwater and groundwater – Southeastern Mediterranean region

very surprisingly, the isotopic composition of rainwater of storms traveling along southwestern trajectories – that is coming from the Mediterranean and crossing the Libyan and Western Egyptian desert – fell on the Global Meteoric Line, similar to the paleo-water under Sinai and the Negev. Such storms were, in many cases, ushered in by heavy dust storms raised by winds blowing from the desert into the cyclonic low along which the storm advanced (Fig. 2). The conclusion was that the paleo-water under the Sinai and the Negev was, indeed, recharged by rainstorms, which, at that time, had prevailing southwestern trajectories.[8]

Based on these findings, Issar and Bruins suggested a new explanation for the processes of the deposition of loess over the Sinai and Negev during the Quaternary. The conventional theory maintained that the loess was deposited throughout the entire duration of the Pleistocene. According to the new explanation, the bulk of the loess was nearly exclusively deposited during the glacial periods. The new model takes into consideration that the climate belts had shifted southward during the glacial periods causing the southwesterly cyclonic lows to become dominant. They became charged with dust and salts while passing over the deserts. The rains that immediately followed the dust storms then deposited the dust as loess. Simultaneously, the rainstorms recharged the Nubian Sandstone aquifer with water of a special isotopic signature, i.e. lighter hydrogen and oxygen isotopes as well as with the salts.[9]

While the anomaly of high calcium sulfate and sodium chloride in the paleo-water can easily be explained by the dust storms composition over deserts, the relatively high presence of magnesium sulfate demands an explanation. We suggest that during the Pleistocene glaciation the sea retreated and exposed the shallow sea-shelf of the Mediterranean. The bottom of the Suez Gulf were exposed as well and became vast areas of salt marshes, locally named *sabkha*. Sodium chloride and calcium sulfate as well as magnesium sulfate were de-

[8] A. Issar, A. Bein, and A. Michaeli, "On the Ancient Water of the Upper Nubian Sandstone Aquifer in Central Sinai and Southern Israel" *Journal of Hydrology* 17:353–374 (1972).

A. Issar, "Fossil Water Under the Sinai-Negev Peninsula" Scientific American 253/1:104–110 (1985).

A. Issar and R. Nativ, "Water Beneath Deserts: Keys to the Past, A Resource for the Present" *Episodes* 11/4:256–262 (1988).

J.R. Gat and A. Issar, "Desert Isotope Hydrology: Water Sources of the Sinai Desert" *Geochimica et Cosmochimica Acta* 38:1117–1131 (1974).

M. Levin, J. Gat and A. Issar, "Precipitation Flood and Groundwater of the Negev Highlands. An Isotopic Study of Desert Hydrology" *International Atomic Energy Agency Workshop on Arid Zone Hydrology AG-158*, Vienna, IAEA, pp. 3–22 (1980).

A. Issar and J. Gat, "Environmental Isotopes as a Tool in Hydrogeological Research in an Arid Basin" *Ground Water* 19/5:490–494 (1981).

C. Leguy, M. Rindsberger, A. Zangvil, A. Issar and J. Gat, "The Relation Between the Oxygen 18 and Deuterium Contents of Rain Water in the Negev Desert and Air-masses Trajectories" *Isotope Geoscience* 1:205–218 (1983).

[9] A.S. Issar and J. Bruins, "Special climatological conditions in the deserts of Sinai and the Negev during the Late Pleistocene" *Palaeo 3* 43:63–72 (1983).

D.H. Yaalon and J. Dan, "Accumulation and Distribution of Loess Deposits in the Semi-desert and Desert Fringe Areas of Israel" *Zeitschrift der Geomorphologie.* Suppl. 20:91–105 (1974).

D.H. Yaalon and E. Ganor, "The Climatic Factor of Wind Erosibility and Dust Blowing in Israel" *IJES* 15:27–32 (1976).

posited in and along the edges of these marshes. The passing storms swept these salts away, mixed them with the dust, already containing salts, and deposited it as loess, and its mineral holding water recharged the sandstones.

As the climate warmed at the end of the glacial period, the loess deposition nearly stopped and sand dunes began to invade the coastal plain, clearly visible in the northern Sinai and Negev where the sands overlie the loess (Plate 2). Epi-Palaeolithic tools dating to around 15,000 years ago were found in the transition layers between the loess and the sand.

The massive invasion of sands was a result of the reduction of loess deposition and an increased amount of sand supply brought by the Nile as well as the erosion of the marine fringes delta. The changed global climatic pattern involving a warmer climate had caused the strengthening of monsoon rainstorms over subtropical Africa from which the Nile receives most of its water and sediments.

All these observations brought Issar to raise the following question: Is it possible that the invasion of the young sand dunes at the end of the Roman-Byzantine period, synchronous with the Arab conquest, was primarily due to climate change rather than to human intervention? Could it be that what happened during the post Last Glacial period repeated itself on a smaller scale, towards the end of the first millennium C.E.? The correlation between the observations of C. Klein of high level of the Dead Sea during most of the Roman Byzantine periods and a low level after the Arab invasion in the 7th century C.E.[10] convinced Issar that indeed, like during the end of the Pleistocene when a cold and humid period was followed by a warm and drier period, the colder

Plate 2. Post Last Glacial sand dunes overlying layers of loess, border of northern Sinai and Negev (Photo A.S. Issar)

---

[10] C. Klein, "Morphological Evidence of the Lake Level Changes, Western Shore of the Dead Sea" *IJES* 31:67–94 (1982).

and more humid Roman-Byzantine period was followed by the Arab warm and dry period. The desert fringes of the Levant bloomed during the first period and were abandoned during the second. This process was most obvious in the Negev Desert where Issar came to live at the mid 1970's at the campus of the Institute for Desert Research of the Ben-Gurion University of the Negev. Just a few kilometers from this campus is the ancient Nabataean-Byzantine city of Avdat (Plate 3). This city, like others in the region, was already more or less deserted when the young sand dunes advanced into the coastal plain some time after the Arab invasion The abundance of abandoned agricultural terraces and farms and agricultural installations, such as wine presses (Plate 4), flour mills and olive presses, raised again and again the question aboutthe impact of climate change on the ancient socioeconomic systems. Could not climate change, indeed, be the primary reason for desertification of the region? We mentioned already that this notion stood in clear contrast to the then dominant paradigm supported by the investigations of M. Evenari and his team in the same region, and by the local archaeologists noted earlier. They claimed that the desert cities were doomed when the Byzantine administration collapsed and the trade on the spice routes between Arabia Felix and Byzantium was interrupted, not withstanding the fact that the isohyet line of 200 mm/year eperates between the region of deserted cities to those which remained inhabited. South and south east of this 200 mm line were all the cities which were deserted. For instance, Beer Sheva which was a flourishing Episcopalian city during Byzantine times was almost totally deserted whereas Hebron, situated in a more humid area only 50 km away, remained inhabited. This example is true for most of the towns to the eastern side of the mountainous backbone of this country, which creates the shadow side of the rains. Others, such as Sussia in south Judaea, were deserted only later around 800 C.E.

The deterministic paradigm gained an edge with publication of the results of a multidisciplinary investigation of a core drilled at the bottom of the Sea

Plate 3. Partly reconstructed eastern wall of Avdat and its acropolis (on the *right*). (Photo A.S. Issar)

Plate 4. Wine press at Shivta/Subeita (Photo A.S. Issar)

of Galilee. The report presented the time series of a collection of proxy data including ^{18}O and ^{13}C data as well as pollen dated by ^{14}C. Even though there were pronounced changes in the isotopic and pollen assemblages during the 5000 years in which the layers in the lake were deposited, the scientists did not suggest climate change as a reason for these changes. The changes in the pollen assemblage from that of Mediterranean woodland to olive during the Roman-Byzantine period and the major reduction in olive pollens and return of the natural vegetation during the Moslem period were explained entirely by economic reasons. The authors reasoned that the demand for oil during the Roman Empire brought the people to cut the natural vegetation and plant olives, and an economic crisis caused the neglect and abandonment of the plantations and the return of the natural vegetation. Yet, in the same time-series, the heavy isotopes of ^{18}O were depleted during the period in which the olive flourished – denoting a colder and more humid climate, and when the olives diminished, the isotope composition became heavier. Moreover, the same proxy data showed a similar change in the pollen and isotopes around 2400-2000 B.C.E. – a period of general crisis in the Near East and other parts of the northern hemisphere which will be discussed more in detail later on. Also in this case, the anthropogenic paradigm of over-irrigation, soil salinization and economic disaster, was preferred with no consideration of climate change.[11]

---

[11] M. Stiller, A. Ehrlich, U. Pollinger, U. Baruch and A. Kaufman, "The Late Holocene Sediments of Lake Kinneret (Israel) – Multidisciplinary Study of a Five Meter Core" *GSI Current Research Report* pp. 83–88 (1983–84).

Based on the isotopic and pollen proxy data, Issar suggested a conceptual model in which the changes in the time series can be explained primarily by climate change taking into consideration the simultaneous changes in the assemblages of the pollen and the isotopes, as well as the hydrological regime of the Sea of Galilee. This lake is part of the drainage basin of the Jordan River, which feeds and drains the lake southwards to the Dead Sea. About 75 percent of the water feeding the upper part of the Jordan and flowing into the Sea of Galilee originates from springs draining the Lebanon and Anti-Lebanon mountains built mainly of highly permeable Jurassic-age limestone. The precipitation, much in the form of snow, infiltrates into the solution channels in the limestone, fills the saturated part of the subsurface, and flows out as springs. Because of the high permeability of the aquifer's layers feeding the springs of the Jordan River, their flow is immediately influenced by the quantity of precipitation in the Lebanon and Anti-Lebanon mountains. As the route of the water from the recharge on the surface of the drainage basin to its emergence at the outlet of the springs is rather short, the isotope composition of the springs' water represents that of the precipitation on the high mountains during a period of a few years to a maximum of a few decades, Therefore, the rains as well as the water of the springs will be depleted in ^{18}O and ^2H during cold and rainy periods and enriched during warm periods. The water containing CO_2 dissolves the carbonate rock while flowing through the limestone and is subsequently deposited in the lake and contains the oxygen isotopic "fingerprint" of the precipitation.

Another isotope, which serves as an environmental indicator, is ^{13}C. Whereas areas covered by grasses and bushes are characterized by a higher ratio of the isotope, soil moisture in forested areas is relatively depleted. Thus, under normal Mediterranean climate conditions, the vegetation during colder and wetter conditions is that of woodlands, and the water percolating into the subsurface is depleted in both ^{13}C and ^{18}O. The opposite is seen in warm and dry periods.

The other source of water are floods draining the eastern part of the watershed of the Galilee and of the western Golan Heights. Springs supply water along the shores of the lake, some of which are saline and warm and were known to have been spas in ancient times. The ratio of flood water in the water balance of the Sea of Galilee is relatively small compared with that flowing from the springs recharged on the mountains of Lebanon and Anti-Lebanon which, due to their height, receive a high share of the precipitation. In addition, they are built of highly permeable limestone which facilitates a high percentage of percolation (up to 30%) and low surface flow percentage (3–5%). On the other hand, the surface water basins are spread over areas that get lower amounts of precipitation, because of their location in the rain-shadow of the Galilee. At the same time, a major part of the surface basin of the Golan heights is built of basalts and covered by black soils, which donate to the floods washed-away soil sediments rather than dissolved carbonates. Thus, due to the high percentage of spring water, the water reaching the sea sediments of the lake contain mainly the "fingerprints" of environmental isotopes ($^{18}O,^{13}C$) of the precipitation falling on the high mountains (reaching an altitude of around 2800 m), and partly of the floods and small springs of its drainage basin. On the other hand, the pollen assemblage represents the vegetation of the basin draining to the lake.

As most of the surface drainage basin is in the shadow of the rain, it receives on the average at present about 500 mm of rainfall per year. A 20 to 30% increase would bring the annual amount of precipitation to about 600 mm – optimal for olives. A decrease in this amount, however, would reduce the rainfall to 400 mm and below – the lower limit the olive tree tolerates for bearing sufficient fruit to justify its cultivation. On the basis of these data, one could reconstruct the paleo-climatic scenario for the post Roman-Byzantine Period, i.e. that of the penetration of the sand dunes and the desertion of the cities in the Negev, as follows: A series of dry years with an annual average precipitation of less than 400 mm over the surface drainage basin caused the olive crop to diminish to a level where it was no longer economical to cultivate. A few consecutive years of drought may even have caused many trees to dry up and die. Thus, the olive plantations on the terraces were abandoned and the natural vegetation of dwarf oaks, pistachio, etc., returned to dominate the landscape. With the return of the natural *maqui* type vegetation, the permanent inhabitants could shift to an economy of raising goats and sheep, and in good years to sow the terraces for grains and fodder. This was the general socio-economic system that characterized the semi-arid Levant a century ago, and still characterizes large parts of it even today. The more arid the terrain, the higher is the population of nomadic bedouin tribes who inhabit it in relation to the farming sedentary population.

As mentioned earlier, the increase and decrease in olive pollen parallels the increase and depletion in the composition of ^{18}O and ^{13}C that occurred during the third millennium B.C.E., and indicates that during this period the cultivation of olives in this region responded to climate change. If this model is applied to Mesopotamia for the same period, then the blame placed on the Mesopotamians for salinization of their soils during this time, as will be discussed later, should be reconsidered.

The preceding discussion has illustrated the constraint of using pollen as proxy data for reconstructing paleo-climates, particularly with the development of agriculture and human interference with the natural vegetation. These data, therefore, should be compared with other time-series of proxy data before arriving at conclusions about paleo-climatic scenarios. One source of these data is Lake Van in eastern Anatolia[12] (Fig. 3). Like the Dead Sea, Lake Van is a ter-

---

[12] E.T. Degens, H.K., Wong, S. Kempe and F. Kurtman, "A geological study of Lake Van, Eastern Turkey" *Geologische Rundschau* 73:701–734 (1984).

M. Schoell, "Oxygen Isotope Analysis on Authogenic Carbonates from Lake Van Sediments and Their Possible Bearing on the Climate of the Past 10,000 Years" in *The Geology of Lake Van*, E.T. Degens, and F. Kurtman (eds.) The Mineral Research and Exploration Institute of Turkey, Ankara, pp. 92–97 (1978).

G. Lemcke and M. Sturm, "d18O and Trace Element as Proxy for the Reconstruction of Climate Changes at Lake Van (Turkey): Preliminary Results". In: *Third Millennium B.C. Climatic Change and Old World Collapse*. G. Dalfes et al. op. cit. (1997).

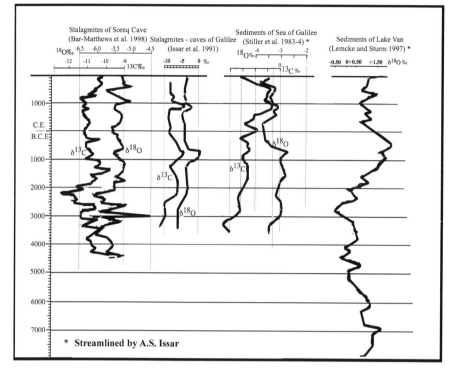

Fig. 3. Isotopes in stalagmites and lake deposits

minal lake. With an area of 3522 square km it is the fourth largest terminal lake in the world. Unlike the Dead Sea, Lake Van is located at a high altitude – more than 1720 m above mean sea level. Because of the high evaporation and hydrothermal activity in the vicinity, the lake is saline alkaline. The precipitation regime feeding the rivers that flow into the lake is influenced by the Mediterranean climate system: in winter, the westerlies driving cyclonic depressions bring rain, and in summer the predominant subtropical high-pressure belt has a drying influence. Today the precipitation on its southern and western water province is about 1000 mm while the northern and northwestern areas receive about 400 mm. The dissolved salts – mainly carbonates – precipitate during the summer forming laminar varves, which can be counted and enables dating of the layers. A multidisciplinary group studied the lake sediments in the early 1980's. Analysis of their $^{18}O/^{16}O$ ratios showed a sequence of changes quite similar to that of the Sea of Galilee. Complicating the picture is that part of the water flowing into the lake originates from springs that run into the river and discharge into the lake. The storage time of the groundwater in the subsurface must, therefore, should be taken into consideration as a retardation factor (see Appendix I) when investigating the isotopic character of the sediments.

Investigations on the ancient levels of the Dead Sea supplied additional information about the climate changes during historical periods. For example,

Klein showed that around 2000 years ago the fluctuating level of the Dead Sea reached a peak of about 70m higher than today.[13] In the same period, the level of the Nile fell, probably due to the weakening monsoon system.[14] Frumkin (Fig. 3a) in a series of studies provided additional support for significant climate changes during historical periods. He and his colleagues investigated the salt caves of Mount Sedom (Sodom) – a huge salt plug emerging as an elongated hill along the southwestern shore of the Dead Sea. They discovered that when the sea level rose, the water entered the caves and left deposits along the ancient high shorelines. These deposits contained driftwood and could be dated by [14]C. When the sea retreated, rivulets began draining the water, which had entered through the caves' chimneys and cut deep canyons in the salt. The highest water level occurred in the first half of the third millennium B.C.E. and was followed by a severe and abrupt decline, falling to below the altitude of the shallow bottom of the southern part of the Dead Sea that thus dried up. The sea level was high again during the last centuries of the first millennium B.C.E. and the first and middle centuries of the first millennium C.E., i.e. the Iron Age and the Roman and Byzantine period. In the Moslem period, namely the second half of the first millennium C.E., it again fell to an extreme low level. We can state in a general way that the high levels of the Dead Sea interpreted as cold and humid periods corresponded with periods in which the desert became moist and green, whereas low levels correlated with warm and dry periods and consequently with desertification.[15]

The investigations carried out on the Holocene levels of the Dead Sea by R. Bookman (Ken-Tor), Y. Enzel, A Agnon and M. Stein supplied additional detailed data on the Dead Sea levels since 2000 B.C.E. Palynological investigations carried out on the sediments of the Dead Sea that the environmental history during the last 10K years was strongly by both climatic and human impact. [16]

[13] Klein (1982) op. cit.

[14] S.H. Nicholson and H. Flohn, "African Environmental and Climatic Changes and the General Circulation in the Late Pleistocene and Holocene" *Climatic Change* 2/4:313–348 (1980).
A. Migowski, A. Agnon, J.F.W. Negendank and M. Stein, "Holocene climatic record in laminated sediments from the Dead Sea – comparison with the developments of the Near East Civilizations" *Israel Geological Society Annual Conference Abstracts* p. 85 (2002).
Y. Henzel, R. Ken-Tor, Y. Enzel, D. Sharon, H. Gvirtzman, U. Dayan, M. Stein, "The Holocene Dead Sea Level Variations and the Hdrochlimatology of the Near East" *UNESCO – Hebrew University of Jerusalem Workshop on Science for Peace in the Middle East*, Program and Abstracts, Y. Becker (ed.) (2002).

[15] A. Frumkin, "The Holocene History of Dead Sea Levels" in *The Dead Sea, The Lake and its Setting*, T.M. Niemi, Z. Ben-Avraham, J.R. Gat (eds.). Oxford University Press, Oxford, pp. 237–248 (1997).
A. Frumkin, G. Kadan, Y. Enzel, Y. Eyal, "Radiocarbon chronology of the Holocene Dead Sea: Attempting a regional correlation" *Radiocarbon* Vol. 43/3:1179–1189, (2001).
A. Frumkin and Y. Elitzur, "Historic Dead Sea Level Fluctuations Calibrated with Geological and Archaeological evidence" *Quaternary Research* 57:334–342 (2002).

[16] R. Bookman (Ken-Tor) et al. op. cit.
R. Bookman (Ken-or), Y. Bartov, Y. Enzel, M. Stein, "Quaternary lake levels in the Dead Sea basin: Two centuries of rsearch". *Geological Society of America, Special Paper* 401:155–170 (2006).
T. Kitt, Holozaene Vegetation und Klimageschichte im Einzugsgebiet des Toten Meers (Naher Osten), *Ber d. Renh-Tuxen Ges.* 17, Hannover pp. 85–92 (2005).

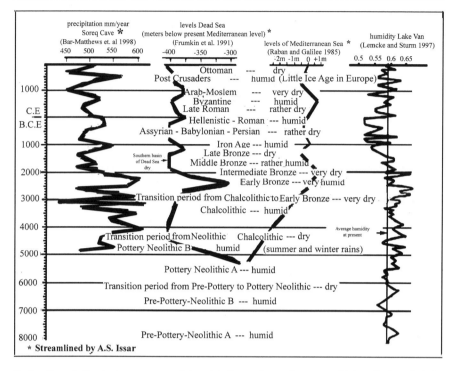

Fig. 3a. Correlation between proxy-data related to palaeo-climates and historical periods in the Near East

Although information derived from ancient lake and sea levels is of great importance in establishing paleo-climates, its value has limitations in that its markers are confined to extreme points, i.e. highest or lowest levels. Traces of intermediate levels have a high chance of being abraded by the water mounting or retreating to the extreme levels. As with the pollen time series, these data must also be compared with other time-series of proxy data.

Comprehensive on- and offshore research along Israel's Mediterranean coast showed fluctuations in sea level over the last 10,000 years that *grosso modo* confirmed the paleo-climate hypothesis, only in reverse correlation with that of the Dead Sea (Figs. 3a & 3b); the reason being that during cold periods sea levels fall as a result of water accumulating in the glaciers whereas at the same time, the level of the Dead Sea rose because of a more humid climate – and *vice versa* in warm periods. Climate fluctuations usually influence the level of the groundwater table, which rises during humid periods and falls during dry periods. This change may be compromised in the groundwater table in coastal plains by the rise and fall of the sea level to which the groundwater flows. Nevertheless, results of research carried out in ancient wells in Israel's coastal plain were in accord with Issar's hypothesis that during the cold humid parts

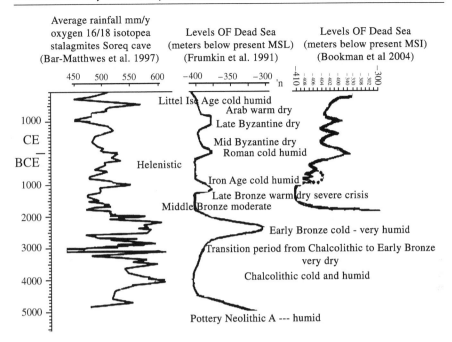

Average rainfall mm/y
oxygen 16/18 isotopea
stalagmites Soreq cave
(Bar-Matthwes et al. 1997)

Levels OF Dead Sea
(meters below present MSL)
(Frumkin et al. 1991)

Levels OF Dead Sea
(meters below present MSl)
(Bookman et al 2004)

Littel Ise Age cold humid
Arab warm dry
Late Byzantine dry
Mid Byzantine dry
Roman cold humid
Helenistic
Iron Age cold humid
Late Bronze warm dry severe crisis
Middle Bronze moderate
Early Bronze cold - very humid
Transition period from Chalcolithic to Early Bronze very dry
Chalcolithic cold and humid
Pottery Neolithic A --- humid

Fig. 3b. Correlation between proxy-data related to palaeo-climates and historical periods in the Near East

of the Roman and Byzantine periods, the groundwater table was high while it dropped during later periods.[17]

Additional support for higher humidity during the Roman period came from the ramp built by the Romans in the siege of Masada, also in the Dead Sea vicinity (Plate 5). This ancient fortress, built by king Herod around 30 B.C.E., was situated in the eastern extreme of the Judaean Desert on a cliff overlooking the Dead Sea. It was the last Jewish fortress in Judaea to resist the Roman army after Jerusalem was conquered in 70 C.E. According to Josephus, the siege lasted three years and ended with the defenders committing suicide when the Romans prepared to storm the fortress after breaking through its western wall. The Romans built a siege wall and army camps surrounding the fortress. They also built the ramp for advancing their battering ram and to allow the legionnaires to storm the walls. For this purpose they cut trees in the area,

---

[17] A. Raban and E. Galili, "Recent Maritime Archaeological Research in Israel – A Preliminary Report" *The International Journal of Nautical Archaeology and Underwater Exploration* 14/4:321–356 (1985).
E. Galili, M. Weinstein-Evron and A. Ronen, "Holocene Sealevel Changes Based on Submerged Archaeological Sites off the Northern Carmel Coast in Israel" *Quaternary Research* 29:36–42 (1988).
Y. Nir and I. Eldar, "Ancient Wells and Their Geoarchaeological Significance in Detecting Tectonics of the Israel Mediterranean Coast Line Region" *Geology* 15:3–6 (1987).
D. Sivan, K. Lambeck, R. Toueg, A. Raban, Y. Porath, B. Shirman "Ancient coastal wells of Caesarea Maritima, Israel, an indicator for relative sea level changes during the last 2000 years" *Earth and Planetary Science Letters* (2004).

placed them as groundwork and covered them with soil (Plate 6). The trees – tamarisk and acacia – are still dominant in the area. This shows that the climate was still desert then; today the average annual precipitation in this region is 50 mm. A group of botanists analyzed the various trees buried in the ramp and found a higher ratio of tamarisk to acacia than is seen in the region today. As tamarisk trees need a rather shallow groundwater table to thrive, this meant there were more floods recharging the streambeds along which these trees grow. The botanists thus concluded that the climate was more humid at the time of the siege. This conclusion was further supported by other research using the tools of environmental carbon and oxygen isotopes. Based on preliminary results of work carried out at the Institute for Desert Research at Sede Boker in collaboration with the Isotope Department of the Weizmann Institute at Rehovot and the Institute for Hydrology of Neuherberg Germany, the study aimed to use isotopes to better understand the water regime of tamarisk trees in the desert. It was part of a more general research project in which Issar participated in order to investigate the hydrology of forested sand dunes and stream terraces in the desert.[18]

We shall take this opportunity to elaborate on the principles of the method used in this research, because this method is an example of interdisciplinary

Plate 5. Masada viewed from the west and the Roman siege ramp (Photo A.S. Issar)

[18] Y. Yadin, *Masada: Herod's Fortress and the Zealot's Last Stand.* Weidenfeld and Nicolson, London (1966).

N. Liphschitz, S. Lev-Yadun and Y. Waisel, "Dendochronological Investigations in Israel (Masada)" *IEJ* 31:230–234 (1981).

D. Yakir, A. Issar, J. Gat, E. Adar, P. Trimborn and J. Lipp, "13C and 18O of Wood from the Roman Siege Rampart in Masada, Israel (AD 70-73): Evidence for a Less Arid Climate for the Region" *Geochimica et Cosmochimica Acta* 58/16: 3535–3539 (1994).

A.S. Issar and D. Yakir, "The Roman Period's Colder Climate" *Biblical Archaeologist* 60:101–106 (1997).

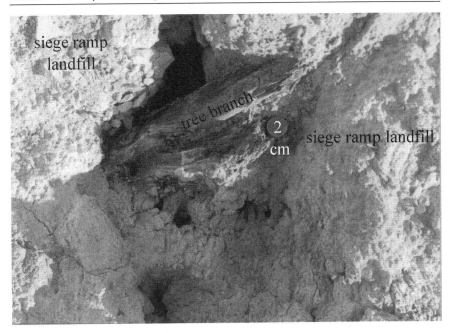

Plate 6. A stump of tamarisk tree (sampled for isotopes) protruding from the land-fill of the Roman siege ramp (Photo A.S. Issar)

endeavor in which the combined results of detailed research in archaeology, hydrology, botany, plant physiology and isotope chemistry have yielded the most interesting results. The idea to use plant remains as climatic indicators was based on observations by the above-mentioned team that the concentration of naturally occurring, non-radioactive isotopes such as ^{13}C in plant organic matter is influenced by the availability of moisture to the tree. All of the carbon in plants comes from the carbon dioxide (CO_2) in the atmosphere. The leaves diffuse the CO_2 through minute pores called stomata. Inside the leaves a certain enzyme fixes a fraction of the gas by using water molecules and energy supplied by solar radiation to produce carbohydrates (e.g. sugars, cellulose). The enzyme preferentially fixes CO_2 molecules containing ^{12}C, the lighter and dominant (99%) isotope. Thus, under normal conditions, wood is depleted in ^{13}C relative to CO_2 in the atmosphere (by a few tenths of a percent). The enzyme's preference for the lighter carbon isotopes is satisfied due to the rapid mixing with the CO_2 in the atmosphere, through the open stomata. The second task of these tiny pores is to transpire the water brought up by the roots (together with the dissolved nutrient minerals). The mixing process that maintains the enzyme's preferable ratio of the lighter carbon isotopes may change if the plant is under stress, which may occur if the rate of evaporation exceeds that of water intake by the roots, because of high temperatures or drought. Under such conditions, the tree tries to reduce transpiration by closing the stomata. As a result, the exchange of CO_2 between the leaf and the atmosphere is reduced, trapping

$^{13}CO_2$ and thus increasing its concentration in the cellulose. Thus heavier cellulose, isotopically speaking, indicates periods in which the tree was under stress, while light cellulose indicates periods of optimal water supply.

The team tested their conclusion using a series of isotopic analyses of wood from tamarix trees growing under more humid conditions in central and northern Israel and under arid conditions in the Negev and Judaean Deserts. As expected, the ^{13}C content of the wood from the desert regions was significantly higher than that of the wood from central Israel. With these results in hand, Issar initiated sampling and analysis of the wood extricated from the siege ramp at Masada. The results (Fig. 4), indeed, showed that these trees grew under more humid conditions than prevail today in this region. The isotopic ratios are rather close to that of trees growing today in the more humid northern Negev region, closer to the Mediterranean Sea which receives about twice as much annual rainfall than the Masada region.

Also important to this discussion is the dendro-chronological method for identifying the types of trees found in archaeological excavations. The remains are dated using radiometric methods and the type of wood is identified by comparing the microscopic structure of the woody material of the ancient sample with that of modern trees. Such investigations were carried out on remains of trees found in the Negev Desert. Pistachio, tamarix and olive trees which grew during the Last Glacial Period (27,000–18,000 years ago) and were cut for firewood in the Upper Palaeolithic period. Pistachio and olives grew ca. 6700 B.C.E. (PPNB) – a relatively humid period – and were used for the same purpose (See Chap. 4). Dendro-chronological studies for obtaining paleoclimate information arrive at time series by investigating the number,

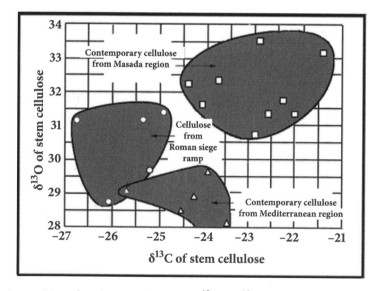

Fig. 4. Composition of environmental isotopes (^{13}C and ^{18}O) in cellulose of tamarisk trees from the Roman siege ramp at Masada compared with that of contemporary tamarisk trees

density and width of the rings, as well as their isotopic analysis for radiocarbon dating and their carbon and oxygen environmental isotopes. Relative to other proxy data time series, the dendro-chronological data apply mainly to the last millennium. Considering that the history of the Near East stretches over five millennia, not to mention the proto-history and prehistory, this tool has a limited implication.[19]

Further support for Issar's suggested conceptual model of climate change came from results presented at a NATO-sponsored international workshop held at Kemer, Turkey in 1994. It dealt with the reasons for the historical calamity affecting the Near East towards the end of the second millennium B.C.E. Along with other data, a detailed isotope curve of Lake Van, constructed by Lemcke and Sturm, and interpreted in values of humidity corresponded generally with the time series discussed above (Fig. 3a).[20]

The workshop represented a milestone in the progress towards a less anthropogenic and more deterministic paradigm. It concentrated on the end of the third millennium B.C.E. just one of the historical events in which climate change played a major role. Some senior scientists were reluctant to abandon entirely their former worldview and to adopt a more deterministic conceptual model, which was in part due to the difficulty in uncovering direct evidence for climate change in the remains of an archaeological site itself. Unlike evidence of destruction by war, which can easily be detected in a stratigraphical excavation, evidence of environmental change is found at the bottom of a lake or in the stalagmites of a cave and, often, far from the site. Moreover, these data constitute a totally different body of evidence than that which archaeologists are accustomed to work with, such as comparative studies of architecture, pottery and other artifacts. It appears that the very simplicity of the climate change explanation, compared with the complexity of the anthropogenic paradigm, instills a reluctance to accept it. Butzer, who participated in this workshop, reflects these reasons in the abstract of his article:

"The local evidence for an incisive and protracted 'dry' anomaly is inconclusive and examination of a wide range of paleoclimatic proxy records fails to support an abrupt climatic shift to greater aridity affecting the Near East and the Aegean world between 2400 and 1900 B.C. Settlement records to the east, west and northwest of Leilan are inconsistent; in two cases the settlement network expanded rapidly after the Leilan collapse, in another, sites were abandoned at about the same time. This is incompati-

---

[19] N. Liphschitz, and Y. Waisel, "Dendrochronological Investigations in Israel: Central Negev-Nahal Zin", Appendix in: *Prehistory and Palaeoenvironments in the Central Negev*, E.A. Marx (ed.) Southern Methodist University, Dallas. pp. 355–356 (1977).
More information about dendrochronological methods can be found in:
R.S. Bradley, "Palaeoclimatology. Reconsructing Climates of the Quaternary", *Academic ress – International Geophysics Series*, V. 64:397–436 (1999).
W. Bircher, "Dendrochronology Applied in Mountain Regions" in: *Handbook of Holocene and Palaeohydrology*. B.E. Berglund (ed.). John Wiley & Sons, Ltd., Chichester, pp. 387–403 (1986).
[20] H.N. Dalfes et al., op. cit., pp. 654–678.

ble with a major climatic shift and demands exploration of more complex hypotheses to explain sociopolitical devolution."

This statement is in contrast to another of Butzer's conclusions in the same article, as follows:

"Many, if not most urban sites from the Balkans to Mesopotamia and Palestine were abandoned, destroyed or much reduced in size. Early states such as the Akkadian Empire and Old Kingdom Egypt collapsed around 2230 B.C. Troy II was destroyed and the Indus Valley Civilization came to an end."[21]

We ask how it is possible that the same sociopolitical devolution befell all these regions at the same time, when logic prohibits the assumption that the same sociopolitical developments occurred simultaneously all over the world?

Thus, the present authors raise the question: Even with all complexities involved in socioeconomic systems, is it not the role of the scientist to find the *causa sine qua non*, which may have initiated the documented collapse in the region? In other words, does an a priori decision that the model must be complex – because it is only *"one potential variable within a multi-dimensional, interactive system"*[22] impede the correct conclusion, when climate change happens to be the paramount variable? This reluctance to adopt "simplistic" explanations was observed also by Moore in his discussion of the current general trend in archaeology to explain the origins of agriculture. After describing his own rather neo-deterministic model to explain this process he describes the conventional current trends as *"growing awareness of the complexity of the whole process"* and that *"archaeologists no longer accept as sufficient those explanations that rest on one or two prime movers."* At the same time, *"archaeologists seem reluctant to offer new, more satisfying explanations"*.[23]

It is indubitable that climate change may have a variety of effects at various sites and on different communities. For example, communities situated near a perennial water source and fed by a regional aquifer would less likely be affected by a climate change, and if affected, they would more readily be resettled than others less fortunate. Moreover, factors related to other sources of income besides agriculture, such as location on trade routes, mineral resources or technology, and so forth, may have significantly complicated the picture, concealing the primary mover. Yet it is argued that its exposure was made more difficult, if not prohibited, by the argument that climate change is too simple a hypothesis to explain the variety of observations of archaeologists and historians. The same is true regarding resettlement, particularly concerning the question of why certain sites were resettled, and even flourished, while

---

[21] K.W. Butzer, "Sociopolitical Discontinuity in the Near East C. 2200 B.C.E. – Scenarios from Palestine and Egypt" in Dalfes et al. op. cit. p. 245 (1994).

[22] Butzer op. cit., p. 287 (1994).

[23] A.M.T. Moore, "The Development of Neolithic Societies in the Near East" in: *Advances in World Archaeology*, vol. 4, F. Wendorf and A.E. Close (eds.), Academic Press, London pp. 1–70 (1985).

others remained in ruins. Examination of the paleo-climate time series on the one hand and the physical conditions in typical sites on the other may explain such cases. It will facilitate, for example, understanding of the renaissance of the urban centers, which characterizes the Middle Bronze Age in the second millennium B.C.E.

In 1968 a cave was discovered after detonating a wall in a limestone quarry in the cliffs overlooking Nahal Soreq west of Jerusalem. It is a large hall formed by dissolution of the rock by water percolation (a karst cave) and contains various types of stalactites, stalagmites, and curtains of flowstones, still being formed by the constant dripping of water from the roof of the cave. On the recommendations of the geological team, the area was proclaimed a protected natural reserve named Soreq Cave. The cave is situated on the western slope of the Judaean anticlinorium, about 40 km inland of the Mediterranean Sea, in an area receiving about 500 mm of rain per year. It is about 400 m above sea level. The rocks that build this region are crystalline dolomitic limestones (low magnesium calcites) of Cenomanian – Cretaceous age. The rocks were dissolved during the primary phase of the cave formation, which may have already occurred during the Neogene. During this period the cave was beneath the groundwater table with water flowing through it directly to the nearby sea. At the end of the Neogene the area was lifted up, becoming part of the non-saturated zone and speleothems began to form. The formation started by the dissolution of the rock by water, which has become slightly acidic from dissolving atmospheric and soil CO_2. As soon as the subsurface water start dripping into the cave, and because the pressure of this gas in the cave is lower, the CO_2 diffuses and the dissolved carbonates precipitates.

Because the stalagmites were still being formed by the dripping water it was possible to analyze the chemical and isotopic composition of the rainwater and compare it with that of the cave water and precipitated carbonates. A joint team of geologists and hydro-chemists headed by M. Bar-Mathews undertook a comprehensive study of the cave. A reverse correlation was found between the intensity of the rain, the temperature and the ^{18}O, and deuterium composition of the cave dripping water. This correlation and the isotopic composition of the rainwater showed that the water represented the average oxygen and hydrogen isotopic ratios characterizing the Mediterranean meteoric-line. Petrographic examination of the stalagmite crystals showed that they did not go through a process of dissolution and re-crystallization which lead to the conclusion that the stalagmites reflect the average oxygen isotopic composition of the groundwater and thus reflect the climatic conditions. The result was an intensive and very accurate analysis of the succeeding rings of a group of 20 stalagmites. The samples were dated by the ^{230}thorium-uranium method, which allows calculation of the dates with an uncertainty range of not more than a few decades. The residence time (retardation factor) of the water in the subsurface was found to be not more than a few decades. The $^{18}O/^{16}O$ and $^{13}C/^{12}C$ isotopic ratios were analyzed very precisely, allowed to draw conclusions about climate changes in this region over the last 60,000 years and particularly to calculate the annual average precipitation for the different periods within the last 7000 years, by correlating isotopes of the dated sections of the stalagmites to the ratios in contemporaneous rainwater.[24] The present authors adopted this part of the curve as the type-section for the climate changes, which had taken place during the Holocene. The term type-section is, in fact, a geological term denoting a profile, based either on an outcrop or a borehole, that contains enough stratigraphical, lithological and paleontological information to be used as a key reference for other outcrops or boreholes. This curve will often be referred to in the following chapters, together with other time-series of proxy data. The second stage will be to draw conclusions with regard to the paleo-environment. After the ancient environment has been reconstructed, we will investigate the historical scenario and correlate between the two, doing our best to avoid "circular" argumentation in historical scenarios.

A similar record to that of the Soreq Cave, but for a longer period, starting ca. 170K B.P. namely the glacial period of Marine Isotopic Stage 6 (penultimate glaciation), was obtained from a stalagmite from a cave east of Jerusalem. As in the Soreq Cave the precipitation during the glacial periods is heavier in its oxygen 18 compositon than the interglacial period, although the source of the precipitation was the westerlies system coming over the eastern part of the Mediterranean Sea.[25]

The results conformed in a near perfect manner the other time-series discussed in this chapter (Figs. 3, 3a & 3b) and with the major historical events interrelated with climatic changes and their impact on the hydrogeology of the region and in some instances with Biblical events related to occurrence of major droughts.[26] Agreement was also found with the archaeological investigations of Zohar who had reached similar conclusions with regard to the reasons for the so-called Dark Ages – non-urban intermediate phases between periods of

---

[24] A. Kaufman, G.J. Wassetburg, D. Porcelli, M. Bar-Matthews, A. Ayalon, and L. Halicz, "U-Th Isotope Systematics from the Soreq cave, Israel and Climatic Correlations" *Earth Planet Science Letters* 156:141–155 (1998).

A. Ayalon, M. Bar-Matthews, and E. Sass, "Rainfall-recharge Relationships Within a Karstic Terrain in the Eastern Mediterranean Semi-arid Region, Israel: 18O and D characteristics" *Journal of Hydrology* 207:18–31 (1998).

M. Bar Matthews, A. Matthews and A. Ayalon, "Environmental Controls of Speleothems Mineralogy in a Karstic Dolomitic Terrain (Soreq Cave, Israel)" *Journal of Geology* 99:189–207 (1991).

M. Bar-Matthews, A. Ayalon, A. Matthews, L. Halicz and E. Sass, "The Soreq Cave Speleothems as Indicators of Palaeoclimate Variations." *GSI Current Research* 8:1–3 (1993).

M. Bar-Matthews, A. Ayalon, A. Matthews, E. Sass and L. Halicz, "Carbon and Oxygen Isotope Study of the Active Water-carbonate System in a Karstic Mediterranean Cave: Implications for Palaeoclimate Research in Semiarid Regions" *Geochimica et Cosmochimica Acta* 60:337–347 (1996).

M. Bar-Matthews, A. Ayalon, A. Kaufman and G.J. Wassetburg, "The Eastern Mediterranean Palaeo-climate as a Reflection of Regional Events: Soreq Cave, Israel" *Earth and Planetary Science Letters* 166:85–95 (1998).

M. Bar-Matthews, A. Ayalon and A. Kaufman, "Middle to Late Holocene (6500-Year Period) – Palaeo-climate in the Eastern Mediterranean Region from Stable Isotopic Composition of Speleothems from Soreq Cave, Israel" in Issar and Brown op. cit. pp. 203–214 (1998).

[25] A. Frumkin, D.C. Ford and H. Schwartz, "Continental Oxygen Isotopic Record of the Last 170,000 years in Jerusalem" *Quaternary Research* 51:317–327 (1999).

[26] A.S. Issar *Water Shall Flow from the Rock – Hydrogeology and Climate in the Lands of the Bible.* Springer Verlag, Heidelberg (1990).

A.S. Issar, "Climate Change and History during the Holocene in the Eastern Mediterranean Region" in Issar and Brown op. cit. pp. 113–128 (1990).

"advanced," or urban, civilizations, which recurred with a certain periodicity during the pre- and proto-history of the ancient Near East. These were periods of fundamental ecological, religious and social changes, often accompanied by migrations and changes in the ethnic composition of the population in the area.[27]

He established that although each intermediate age had its own characteristics and course of history, all of them nevertheless shared some important cultural elements in common. The growing importance of pastoral nomad societies, similarities in settlement patterns, subsistence economy, architecture, funerary customs and even details in pottery seemed to re-surface during these obscure periods, but were absent in the intervening urban phases. It became clear that the periodic collapses of certain urban cultures – such as the Early Bronze III Age in Palestine, the disintegration of the Sumerian city-state system in Mesopotamia or the decline of the Old Kingdom in Egypt during the second half of the third millennium B.C.E. – had the same primary cause, i.e. unlimited population growth exceeding the carrying capacity of a given area, and therefore, environmental overexploitation as the result of climate changes, leading to collapse during severe and prolonged dry periods. It must be remembered that the desiccation extending over centuries was the primary cause of collapse, subsequently aggravated by human degradation of the natural environment; not the reverse. A similar picture appears towards the end of the second millennium B.C.E., with tremendous cultural changes in all areas of the Near East and practically all over the northern hemisphere.

The rhythmic regularity of collapse continued in later historical periods, then, however, with the rise of each new civilization, the influence of the urban cultures and their improved technologies expanded and was able to more deeply penetrate the peasant-pastoral lifestyle and to mitigate the adverse effects of droughts. Consequently, the original pristine character of the earlier "Dark Ages" was lost, but without entirely ceasing to exist as a general historical phenomenon to which we shall return later.

Since Edward Gibbon's monumental *"The Decline and fall of the Roman Empire"* in 1765, historians and archaeologists alike have been fascinated by the demise of civilizations. Libraries became filled with books dealing with the question of how only ruins remained of once-great cities. It should be stated at the outset that, despite certain similarities and outward appearances, no collapse is like any other; history does not repeat itself, even if might sometimes seem so. Most recent years have seen a growing consensus among scholars that climatic and environmental changes were often the root of a chain reaction leading to economic recession, resource depletion, overpopulation, conflict and invasions and were, therefore, ultimately responsible for the decline and collapse of the ancient Near Eastern civilizations.[28]

---

[27] M. Zohar, *Early Transition Periods in the Archaeology of Syria-Palestine*, Ph.D. thesis Hebrew University, Jerusalem (1994).

[28] A. Tainter, "States: Theories of the Collapse of States" in: *The Oxford Companion to Archaeology*. B.F. Fagan (ed.) Oxford University Press Oxford, UK pp. 688–690 (1996).

In contrast, fewer scholars bothered to scrutinize the conditions leading to the genesis and flowering of these very same civilizations. The decline of a culture can only be understood when the reasons and the circumstances of its birth are understood. Needless to say, climate, geology and other material and physical conditions were often of no great concern to most archaeologists. The humanistic approach prevailing in Old World archaeology appears to have conditioned students in these fields to accept as fact that the human race is born with an innate desire for loftier civilization achievements, a rather bizarre notion in our age. This approach spawned even quainter ideas such as 'diffusionism', i.e. the notion that all higher civilizations had but one hearth of origin and carried by superior beings all over the world. When these and similar preconceptions were overcome, the next pitfall became, comparable to all fields of knowledge, the large amount of knowledge accumulated over the last century led to specialization to such an extent that most active archaeologists – with some outstanding exceptions – prefer to stay "within their square".[29]

The most surprising result for the researcher was on the simple human plane: The skeletons found in excavations dated to the Dark Ages looked healthier and better fed, with less evidence of disease than those of the civilized phases.[30] The so-called Dark Ages began to look anything but bleak or "dark" for the people of their time. As archaeologists, we deplore the absence of monumental temples or palaces, of sumptuous tombs filled with fanciful art objects or historical records, yet we must appreciate these ancient peoples who, despite often severe environmental restrictions, managed to develop their specific cultures, probably without kings or aristocrats, priests or bureaucrats.

Another example of reaching opposite conclusions from the same evidence can be seen in mapping the settlement pattern. For archaeology students, the position of many large tells in the Shephela of Palestine, such as Lachish, Gezer, Tell Aphek/Antipatris, Megiddo, Hazor, etc., in a line running from southwest to north-northeast is attributed overwhelmingly to the famous *Via Maris* – an international trunk road system connecting Egypt with Syria-Mesopotamia. To a certain extent this is undoubtedly true, as trade usually brings prosperity. Of equal or even greater importance, however, is the fact that each of these cities

---

N. Yoffee and A. Cowgill (eds.), *The Collapse of Ancient States and Civilizations*. University of Arizona, Tucson (1987)

[29] S. Richard, "The Early Bronze Age: The Rise and Collapse of Urbanism" *Biblical Archaeologist* 50:22–43 (1978).

W.G. Dever, "From the End of the Early Bronze Age to the Beginning of the Middle Bronze Age" In: *Biblical Archaeology Today, Proceedings of the International Conference on Biblical Archaeology, Jerusalem, April 1984*, J. Amitai (ed.), The Israel Academy of Science and Humanities, Jerusalem, pp. 13–135 (1985).

B. Fagan, Floods, *Famines and Emperors – El Niño and the Fate of Civilisations*. Basic Books, New York (1999).

[30] P. Smith, The Physical Characteristics and Biological Affinities of the MBI Skeletal Remains from Jebel Qa'aqir. *BASOR* 245:65–73 (1982).

Ibid. The Skeletal Biology and Palaeopathology of Early Bronze Age Populations in the Levant. in: *L'Urbanisation en Palestine à l'âge du Bronze Ancien. Bilan et Perspectives des Recherches Actuelles, Actes du Colloque d'Emmaus 20–24 Octobre 1986*. ed. Miroschedji. (London, BAR International Series 527 (ii), 297–313 (1989).

had a secure and abundant water supply and arable hinterland, which enabled their establishment, and allowed them to function and flourish throughout history.

We think that applying Occam's Razor, in this case down-to-earth natural environmental factors such as climate conditions are as important for understanding the history of civilizations as complex socio-economic deliberations which must be viewed in proper proportion.

An understanding of the impact of the natural environment on the history of human society, and *vice versa*, requires an interdisciplinary and holistic approach. The team of investigators must include specialists in the archaeological, historical and anthropological sciences, as well as botanists, hydrologists, isotope-chemists, etc. Once this approach is adopted, there is a good chance that the anthropogenic paradigm will become more balanced, reducing the burden of blame on human society.

CHAPTER 3

The Near East: A Bridge from the Garden of Eden to the Fields of Toil

From the Palaeolithic to the end of the Pre-Pottery Neolithic Period (From ca. 2 million to ca. 8000 years ago)

"And unto Adam he said, Because thou hast hearkened unto the voice of thy wife, and hast eaten of the tree, of which I commanded thou saying, Thou shalt not eat of it; cursed is the ground for thy sake; in sorrow shalt thou eat of it all the days of thy life... Therefore the Lord God sent him forth from the garden of Eden, to till the ground from whence he was taken."

(Genesis, 4; 17,10,23)

3.1
The Password 'Climate Change' (Table 2, Fig. 5)

Since the very beginning of the human race, the Near East served as a bridge between Africa – where our species first evolved – and the rest of the world. The passage over this bridge opened and closed with the global fluctuations of climate. The first glacial periods at the beginning of the Quaternary caused the greenhouse of equatorial Africa to become less hospitable, while making the desert belt of the Near East more humid, green, and thus passable. Flint tools found along the shores of dried up lakes and swamps in the Negev Desert provide evidence that members of the first wave, *Homo erectus*, as well as the second wave, *Homo sapiens*, camped there *en route* to all the other continents.

Before and during the Last Glacial Period this region experienced a series of cold and mainly humid phases when many lakes appeared spreading out over wide areas now desert. For example, in the Jordan-Dead Sea Rift valley existed, prior to that of Lake Lisan, the precursor of the Dead Sea, one stage which deposited the Samra Formation, the type locality described by Picard north of Jericho, where it overlies unconformably the deposits of Lisan Lake. The Samra Formation was dated to have started before 350K and extended to about 63K years B.P.[1] The Lisan Lake, which will be discussed in more detail

---

[1] L. Picard, "Structure and Evolution of Palestine with comparative notes on neighbouring countries" Hebrew University, *Geology Department Bulletin* 4 (1943).
A. Kaufman, Y. Yechieli and M. Gardosh, "Reevaluation of the lake-sediment chronology in the Dead Sea basin, Israel based on new ^{230}Th/U dates" *Quaternary Research* 39:292–304 (1992).

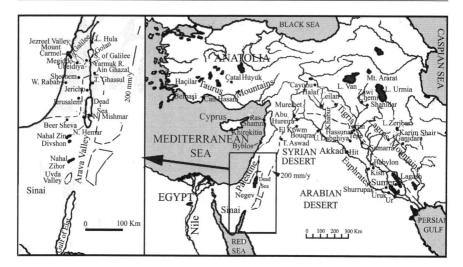

Fig.5. Map of epi-Palaeolithic, Neolithic and Chalcolithic archaeological sites

later on, extended during the Last Glacial Period from the northern part of the Arava Valley, south to the Dead Sea, to the Sea of Galilee and even beyond into the upper Jordan gorge.[2]

In Jordan deposits of Quaternary lakes and swamps were observed and mapped in the El Jafr depression and Faynan area in southern Jordan, in the El Azraq basin in eastern Jordan as well as in the Wadi Hisma, along the southern edge of the Jordan Plateau. Deposits of an Upper Quaternary lake were found also in Wadi Hasa the principal drainage of west Central Jordan.[3]

---

[2] J.F. Shrer, Jr. M.P. Bishop, K.J. Cornwell, M. Inbar, "Catastrophic Geomorphic Processes and Betsaida Archaeology, Israel" In: *Bethsaida – A City by the North Shore of the Sea of Galilee*, eds. R. Arav and R.A. Freund, Vol II Bethsaida Excavatins project, Truman State University Press (1999).

[3] R. Huckriede and G. Wieseman, "Der jungpleistozäne Pluvial-See von El Jafr und weitere Daten zum Quarter Jordaniens" *Geologie Palaeontologie*, Marburg, Vol. 2:73–95 (1968).

A.N. Garrard, S. Colledge, C. Hunt and R. Montague "Environment and Subsistence during the Late Pleistocene and Early Holocene in the Azraq Basin" *Paléorient* 14/2:40–49 (1988).

P. Sanlaville, "Considérations sur l'environnement dans le sud du Levant durant les stades iso-topiques 5 et 4" *Paléorient*, 28/1 (2002). The authors disagree with Sanlaville's conclusion that glacial stage 4 (75–55K B.P.) was arid. Like the other cold glacial stages it was also humid. Indeed Sanlaville notes that "the aridification was characterised by irregular intense rainfall . . . that brought out the development of alluvial terraces in valleys". To the author's opinion this in desert regions is characteristic of a humid climate and Sanlaville's conclusion is an example of misinterpretation of data in order to fit it into the paradigm of "cold climate equates dryness and vice versa". This wrong, yet deep rooted paradigm, will be discussed later on.

D.O. Henry, "The Middle Paleolithic of Jordan" in: *The Prehistoric Archaeology of Jordan*, D.O. Henry (ed.), BAR, International Series, Oxford, UK, no. 705:23–38 (1998).

J. Schuldenrein and G.A. Clark "Prehistoric Landscapes and Settlement Geography Along the Wadi Hasa, West-Central Jordan. Part II: Towards a Model of Palaeoecological Settlement for the Wadi Hasa" *Environmental Archaeology* Vol. 8/1 (2003).

E.R. Winer, J.A. Rech, N.R. Cinman, "Late Quaternary wetland deposits in Wadi Hasa Jordan and their implications for paleoenvironmental reconstruction" *Geological Society of America, Abstracts of 2006 Philadelphia Annual Meeting* Vol. 38/7:314 (2006).

Table 2. A historical-archaeological timetable of the Near East, from 10,000 to 3000 B.C.E.

years B.C.E.	Egypt	Syria-Palestine	Mesopotamia	Anatolia
3000	Archaic	E.B. I (Transitional)	Jemdet Nasr	Early Bronze IA
4000	Gerzean (Naqada II & III)	Mature Chalcolithic (Ghassulian/Beer Sheva)	Gawra I–III (N)/ Early Late Uruk (S)	Late Chalcolithic
	Amratian (Naqada I)	Early Chalcolithic	Ubaid III & IV (South & North)	Middle Chalcolithic
5000	Badarian	Late Pottery Neolithic (Wadi Rabah)	Halaf (N) Ubaid 0, I, II (S)	Early Chalcolithic
6000	Neolithic: Fayum A Merimde	Munhatta Early Pottery Neolithic (Yarmukian)	Samarra Hassuna	Late Ceramic Neolithic
	Beni Salame	Pre-Pottery Neolithic B	Proto-Hassuna	Early Ceramic Neolithic
7000	?		Pre-Pottery Neolithic B	
8000	Fayum B/ Qarunian?	Pre-Pottery Neolithic A	Pre-Pottery Neolithic A	Aceramic Neolithic
9000	?	Natufian	Natufian	Hallan Çemi Phase
10,000	Epi Palaeolithic	Geometric Kebaran A and B/Harifian		Epi-Palaeolithic

As the glaciers in the higher latitudes melted with the warming climate, precipitation declined, causing the lakes and springs in the lower latitudes to dry up. As a consequence, vegetation and animal and human life retreated to refuge areas along perennial rivers and spring outlets emerging from water-bearing rocks fed by rains falling on more humid terrain. Sedentary epi-Palaeolithic settlements bear witness to mankind's first steps to live year-round in one place. The gradual desiccation of the Near East during this period likely ushered in the so-called agricultural revolution of the Neolithic period, with the first attempts at artificially growing crops outside their natural habitat, and also at domesticating animals.

Yet, artificial seeding and planting, and even animal domestication, could not compensate for the reduced supply of natural food and commodities. The warmer and drier the climate became, the more random was the availability of water and the greater the fluctuations in the food supply. To overcome this instability, agricultural settlements were established near perennial sources of water. With the ability to water the land, and thus modify the natural environment to suit its needs, human society was freed of the constraints imposed by the warmer climate.

G. Barker, O.H. Creighton, D. Gilbertson., C. Hunt, S. McLaren., D. Mattingly, D.C. Thomas, "The Wadi Faynan Project, Southern Jordan: a preliminary report on geomorphology and landscape archaeology" *Levant* 29:19–4 (1997).

3.1.1
Humble Beginnings

All scientific evidence about the origin of man points unequivocally to Africa as the home of the hominids that were to become the ancestors of man[4] and some of our closest relatives still survive on this continent. The details of the evolution of groups of primates into hominids and their descendants' diffusion to other continents are clearly beyond the scope of this work. It is, however, important for us to emphasize the fact that this entire process was interwoven with profound tectonic and climatic changes which repeatedly affected the environment. Plate tectonics in eastern Africa some ten million years ago or more created a developing rift between the African and the Arabian shield, the Syro-African Rift Valley accompanied with the uplift and aridification of eastern Africa starting around eight million years ago.[5] Early primates living in the equatorial zone on both sides of the valley probably had already used stones and sticks as tools and had, for a relatively long time, developed a basic intelligence. Some six to seven million years ago, the incessantly changing climates and environments forced several species of forest dwelling hominids to adapt to live in a very wide range of surroundings which included open grasslands, in particular in the area to the east of the slowly forming Rift Valley. Research on ancient variations of the lakes which had formed on the bottom of the Rift Valley showed alterations between extreme high and low water levels reflecting sharp climatic fluctuations ca. 3 million years ago. Unfortunately, the strong erosion destroyed a large amount of deposits at a crucial time when speciation of the very early hominids began and the brain of some began to show signs of growth and complexity.[6]

Some species of hominids had developed for a long time a taste for meat, first by scavenging and then by active hunting. After some more millions of years the brains of some groups grew further, whether gradually or by leaps and bounds is disputed. Safety from predators was found in numbers and the groups' sizes began to grow, favoring social intercourse and the development of verbal communication[7] starting a new branch on the tree, or perhaps better bush of life. The fossilized bones of several species of these early hominids have been found all over southern and eastern Africa.

Some 2.5 million years ago, their descendents whom, for the time being, we call *Homo habilis* and *ergaster* started to make tools from pebbles of hard but

---

[4] R.E. Leaky, R. Lewin, *Origins*, Dutton, New York (1977).

[5] P. Sepulchre, G. Ramstein, F. Fluteau, M. Schuster, J.-J. Tiercelin, and M. Brunet, "Tectonic Uplift and Eastern Africa Aridification" *Science* 313:1419–1423 (2006).

[6] M.H. Trauth, M.A. Maslin, A. Deino and M.R. Strecker, "Late Cenozoic Moisture History of East Africa" *Science* 309/5743:2051–2053 (2005).
M. Maslin, M. Trauth, and B. Christensen "A Changing Climate for Human Evolution" *Geotimes* September (2005).

[7] A.S. Issar with R.G. Colodny in their book *From Primeval Chaos to Infinite Intelligence* (Aldershot, UK, Avebury, 1995) investigated the evolution of intelligence from the most primitive to the most evolved forms of life. A philosophical conceptual model is suggested which sees an ever-ascending evolution of complexity and thus intelligence along the dimensions of Space-Time-Information. Adding an explanation for the process of emerging evolution to the Darwinian model of adaptation.

brittle rocks, most probably for the primary purpose of crushing and splitting the bones of dead animals. Initially, they simply smashed the rocks and selected suitable pieces. Later hominids produced tools not by simply smashing rocks, but by deliberately hammering the edge of a stone, creating a sharp, sinusoidal cutting edge. After tool making, discovering how to harness fire was the next most important step in human development, and one that paved the way for all later technological advances. The ability to create fire at will allowed the descendants of a tropical animal to warm themselves in harsher climates, offer protection against predators and improve their nutrition. Food prepared by broiling or cooking allows a higher energy output which is crucial for brain growth.

The circumstances of the success, spread and dispersion of the members of this new primate species beyond southern and eastern Africa over an ever far-ranging territory are not known. A combination of intelligence and social structure probably enabled them to seek refuge in new places when the environmental conditions in their native habitat deteriorated. There is little doubt that they were simply following the footsteps of the herds of herbivorous animals and their predators, gradually entering northern Africa. The huge landmass of the northern part of the African continent has seen many dramatic climate changes, with alternating intervals of extreme desiccation and of humidity, during which plants, animals and humans colonized these areas. Whereas during the humid phases the plants, animals, and humans colonized these areas, in dry periods the vegetation decreased or disappeared forcing the animals and the humans out of the region. While most animals and humans flooded back south, others were pushed further north and finally left the continent over land bridges.[8] The Near East was then, and still is the most accessible land bridge between Africa and the rest of the world. There were probably other, less important and intermittent, escape routes, one of which was today's Straits of Gibraltar leading to the Iberian Peninsula. There is little doubt that around two million years ago or even earlier a trickle of small groups or families of ancient toolmakers, scavengers and gatherers began to wander into Asia and then Europe, constituting the first exodus from Africa.

One can assume that the Near Eastern bridge, which enabled migration to the neighboring continents, conditioned the evolution that followed. At present the border zone between the Near East and Africa is desert; however, it is hard to imagine that the early *Homo habilis* or the slightly more evolved *Homo ergaster* or even *Homo erectus* with their still rudimentary intelligence, simple social organization, and equipped with primitive stone tools could traverse this barrier unless the climate was more favorable and the desert more hospitable than now. They could have migrated northward along the Nile Valley or along the coast of Egypt and from there to the Sinai and Arabian Peninsula and the Negev Desert, all of which appeared very different at various times. At that time the topography of the Near East in general and the Negev and Sinai in

---

[8] M. Schuster, P. Duringer, J-F. Ghienne, P. Vignaud, H. Mackaye, A. Likius and M. Brunet. "The Age of the Sahara Desert" *Science* 311:821 (2006).

particular was quite unlike that of today. The Syrian-African rift system is now a dramatic cleft in the earth's crust, with the world's lowest point on land at the Dead Sea, some 450 m below sea level. The rift stretches southward from the mountains of the Anti-Taurus in eastern Anatolia, first as the Valley of the Orontes, then as the Lebanese Beq'a, the Jordan Valley, the Dead Sea, the Arava Valley, the Red Sea, and eventually reaching deep into eastern and southern Africa. When the first humans probably arrived more than two million years ago, the rift was not yet fully developed and was composed of a series of basins reaching from eastern Africa to the foothills of the Anti-Taurus.[9] For example, part of the drainage of the Negev highlands flowed not to the developing Rift Valley but through a series of intermountain valleys to the Mediterranean Sea.[10] Sediments found in the basins of the Rift Valley were once deposited in freshwater lakes of Lower Quaternary age, indicating that the climate was more humid. This meant that more routes were passable, especially during glacial periods when the sea retreated and the coastal areas widened. These areas drained the surface and subsurface flow of floodwater coming from the highlands, which got more precipitation. These paleo-topographical and climate conditions favored the existence of a shallow groundwater table in the coastal regions and intermountain basins. As a consequence, the deserts, and especially their coastal areas, were dotted with swamps of brackish and fresh water, supporting sub-tropical vegetation. Such swamps are found in these regions today at oases along the Rift Valley and at the proximity of the sea. Thus, theoretically, humans could have traveled quite easily over territories that are now deserts.

Investigations in the Negev Desert – an important steppingstone in this bridge – show that this was, in fact, the case. H. Ginat found deposits from one of these ancient lakes in the riverbed of Nahal Zihor in the Central Negev. (Plate 7) Field evidence and hydrological calculations showed that during the lake's existence, precipitation ranged from 200 to 400 mm per annum, whereas today it is about 50 mm. About half of the lake's water came from floods, while the other half was replenished from the high groundwater table in the gravel underlying the riverbeds and from water-bearing limestone layers in the bedrock below the western end of the lake. The relative light isotopic composition of the freshwater chalks deposited in the center of the ancient lake suggests that most of its water was fresh while sediments of the water close to the shore have a heavier isotopic composition indicating evaporation and brackishness. The relatively heavy carbon isotopic composition of the sediments and the pollen assemblage in the sediments show that the local vegetation was composed mainly of grasses, scrubs and some trees showing that at that time the Negev was a semi-arid steppe dotted with oases around lakes and springs. The spring travertines found at the vicinity of the lakes as well as along the fault lines bordering the Arava valley tell us also about

---

[9] M. Petraglia, "Early prehistory in the Farasa Island and the Southern Red Sea" *Abstracts – Seminar for Arabian Sudies* (2006).

[10] Z. Garfunkel and A. Horowitz, "The Upper Tertiary and Quaternary Morphology of the Negev" *IJES* 5:101–117 (1966).

palaeo - lacustrine layers

recent riverbed

Plate 7. The palaeo-lake deposits in Nahal-Zihor Negev (Courtesy H. Ginat)

a more humid climate. The tools found along the shores of the ancient lakes are of Lower Palaeolithic age and are at least 750K years old. The dating of the travertines of the Arava by U-series ranges between around 300K to around 110K Years B.P. The Sr isotopeic ratios of the lacussstrine limestone and that of the travertines in the southern part of the Arava, is similar, which points to a common source of precipitation.[11] Our ancestral emigrants could find food and water at these sites where the concentrations of flint tools found along the shores of these lakes attest to their encampment in the region. The tools are of the same character, albeit slightly more evolved, as those found in the Olduvai Gorge in Tanzania and resmble those found at Ubaiduya, which will be discussed later on.

Issar and colleagues found a similar assemblage of tools further north in the Negev highlands. At present the area is a desert playa with an average annual precipitation of about 80 mm. The tools were found in layers of gravel inter-fingering with reddish clays and lacustrine silt layers, which indicate a shallow lake or marsh. When the ancient people camped along its shores, the marsh was part of a river system draining the Negev highlands to the Mediterranean Sea. At present the upper part of this system is captured by the Dead Sea drainage system.

---

[11] H. Ginat, "Palaeogeography and Landscape Evolution of the Nahal Hiyyon and Nahal Zihor Basins" *GSI report* 19 (Hebrew with an English abstract) (1997).

Ginat, E. Zilberman and I. Saragusti, "Early Pleistocene Freshwater Lake Deposits in Nahal Zihor, Southern Negev, Israel" *Quaternary Research* v. 59:445–458 (2003).

A. Livnat and J. Kronfeld, "Paleoclimatic implication of U-series dates for lake sediments and travertines in the Arava rift valley Israel" *Quaternary Research* 24:164–172 (1985).

L. Enmar, "The travertines in the Northern and Central Arava" Stratigraphy Petrography and Geochemistry" *GSI/1/99* (1999) (Hebrew with English Abstract).

The site richest Lower Palaeolithic tools in Cis-Jordan is Tell 'Ubaidiya, located in the northern Jordan Valley a few miles south of the Sea of Galilee. The age of the layers in which the tools were found – estimated on the basis of geological and faunal correlations – is Lower Pleistocene about 1.4 million years ago. The lithology and layers suggest a marsh or shallow lake and the abundance of fauna with northern affinities indicates a cold climate.[12]

An assemblage of Middle to Lower Middle Palaeolithic arterfacts was found on the other side of the Rift Valley in the Wadi Faynan area, southern Jordan. In this alluvial basin the deposition of a complex assemblage of fluvial, slope and alluvial sediments was investigated. The unit in which the Palaeolithic tools were found (Dahalat Member) is calcreted at its base, the same as that observed at Sede Boker. To the opinion of the present authors this member was also deposited during a colder and more humid phase, namely environmental conditions which enabled hominids to live in an area, which today is arid.[13]

There are not enough data to precisely date the beginning of the cold period in the Near East that enabled hominids to start leaving Africa. A site containing very primitive tools of Oldowayan type was found in the Galilee, covered by a basalt flow dated to around 2.4 million years.[14] Another early find spot was 'Erq el-Ahmar in the northern part of the Jordan Valley which may point to the probability that hominids left Africa already at the *Homo ergaster* or very early *erectus* stage, a contingency now confirmed by the finds of Dmanisi in Georgia. Here the oldest actual skeletal remains of several *homo ergaster* with very close relationship to their gracile African contemporaries was found in close proximity with Oldowayan tools just above and beneath a lava flow dated to 1.75 million years ago (Lordkapanitze 2005, pers. com.). Later finds of a heavy jaw confused the picture: Was this evidence of the variability within the genus of *homo* or another newer type arriving from Africa? The finds of the last seasons will throw more light on the problem.[15]

Issar's research in the 1950s on the geology of Israel's coastal plain, together with deep drillings, shows that two major sea regressions occurred during the Lower Pleistocene. At the time these investigations were carried out it was not possible to say whether these regressions were caused by tectonic lifting of the

---

[12] O. Bar-Yosef and N. Goren-Inbar "The Lithic Assemblages of 'Ubeidiya A Lower Palaeolithic Site in the Jordan Valley" in: HUJI – Institiute of Archaeology *Qedem* 34 (1993).
    O. Bar-Yosef, "Pleistocene connection between Africa and Southwest Asia an archaeological perspective" *The African Archaeological Review* 5:29–38 (1987).
    E. Tchernov, "New Comments on the Biostratigraphy of the Middle and Upper Pleistocene of the Southern Levant" in: *Late Quaternary Chronology and Palaeoclimates of the Eastern Mediterranean*, O. Bar-Yosef and R.S. Kra (eds.) *Radiocarbon*, University of Arizona, Tucson (1994).

[13] A. Issar, A. Karnieli, H.J. Bruins and I. Gilead, "The Quaternary Geology and Hydrology of Sede Zin, Negev, Israel" *IJES* 33:34–42 (1984).
    S. J McLaren, D.D. Gilbertson, J.P. Gratten, C.O. Hunt, G.A. T Duller, G. A Barker, "Quaternary palaeogeomorphologic evolution of the Wadi Faynan area, southern Jordan. *Palaeogeography, Palaeoclimatology, Palaeoecology* 205:131–154 (2004).

[14] A. Ronen, "The Yiron-gravel lithic assemblage artifacts older than 2.4 MY in Israel" *Archäologisches Korrespondenzblatt* 21:159–164 (1991).

[15] L.K. Gabunia, A.K. Vekua, "A Plio-Pleistocene hominid from Damnisi, East Georgia, Caucasus", *Nature*. 373, 50–512 (1995).

Judaean mountains or resulted from global glaciation. Issar later correlated the second regression with the archaeological strata at Tell 'Ubaidiya'. An investigation of the samples of drillings in the off-shore area west to the southern part of the coastal plain of Israel by Adva avital. enabled to correlate between some of the glacial cycles, which occurred during the middle and upper Pleistocene and the transgression and regression mapped by Issar. From correlating Avital's section west (about 0.5 to 2.5 km) parallel to the coast line with that of Issar's along the coast line, it seems that the ingression termend by Issar no. 4, characterized by the warm water foraminifer "Marginopra" now called Amphisorus, was found also by Avital and is corrlable to Interglacial Marine Isotopic Stage (MIS) 7 dated between 250K and 200K B.P. Thus the regression following (Issar's regression no. 4) is correlable with Glacial MIS 6, which occurred about 140K B.P. The following trangressions (termed by Issar Last Oscillation) are the interglacial 3 and 5 (a few terrestrial clay layers are relics of glacial MIS 4 dated about 70K B.P.). The prominent regression representing the Last Glacial dated between 30K to 20K B.P. is found in the two sections by Issar along the shore and by Avital offshore. The transgression following this regression deposited the calcareous sandstone (local name Kurkar) ridge building the present shore cliff.[16]

Evidence of glacial periods during the Lower Pleistocene, as well as later stages is seen in analyses of oxygen 18/16 compositions of microfossil shells from core-hole samples from ocean floor sediments. Heavy composition marks a cold period, due to the process of glaciers capturing the light isotopes, while light isotopic composition shows that the glaciers melted, and the water captured by them was released. The isotopic composition of the samples indicates that throughout the Pleistocene several global glacial events took place.[17] However, it is premature to correlate the glacial and interglacial episodes revealed by these cores with migrations into Asia due to the scarcity of data. It is important to note, however, that the pollen assemblages in deep cores from the Hula and Dead Sea basins correlate with deep-sea cores, and during each cold period the ratio of tree pollen to other plant pollen shows a greater relative abundance of trees which means that during the cold periods the climate in the Near East was

---

[16] A. Issar, Palaeo 3 (1979).
    A. Avital, A. Almogi-Labin, H. Binyamini, "Reconstruction of the stratigraphic section of the coastal aquifer in the southern continental shelf in the Upper Pliocene, Pleistocene and Holocene according to sediments and fauna from marine drillings in the Askalon region", *Abstract volume of symposium on Exploitation and Regenerating of the Coastal Aquifer* – Israel Water Resources Organization and Water Commissioner, Israel. (in Hebrew) (2005).

[17] C. Emiliani, "Pleistocene Temperatures" *Journal Geology* 63:538–178 (1955).
    N.L. Shackleton, M.A. Hall, "Oxygen and Carbon Isotope Stratigraphy of Deep Sea Drilling Project Hole 552A: Plio-Pleistocene Glacial History" in: *Initial Reports of the Deep Sea Drilling Project 81*, D.G. Roberts and D. Schnitcker (eds.), U.S. Govt. Printing Office, Washington, D.C., pp. 599–609 (1984).
    J. Van Donk, "^{18}O Record of the Atlantic Ocean for the Entire Pleistocene Epoch" in: *Investigation of Late Quaternary Palaeo-oceanography*,. R.M. Cline and J.D. Hays (eds.), *GSA. Memoires* 145:147–163 (1976).

more humid and the forests thrived.[18] Once the early humans had crossed the desert stretches, taking advantage of the oases as stepping-stones, they found themselves in an accommodating environment.

The general picture that emerges is as follows: In the Lower to Middle Pleistocene (2.5 to 0.5 million years ago) there were a few cold glacial events, during which the monsoon and tropical rain systems weakened. As a result, the optimal environmental conditions that existed until then in tropical and subtropical Africa deteriorated.[19] These climatic and environmental changes encouraged the ancient hominids, now named *homo erectus*, to migrate further north and to expand their territories of hunting and foraging. The same global cooling caused the westerlies belt from the Atlantic Ocean to move southward, making the Saharo-Arabian landmass more hospitable and the Sinai and Negev Deserts passable. Consequently, the earliest humans were able to cross these areas to reach the more humid lands of the Fertile Crescent, where the road was open to all of Asia and southern Europe. We can conclude, therefore, that even in the earliest history of mankind, climate changes affecting the natural environment of the Near East were significant in determining the fate of humanity.

In the meantime, another and genetically different branch of humans evolved in eastern Africa during the second half of the last million years into modern *homo sapiens*. Unpublished finds of two skulls found on the shores of an ancient lake in southern Ethiopia were dated to 195K years ago suggest that some proto-*homo sapiens* lived alongside other hominid species in the still developing Rift Valley. Cavalli-Sforza suggested in his monumental study that the earliest modern humans migrated out of eastern Africa between 100,000 and 50,000 years ago, and their routes involved an adaptation to an aquatic way of life along the northern coasts of the Indian Ocean from east Africa along the southern coast of the Arabian Peninsula, India, the Indonesian Archipelago, and Australia. The genetic traits of these adventurers may still be found on remote archipelagoes such as the Andamans, and other parts of southeastern Asia, New Guinea and Australia. Another wave moved north and northeast into Asia and Europe, the Near East serving again as the land bridge.[20]

Based on tool typology some previous scholars maintained that the migrations via the Levantine land bridge occurred some 40,000 years ago. Current dating, however, shows that these more modern humans already existed some 100,000 years ago. Recent analysis and dating of fossilized human skeletons found in a cave at Jebel Qafzeh near Nazareth in northern Israel reveal modern features originally considered to be *homo neanderthalensis*.[21] If this will be

---

[18] A. Horowitz, "Continuous Pollen Diagrams for the Last 3.5 M.Y. from Israel: Vegetation, Climate and Correlation with Oxygen Isotope Record" *Palaeo 3* – 72:63–78 (1989).

[19] W.S. Broecker, "Cooling the Tropics" *Nature* 376:212–213 (1995).

[20] L.L. Cavalli-Sforza, *Genes, Peoples and Languages* (translated by M. Seielstad). North Point Press, New York, pp. 92–96 (2000).

P. Forster, S. Matsumara. "Did Early Humans Go North or South?" *Science* 308:965–966 (2005).

P. Mellars. "Going East: New Genetic and Archaeological Perspectives on the Modern Human Colonization of Eurasia" *Science* 313: 796–800 (2006).

[21] F.W. Farrand, "Confrontation on Geological Stratigraphy and Radiometric Dates from Upper Pleistocene Sites" in: *Late Quaternary Chronology and Palaeoclimates* pp. 33–53.

proven to be true, the migration occurred during a warm interglacial period. Most scholars today tend to adopt this earlier date as the time of *Homo sapiens'* arrival from Africa and that the influx of Neanderthal-type people descended from a northern *homo erectus* populations took place later at the height of the glacial period around 50 to 40,000 years ago, driven south by the extremes of the glacial regime in Europe. The Near East saw a few more glacial and interglacial periods, during which *homo neanderthalensis* roamed the plains of the Near East, most probably alongside *homo sapiens.* Whether or how the two groups reacted and interacted with each other, if at all, is not known. Did they, for example, compete for the same resources or did each have its own ecological niche? Did they inhabit the same caves at the same time? Their DNA record seems to indicate that they probably did not mix. It seems to the authors, however, that the final answer to this issue was not yet given.[22]

The studies of the deposits and of the ancient levels of Lake Lisan, the precursor of the Dead Sea, tell a lot about the climate during the upper Pleistocene and especially during the Last Glacial Period. The oxygen 18 composition of the aragonites deposited in this lake, demonstrate that during most of the Last Glacial period the climate was characterized by high precipitation.[23] This conclusion is backed by observations of the lakes' levels, which can be generalized by the equation of glacial cold period spelled a humid climate causing the rise in the level of Lake Lisan. From 70K years the level was rather high until 55K B.P. reaching about 240 m, below present MSL. From about 55K to about 30K the level was rather high except for three short periods of low levels synchronic with north Atlantic Heinrich Events. These events, which were extremely cold caused the influx of cold water from the Atlantic into the Mediterranean. This cooled the surface water of the Mediterranean reducing the amount of evaporation from the surface of the sea and thus the vapor available for forming rainstorms reaching the continent. From 30K to 23K, i.e. the peak of the Last Glacial, the level came up to its highest level reaching 165 m below present compared with its present elevation of about 420 meters below MSL. The lake extended over most of the Jordan Valley from north of the Sea of Galilee to the northern part of the Arava Valley.[24]

---

[22] L.L. Cavalli-Sforza, op. cit (2000).

J.J. Shea "The Middle Paleolithic: Early Modern Humans and Neandertals in the Levant" *Near Eastern Archaeology* 64, No. 1/2:38–64 (2001).

[23] Y. Kolodny, M. Stein, M. Machlus, "Sea-Rain-lake Relation in the Last Glacial East Mediterranean revealed by $\delta^{18}O$-$\delta^{13}C$ in Lake Lisan aragonites", *Geochimica et Cosmochimica Acta* 69/16: 4045–4060 (2005).

[24] L. Picard, 1943 op. cit.

D. Neev and K.O. Emery, "The Dead Sea, Depositional Processes and Environments of Deposition" *GSI Bulletin* 41 (1967).

D. Neev and K.O. Emery, *The Destruction of Sodom, Gomorrah and Jericho – Geological, Climatological and Archaeological Background,* Oxford University Press, New York Oxford (1995).

D. Neev and J.K. Hall, "Climatic Fluctuations during the Holocene as Reflected by the Dead Sea Levels" *Preprint presented at the International Conference on Terminal Lakes, Weber State College, Ogden, Utah, May 2–5* (1977).

B.Z. Begin, A. Ehrlich, Y. Nathan "Lake Lisan – The Pleistocene precursor of the Dead Sea" *GSI Bulletin* 63. (1974).

B.Z. Begin, W. Broecker, B. Buchbinder, Y. Druckman, A. Kaufman, M. Magaritz and D. Neev, "Dead

Lakes existed also in in the arid part of Jordan like the El Jafr Depression and the Azraq Basin as already mentioned. A lake existed also existed in the Palmyra region in the Syrian Desert until 19–18K years ago. Sediments of fresh water bodies were located as in Wadi Feiran of southernmost Sinai between 24,000 to 20,000 years ago.[25]

As already discussed the precipitation falling on the desert areas of the Near East during the cold humid periods recharged the aquifers by water of a charateristic isotopic composition.[26] Summing up, we can state that during the Last Glacial Period the westerlies and the Saharan high-pressure belt had migrated southward, causing many of the cyclonic rainstorms' trajectories to cross the Mediterranean coast of northern Africa and enter the Sahara, where they produced sand and dust storms. These storms normally ended with heavy rainstorms, after which the dust was deposited as loess. These loess soils, therefore, have characteristics of both eolian (windblown) as well as pluvial (waterborne) deposits. At the same time, the monsoon rain system abated. As a result, the Nile flow decreased and the amount of silt and sands carried by the river to the Mediterranean was reduced.

Flint tools found in deposits throughout the Negev and Sinai deserts provide evidence that these now arid lands were hospitable to hunters and gatherers during most of the Upper Palaeolithic period. Charcoal remains of local pistachio, tamarisk and olive trees indicate a relatively humid climate.[27]

All these results stand in stark contrast to the opinion of more than a few scholars of the prehistory and history of the Near East who associated, erroneously, cold climates with dryness and warm climates with humidity. One of the influential but unfortunately misleading papers and rather frequently cited, is O. Bar-Yosef's article focusing on the Natufian culture in the Levant, where he firmly states that "during the Late Glacial Maximum, dated to ca. 20,000 to 14,500 B.P. the entire region was cold and dry". The investigations about the Lisan Lake cited above, as well the Lakes of Sinai etc., show that Bar-Yosef's

Sea and Lake Lisan Levels in the Last 30,000 years, A Preliminary Report" *GSI Report* 29 (1985).

Y. Bartov, *The Geology of the Lisan Formation in the Massada Plain and the Lisan Peninsula*, M.Sc. thesis HUJI, (1999) (Hebrew with English abstract).

Y. Bartov, L.S. Goldstein, Y. Enzel, Y. Stein, "Catastrophic arid episodes in the Eastern Mediterranean linked with the North Atlantic Heinrich events" *Geology* 31/5: 439–442. (2003).

[25] Y. Sakaguchi, "Paleoenvironments in Palmyra District during the Late Quaternary", in: *Paleolithic Site of the Douara Cave and Paleogeography of Palmyra Basin in Syria, Part IV: 1984 Excavations*, T. Akazawa, Y. Sakaguchi, (eds.) University of Tokyo Press, pp. 1–27 (1987).

A. Issar and Y. Eckstein, "The Lacustrine Beds of Wadi Feiran, Sinai: Their Origin and Significance" *IJES* 18:21–27 (1969).

A. Kaufman, Unpublished Age Determinations, Weizman Institute of Science, Rehovot, Israel.

[26] Issar Bein and Michaeli op. cit, (1972).

Gat and Issar, op. cit (1974).

M.A. Geyh, J. Khouri, R. Rajab and W. Wagner, "Environmental Isotope Study in the Hamd Region" Series C, *Geologisches Jahrbuch* 38:3–15 (1985).

M.A. Geyh, "The Palaeohydrology of the Eastern Mediterranean" in: *Late Quaternary Chronology and Palaeoclimates of the Eastern Mediterranean*, eds. O. Bar-Yosef and R.S. Kra, pp. 131–145 (1994).

[27] N. Liphschitz and Y. Waisel, "Dendrochronological Investigations in Israel: Central Negev-Nahal Zin" Appendix in: *Prehistory and Palaeoenvironments in the Central Negev*, A.E. Marx (ed.), Southern Methodist University, Dallas, pp. 355–6 (1977).

conclusion, like his assertions about the cold but dry climate of the Natufian period, are based on casuistry rather than field data. In contrast a recent review based on field evidences of paleoclimates and paleoenvironments in the Levant by Bartov, Bartov et al. and Robinson et al. establish the fact that during most of the Last Glacial Period, except during the short Heinrich Events, the Levant was cool and humid.[28]

As global temperatures rose and the glaciers in the higher latitudes and mountain chains melted, the Near East climate gradually became more like it is during the more recent periods. The dominant trajectory of the cyclonic lows was also similar to today's trajectory – from west or from northwest. The dust-laden storms became infrequent and loess deposits were reduced to a minimum. On the other hand, the Nile began to rise and the quantity of sand transported from Nubia increased, as did the volume of sand and silt reaching the southeastern part of the Mediterranean.

The isotope composition of the Soreq Cave stalagmites shows that the transition from the Last Glacial Period to the Holocene is characterized by a sharp decrease in both $d^{18}O$ and $d^{13}C$, punctured by very high jumps to higher values One of these peaks, occurring around 19,000 B.P., can be interpreted as a warm and dry spell (Fig. 6). This may explain the low level of the Sea of Galilee – 15 meters below the present level – during this time.[29]

An archaeological site on the banks of the lake at this level contained the remains of wooden buildings and floors of an encampment of early hunters and gatherers. Twisted fibers used for fishing nets were also found.[30] The people living at this site belonged to a wide spread culture named Kebaran for a cave at the southwestern edge of Mount Carmel and in the vicinity of a swamp bearing the same name dated to between 16,000 and 18,000 B.P. Another camp of the same period is in Nahal Oren, one of the gorges of the same mountain. The Kebaran tool assemblage is characterized by microlithic tools, i.e. very small stone chips produced by breaking a long blade into small segments; the more recent of these tools – standardized triangular burins and

---

[28] O. Bar-Yosef, "The Natufian Culture in the Levant Threshold to the Origins of Agriculture" *Evolutinary Anthropology* 159–177 (1998).

Bartov op. cit (1999).

Bartov et al. op. cit. (2003).

S.A. Robinson, S. Black, B.W. Sellwood, P.J. Valdes, "A review of paleoclimates and paleoenvironments in the Levant and Eastern Mediterranean from 25,000 to 5000 years B.P.: setting the environmental background for the evolution of human civilization" *Quaternary Science Reviews* 25:1517–1541 (2006).

W. Nützel, "The Climate Changes of Mesopotamia and Bordering Areas" *Sumer* 32:11–24 (1976); H.J. Nissen, *The Early History of the Ancient Near East, 9000–2000 B.C.*, The University of Chicago Press, Chicago (1988).

S. Pollock, *Ancient Mesopotamia*. Cambridge University Press, Cambridge, UK (1999).

D.A.E. Garrod, "Primitive man in Egypt, Western Asia and Europe in Palaeolithic Times" in: *The Cambridge Ancient History*, Cambridge University Press, Cambridge, UK, pp. 54–57 (1980).

Garrod, maintains that during the early Last Glacial Period the climate was cold and humid, while during its later part it was cold and dry.

[29] Bar Matthews et al. op. cit p. 159 (1998).

[30] D. Nadel, A. Danin, E. Werker, T. Schick, M.E. Kislev, and K. Stewart, "19000-year-old Twisted Fibres from Ohalo II" *Current Anthropology* 35/.4:451–457 (1994).

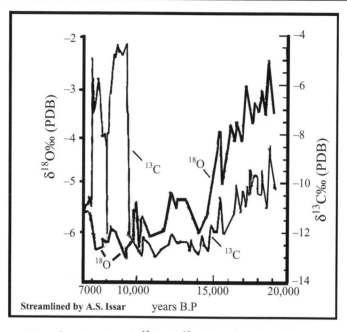

Fig. 6. Composition of environmental (^{13}C and ^{18}O) in stalagmites of the Soreq Cave from 20,000 to 7000 years B.P. (Bar-Matthews et al., 1998)

small crescent-shaped (lunate) blades – prompted scholars to call the final phase of this culture "Geometric Kebaran". These microliths were not used alone, but were placed on the tip of arrows or in rows in handles of wood, bone, or antler. There is little doubt that some were used as sickles to cut wild cereals. Microliths were first developed in Africa during the Palaeolithic period, appearing originally in the south of the continent, then the east, and later in the northeast, including the Nile Valley. The Kebaran people had a rich diet based on wild varieties of Mediterranean type grains, nuts, fruits, game and fish. Whether they had already started planting some of the seeds they had gathered cannot be ascertained and some recent finds are debateable.[31] However, since wild cereal grains comprised a major part of their diet, they may have experimented with seed planting long before the accepted date of the agricultural revolution.[32] The remains of grain seeds found at Gilgal included rye and oats, typical grains for temperate to cold climates, which may well have indicated a cold spell during the turbulent and changeable climate of the time.[33]

---

[31] M.E. Kislev, A. Hartmann, and O. Bar-Yosef. "Early Domesticated Figs in the Jordan Valley" *Science* 312:1372–4 (2006).

[32] H. Pringle, "The Slow Birth of Agriculture" *Science* 282:1446–145 (1998).

[33] E. Weiss, M.E. Kislev and A. Hartmann, "Autonomous Cultivation Before Domestication" *Science* 312:1608–1610 (2006). S. Lev-Yadun, A. Gopher and A. Shahal, "A. How and When was Wild Wheat Domesticated" *Science* 313:296 (2006).

One of the most interesting archaeological sites in this context is Wadi Kubbaniyya, between Kom Ombo and Aswan in Upper Egypt. The findings at this site illuminate some aspects of the earliest processes of a sedentary way of life between 19,000 and 15,000 years B.P.[34] Besides fishing and hunting, there is evidence of intensive gathering and, perhaps, intentional planting of tubers of the wild nut-grass *Cyperus rotundus*, a plant rich in carbohydrates and fibers. These finds could well represent the earliest cultivation of typical African tuber plants, such as yam and others.

Evidence of cultures similar to the Levantine Kebaran has been uncovered in most other regions of the ancient Near East, from the southwest Anatolian site of Belbashi near Antalya to the Zarzian and other localized epi-Palaeolithic cultures in the valleys of the Zagros Mountains and northern Iran.

3.1.2
The End of the Last Ice Age

The climate in the Near East during this period varied greatly, alternating between cold/humid and warm/dry. This can be deduced from Bar-Matthews' et al.'s findings that the speleothemes dated from 17,000 to 15,000 B.P. developed irregular alternating thick and thin laminas with variable colors (from white to dark red) and higher detrital components.[35] This variability did not favor sedentary life in the drier parts of the Near East as can be seen from the small size of the few and far between campsites of presumable hunting expeditions.[36] In contrast, in the more humid areas permanent and larger encampments are found near perennial springs, lakes and marshlands.

Our data suggest that after ca. 15,000 years B.P. the climate was increasingly becoming similar to the present Mediterranean regime of cold humid winters and warm dry summers. The winter and spring storms gathered force between 14,500 to 13,000 B.P. due to the southward shift of the westerlies, and probably affected the southern and more arid parts of the Levant, where many small sites of Geometric Kebaran, a late phase of the Kebaran, were found throughout the Sinai Peninsula and the Negev.[37]

The percentage of heavy oxygen and carbon isotopes in the stalagmites rose abruptly from 13,500 to around 11,500 B.P. The ratio between the two isotopes is similar to that of the immediate post-glacial period and different from that of the Holocene, characterized by summer precipitation. At Lake Van in Turkey also both isotopes in the sediments show heavier values, interpreted as indicating low humidity from around 13,300 to 10,500 B.P, whilea short humid interval at 11,000 B.P. is suggested to be correlated with the cold spell

---

[34] F. Wendorf, R. Schild and A.E. Close, (eds). *The Prehistory of Wadi Kubbaniya, vol. 3: Late Palaeolithic Archaeology*. Southern Methodist University Press, Dallas, pp. 820–821 (1989).
T. Shaw. P. Sinclair, B. Anda and A. Okpopo (eds.) *The Archaeology of Africa: Food, Metals and Towns*, Routledge, London (1993).

[35] Bar-Matthews et al. op. cit. (1996).

[36] N. Goring-Morris, *NEAEHL* p. 1039.

[37] O. Bar Yosef, *NEAEHL* p. 1105.

of the Younger Dryas in Europe.[38] Cores in ocean sediments confirm these observations, as they show the opposite trend, characterized by lighter values of ^{18}O, a result of melting of the sea ice.[39] One would thus conclude that the retreat of the deciduous oak forest in the Near East was due to warm and dry conditions rather than cool conditions contrary to what Baruch and Bottema have suggested. The warm, dry climate must have caused Lake Lisan to shrink to the level of the Dead Sea today.

By 12,000 B.P. permanent settlements of the Natufian culture (named after the D. Garrod's excavations in Wadi Natuf in the western Judaean hill country) proliferated near swamps and springs. One such settlement was excavated near the now drained and dry Lake Hula, not far from the spring of 'Enan in northern Israel. The yellowish calcareous soil at the sites is quite different from the underlying heavy black clays.[40] The black clays may have been deposited during the Last Glacial Period when the climate was more humid and corresponds with the Upper Palaeolithic tools found there. On the other hand, during the Natufian period the climate was not too humid, and the Hula Lake was not as high as it was in the pre-Natufian period.

The excavators found the remains of some fifty permanent round houses with roofs supported by pillars. Inside were hearths, one with an infant buried in it. Near the buildings were the first small cemeteries. The dead were buried wearing ornaments of shells and stones. One young adult was buried with a hand resting on a pup. One cannot say whether the dog was domesticated or adopted from the wild, but it is clear that people had taken a step towards domestication by starting to keep animals in their camp.

Weighing stones of nets and various implements, such as mortars, pestles and mallets for preparing food were also found. Whether the people of 'Enan had begun sowing some of the wild cereal seeds they gathered remains an open question. *"The various innovations and achievements supposed partly or wholly to have developed in the civilizations of the historical era are shown to have been foreshadowed by the accomplishments of prehistoric cultures".*[41]

Natufian settlements overlying the Kebaran layers were also found at Nahal Oren. Here the occupants inhabited both the cave and the slope in front, where they built terraces. Mortars and other utensils found on the terraces, indicate the area was used for preparing tools and food. The lower terrace served also as a cemetery. Sickles made of bone and microlithic sickle blades with the characteristic silicon sheen for cutting grass are evidence that cereals were

---

[38] Lemcke and Sturm, op. cit. (1997).

[39] J.C. Duplessy, "Global Changes during the Last 20,000 Years: The Ocean Sediment Record" in: *Global Changes of the Past*, ed. R.S. Bradley, UCAR/Office for Interdisciplinary Earth Studies, Boulder, USA, pp. 341–357 (1989).
Proceedings of Symposium *The Transition from Foraging to Farming in Southwest Asia* . Groningen (1998).
C.S. Larsen, "The Transition from Foraging to Farming and the Impact on Human Health: The Bioarchaeological Record", in Leroy and Stewart (2002).

[40] J. Perrot, "Enan" *NEAEHL* 2, pp. 389–393.

[41] R. Rudgley, *The Lost Civilizations of the Stone Age*, The Free Press, New York (1999).

harvested. More than 80 percent of the bones were gazelle, suggesting that the people had discovered how to concentrate and cull herds of these caprids.[42]

El-Wad (Me'arat ha-Nahal), a cave on the eastern edge of Mount Carmel, was another Natufian permanent settlement. Here, too, man-made terraces extended from the cave. An abundance of tools, including sickles, and several art objects were found. This settlement also had a cemetery, with the main burial site on the terrace.[43] Here, too, the skeletons had shell ornaments. The abundance of tools, hearths, and graves, and especially the construction of the massive terraces provide substantial evidence that the site was a permanent dwelling place.

Additional similar sites were found in the central part of the Galilee, often far from a permanent water resource or marsh. At HaYonim, for instance, gazelles were the most hunted animals and about 30% of those killed were fawns, a percentage characteristic of a herd culled on a yearly basis. Moreover, bones of house mice and rats also indicate the place was a permanent habitat.[44]

The lowermost excavated layer at Jericho in the Jordan Rift Valley, near the perennial spring of 'Ein es-Sultan the water source of Jericho, also contains implements of Natufian characteristics dating from 9687 ($\pm$107) to 7770 ($\pm$210) B.C.E.[45] A rectangular structure enclosing a platform constructed of mud with small saucer-like basins on the upper surface was found in the same layers. It was suggested that this might have served as an altar on which the hunter-gatherers, and maybe incipient farmers, living at the site presented offerings. If this is true, this humble artifact may represent the earliest manifestation of a spiritual development eventually leading to the ritualism of later periods.

The Natufian culture evolved directly from the preceding Kebaran and appeared in many parts of the Fertile Crescent. Permanently inhabited sites in Syria were excavated at Tell Abu Hureira and Mureibet both situated on the bank of the Euphrates east of Aleppo and show a high level of sophistication in its architecture, art objects and tools. Among the plant remains were grains of rye with the plump features of a cultivated variety. If the dating of the grains is confirmed, it may verify that the agricultural "revolution" of the Neolithic period was, in fact, a process that slowly evolved over several millennia. It may well have started in the last phase of the Pleistocene after the Last Glacial Period and developed through various phases until it peaked during the Neolithic period.[46]

As metioned, the Natufian Period is contemporary with the Younger Dryas (from 13K to 11.5K years B.P.), which was a cold spell during the mostly warm post Last Glacial Period. Based on the fact that most of the Last Glacial Period was humid we conclud that the Younger Dryas also brought humidity to the Middle East and that the socio-economic evolution of the Natufian Period was a result of this humid phase. Contrasting this view is the evidence of a salt layer

---

[42] T. Noy, "Oren Nahal" *NEAEHL* 3, pp. 1166–1170.

[43] M. Weinstein-Evron, "Wad, Cave" *NEAEHL* 4, pp. 1498–1499.

[44] O. Bar Yosef, Hayonim Cave, *NEAEHL* 2, pp. 589–591.

[45] K. Kenyon, Jericho, *NEAEHL* 2, pp. 674–697.

[46] Pringle, op. cit. (1998).

deposited during this period in the Dead Sea which speaks for a dry period. [47] On the contrary, Natufian settlements spread into the Negev Desert.[48] Indeed, Bar-Yosef, being devoted to his paradigm that cold equals dryness, explains the abundance of settlements in the desert by adaptation of the Natufian people to the arid conditions. Applying Occam's Razor we suggest a simpler explanation: the cold and dry periods were an exception such as the Heinrich Events occurring during extreme cold peaks and only were short dry intervals, This will also answer the theoretical conclusions, based on GCM modeling that glacial periods brought aridity to the Middle East and vice versa.[49]

In conclusion it is suggested that during most of the Younger Dryas the climate was cold and humid which caused the deserts of the Middle East, including the Negev and Sinai Deserts, to flourish, enabling the Natufian people to prosper. Only during the peak, which may have lasted a few centuries, a dry spell caused the fall of the level of the Dead Sea and the deposition of salt. This may have caused a temporary migration of the Natufian people northward and their coming back when the dry spell was over . . .

Related sites were discovered in southern Turkey and northeastern Iraq (Kurdistan). This region seems to have followed a unique path towards the agricultural revolution. At Hallan Çemi in eastern Anatolia a permanent village dated to the ninth millennium B.C.E. was excavated.[50] The inhabitants were hunter-gatherers descended from the earlier Zarzian culture experimenting with domestication of pigs, but showed no signs of intensive exploitation of cereals growing wild in the vicinity of the site. At Göbekli Tepe some of the oldest monumental art was discovered: animals – foxes, boar, ducks, birds of prey – and what looks like mythological beings were carved in high relief on soft lime stone blocks and defy explanation.

Unfortunately, we do not have more information about this period from other regions of the ancient Near East. Intensification of research in what is now the eastern Sahara, however, leaves little doubt that during the 11th millennium B.P. the periodic aridity of the end of the ice age gave way to a shift of the tropical monsoon belt which turned northern Africa and the Arabian peninsula into a well-watered and fertile savanna. Animals and people from the south soon moved in whereas the Nile valley was then a largely uninhabited swamp land. As in the following millennia the monsoon began to move south, the desert expanded, forcing animals and people to aggregate around the dwindling lakes and into the Nile Valley where the foundations of the sequence of Egyptian pre-Dynastic cultures were laid, as we shall see below.[51]

---

[47] Neev and Emery 1967, op. cit.

Y. Yechieli, M. Magaritz, Y. Levy, U. Weber, U. Kafri, W. Woelf, G. Bonani "Late Quaternary geological history of the Dead Sea area, Israel" *Quaternary Research* 39:59–67 (1993).

[48] See references in Bar-Yosef op. cit.

[49] W.F. Ruddiman and W.L. Prell. "Introduction to the uplift-climate connection". In: Ruddiman, W.F. (ed.) *Tectonic Uplift and Climate Change*. Plenum, New York, pp. 3–19 (1997).

[50] M. Rosenberg, "Cemi Hallan" in: *NeolithIc in Turkey – New Discoveries*. M. Ozdogan and N. Basgelen, (eds.) *Arkeoloji ve Sanat Yayyinlari*, Istanbul, pp. 25–33 (1999).

[51] R. Kuper, and S. Kroepelin, "Climate-Controlled Holocene Occupation in the Sahara: The Motor of Africa's Evolution" *Science* 313:803–807 (2006).

It appears to be that once a perennial supply of water and food was secured, a sedentary habitat was adopted with all its consequences. A sedentary way of life was undoubtedly the single most important element leading to the early phases of domestication of plants and animals, then full-fledged agriculture and animal exploitation, and, even more importantly, to profound social, cultural and spiritual changes. Farming skills were thus the consequence of the need to survive in times of abrupt and extreme climatic, and thus environmental, fluctuations from a pluvio-glacial to a Mediterranean climate regime. Hunting and fishing expeditions ventured into extended areas away from their permanent settlement as the wild animal population in the vicinity dwindled due to over-hunting and probably influenced the increasingly vegetal component in the diet. Settling for long periods at one site must have required innovations for maintaining an adequate food supply, especially during the dry seasons and drought years. It is difficult to imagine that people who could produce specialized tools and art objects and develop social and religious customs would fail to recognize that putting grains in the soil during the dry season would produce sufficient grains during the humid season to sustain them. Thus, although no concentrated effort was apparently invested to select grains from the highest yielding plants, still the re-seeding and re-planting perpetuated mutations of cereals, which did not disperse. Their seeds were harvested immediately and part was stored to be seeded, first by chance and later by design.[52]

3.1.3
The Start of Settled Life

Our data indicate a period of abrupt climate changes alternating between warm and dry and cold and humid after 11,000 years ago. A rather cold and humid shift was followed by warm spells that brought droughts in the central and northern parts of the Near East, and possibly also monsoon rains to the desert regions in the south. There were also annual fluctuations in the temperature, and thus in precipitation, during this period by the millennium, century and decade, and, of course, from year to year. The decrease in rain caused the lakes and springs in the lower latitudes to dry up. The vegetation, animals and humans were forced to find refuge along perennial rivers and springs emerging from water-bearing rocks fed by rains falling on more humid terrain. Natufian Jericho and the sites on the banks of the Euphrates bear clear witness to this development. The introduction of irrigation methods may have started already during the Natufian period at places like Jericho. There, by imitating nature, the inhabitants could dig a channel, create an artificial a rivulet and water a dry field. Jericho is situated in the Jordan Rift Valley about 10 km north of the present northern edge of the Dead Sea at about 300 meters below sea level. The

---

[52] D. Zohary and M. Hopf, *Domestication of Plants in the Old World*, Clarendon Press, Oxford (1994). D. Zohary and M. Hopf, "Domestication of the Neolithic Near Eastern Crop Assemblage" in: *Prehistory of Agriculture, New Experimental and Ethnographic Approaches*, Monograph,. P.C. Anderson (ed.), University of California, Los Angeles (1999).

valley is arid due to its location in the shadow of the rain. The spring providing the source of water of Jericho is replenished from a wide, high-altitude area part of the mountains of Judaea and Samaria to the west. The precipitation in the area infiltrates the rocks to form a regional groundwater table, part of which flows eastward towards the Rift Valley. Due to the vast storage capacity of the limestone rocks the flow is rather steady throughout the seasons and the years; only an extended period of several arid years may diminish its flow. Today its annual flow is about 10 million cubic meters and is still in use.

The regularity of this water source prompted a group of people to settle at the spring. However, a food source for a growing population could not be guaranteed unless the people could learn to overcome the consequences of the random nature of climate, especially in such a dry area. Thus, irrigation for agriculture was born of the need to ensure a stable food supply. Jericho was, therefore, among the sites of one of the earliest agricultural societies because the climatic conditions of the area were conducive to the invention of irrigation.

Our conclusions concurs with the principles of Childe's oasis hypothesis and his conceptual model of postglacial aridization[53] as well as with Garrod's suggestion that the Natufian people in Palestine were the initiators of agriculture and with Braidwood's and Howe's model of a gradual evolution from hunter-gatherer to farmer during the Upper Pleistocene.[54] Also, we agree with the principle of a "nuclear zone," which, in our terms, is the area of the oasis created by a spring. Whether this emerged at the foothills of the Zagros, the Taurus or the Judaean mountains will be established when more data become available. It would be reasonable to assume that this development took place in a rather uneven fashion, e.g. while plant cultivation progressed in the Levant, caprines were being herded and domesticated along the foothills of the Zagros. The archaeological evidence indicating that Jericho had already started to function as a settlement during the Natufian supports the hypothesis that the agricultural "revolution" was more a process of evolution that started in refuge areas – in this case at an oasis situated near a perennial spring. The site, although small, enabled attempts to sow grain outside its natural habitat.

Another neo-determinist model explaining the transition from a hunting-gathering society to one of incipient agriculture during the Natufian period was proposed by A.M.T. Moore. He attributes the shift to an increase in the population during this time that may have resulted from more hospitable environmental conditions – a rise in temperature and humidity, followed by expansion of the forests – beginning around 15,000 years ago. The population increase reduced the area of hunting and forced sedentary life, which led to the formation of new social hierarchies. The need for food from local resources led to the beginnings of agriculture during this period. Thus, Moore's Archaic Neolithic around 10,500 to 8000 years ago (i.e. Early Pre-Pottery Neolithic),

---

[53] G.V. Childe, *New Light on the Most Ancient East.* Kegan Paul, London (1934).

[54] D.A.E. Garrod, "The Natufian Culture: the Life and Economy of a Mesolithic People in the Near East" in: *Proceedings of the British Academy* 43:211–227 (1957).
J.R. Braidwood, B. Howe, H. Helbaek, F.R. Mason, C.A. Reed and H.E. Wright, Jr., "Prehistoric Investigations in Iraqi Kurdistan" *Studies in Ancient Oriental Civilization* 31, Chicago University Press, Chicago (1960).

village life and agriculture had become the dominant way of life. During this period there was deterioration in the environment as temperatures rose and rainfall became more seasonal. Agriculture became slowly more important for subsistence, enabling the population to grow, and the society to evolve.[55] Without going into the details of Moore's model (as, for example, whether the climate changes happened as he describes them or in some other way, as revealed by isotopic time series data), his emphasis on the interaction between climate, environment and human society is rational. In the present authors' opinion, it lacks, however, appropriate emphasis on the question of *causa sine que non*, namely the essential cause of the shift from hunting and gathering to agriculture when people were forced into niches where survival depended on the development of irrigated agriculture.[56]

The furthest observed strides were taken in refuge areas such as Jericho in Palestine and Tel Aswad in Syria, located near perennial springs, or Tell Abu Hureira, situated on the banks of the Euphrates, in northern Syria and Nevali Cori on the same river across the border in Turkey. This is demonstrated by the findings of occupation layers containing remains of einkorn wheat, emmer wheat, barley, peas and lentils – all domesticated as early as the tenth millennium before present.[57] So, when man had not yet learned to build a kiln to fire his clay vessels, he knew how to sow and irrigate his fields.

The school opposing Childe's, Moore's, and the present author's neo-deterministic model prefers the anthropogenic hypothesis, minimizing the influence of climate change and putting emphasize on population density, social stress and competition for leadership among families, clans, or tribes.[58] These processes are undoubtedly of great importance in understanding the subsequent development of the Neolithic period, but are secondary manifestations and do not represent the primary cause. After all, this period lasted nearly half a millennium, and much can happen in such a long time, even if history remains absolutely silent about the events and peoples that shaped it. Trade in luxury items such as obsidian and sea shells is a sure sign of a developing social stratification, indicating the growth of a class of people who use these items to enhance and display their social status.

The steady desiccation at the end of the Pleistocene urged some human groups to improve upon nature by inventing agriculture and animal domesti-

---

[55] Moore, op. cit. (1985).

[56] N. Brooks. "Climate Change May have Sparked Civilization". *Environment New Service*, Sept. 11, (2006).

[57] Zohary and Hopf, op. cit. (1994, 1999).

[58] L.R. Binford, "Post-Pleistocene Adaptations" in: *New Prespectives in Archaeology*, S.R. Binford, and L.R. Binford, (eds.) Aldine, Chicago, pp. 313–341, (1968).
B. Hayden and R. Gargett, "Big Man Big Heart? A Mesoamerican View of the Emergence of Complex Society" *Ancient Mesoamerica* 1:3–20 (1990).
A. Gofer, "Early Pottery-bearing Groups in Israel – The Pottery Neolithi Period" in: *The Archaeology of Society in the Holy Land*. T.E. Levy (ed.) Leicester University (1998).
J. Cauvin, *Le Premiers Villages de Syrie-Palestine du IXème au VIIème Millénaire avant J.C.* Maison de l'Orient, Lyon (1978).
K.V. Flannery, "The Origins of Agriculture" *Annual Review of Anthropology* 2:271–310 (1973).

cation, along with the subsequent social and cultural changes. All these innovations enabled the agricultural "revolution" of the Neolithic cultures to take off. It should be emphasized that, in our opinion, an important element in this evolution was the conscious manipulation of water resources.

3.2
The Establishment of Agricultural Villages – The Pre Pottery Neolithic – (6000–8000 B.C.E.)

Tell Aswad and Jericho exemplify the refuge-oasis model as both depended on water issuing from large perennial springs fed by limestone aquifers. While Jericho is situated close to the outlet of a perennial spring, Tell Aswad is about 30 km to the east, where the Barada River, formed by the spring, flowed into a basin and created a lake. The lake existed during the Upper Pleistocene and covered a large area of the basin, where the city of Damascus now lies. Today, the area receives less than 200 mm of rainfall per year. Excavations by H. de Contenson[59] showed that the site was inhabited from 7800 to 6600 B.C.E. it consisted of small round huts made from mud bricks, with sunken floors. Flint tools, clay figurines, and bone implements were found but no pottery. Paleo-botanical examinations proved that crops were cultivated right from the establishment of the settlement and they were most probably irrigated.[60]

Detailed knowledge about Jericho's Neolithic culture is owed to the British archaeologist Kathleen Kenyon, who excavated the site from 1952 to 1958.[61] The Pre-Pottery Neolithic A culture (P.P.N.A) survived about a thousand years in Jericho. During this time, the inhabitants built a tower 10 meters in diameter and 8.5 meters high with a 22-step staircase – the world's oldest evidence of monumental architecture. They also built a set of perimeter walls and a ditch. Kenyon described these walls as fortifications, while Bar-Yosef suggests that they were built as protection against floods, whereas Ronen and Adler suggest a ritual purpose.[62] Anyhow, construction of such monumental structures attests to a social organization capable of mobilizing the manpower needed to build the walls and the tower, which served either as a watchtower or as a cult site, and was perhaps a precursor to the ziggurats of the Mesopotamian plain. The inhabitants of Jericho had commercial relations with the Mediterranean and the Red Sea from where they obtained malachite and ornamental seashells as well as with more distant places such as Anatolia where their obsidian had originated arriving here by a chain of exchange. The skulls and other osteological remains found at Jericho show that the people who lived there during the

---

[59] H. de Contenson, "Tel Aswad (Damascène)" *Paléorient* 5:153–156 (1979).

[60] W. Van Zeist, and J.A.H. Bakker-Heeres, "Some Economic and Ecological Aspects of the Plant Husbandry of Tel Aswad" *Paléorient* 5:161–169 (1979).

[61] K.M. Kenyon, *Digging up Jericho*, Benn, London (1957).

[62] O. Bar Yosef, "The Walls of Jericho: An Alternative Interpretation" *Current Anthropology* 27/2: 157–162 (1986).
A. Ronen and D. Adler, "The walls of Jericho were magical" *Archaeology, Ethnology & Anthropology of Eurasia* 2/6:97–103 (2001).

PPNA had the same features as those of the Natufian. Their flint tools were also of the Natufian type. Continuity seems to be seen in excavations at sites of the same age on the Euphrates, such as Nevali Cori and Jerf el-Ahmar where the ongoing French excavations exposed a site of ca. 1000 square meters containing 10 occupational layers all dated to the PPNA (Danielle Stordeur, directeur de recherche C.N.R.S.). The traditional Natufian-style circular mud house plan was changed over time into larger rectangular structures partially surrounding a round "piazza", obviously for some public function. The change implied that the houses were build with wooden beams, plastered and on stone foundations. The flint material for tools can be traced to Anatolia and continued Natufian work traditions. Some flat stones were incised with geometric scratchings of unknown purpose. The large amount of charred wild cereals (probably wild weat) and lentils as well as the bones of gazelles, hermiones (wild ass), aurochs, and birds, make it clear that the inhabitants were sedentary hunter-gatherers. Sedentary village life, cultivating crops, and finally domestication of plants and animals once believed to have formed a "package deal" called Neolithic Revolution is now seen as a long and slow process where each element developed at different places and at various periods without necessarily having been linked.

The PPNA culture at Jericho ended abruptly sometime in the first quarter of the eighth millennium B.C.E. While there are no signs of destruction by earthquake, foe or fire, there are signs of flooding on top of the layers from this period. The proxy data show an abrupt decrease in ^{18}O around 10,000 years ago as well as a simultaneous increase in the ^{13}C compositions. The most likely interpretation suggests a strengthened westerlies storm pattern, which resulted in a colder climate stretching the rainy season into the summer, enabling grasses and legumes (C3 and C4 type vegetation) to survive most of the year. In any case, the isotopic evidence clearly shows a climatic change that most likely involved flooding of Wadi el-Kelt – a seasonal stream that drains a vast area of the highlands to the west and enters the plain nearby. Kenyon also maintains that this was a period of floods and subsequent erosion.

The new people who settled in Jericho around 9000 years ago were seemingly of a different stock with longer skull features[63] and brought a new cultural tradition which Kenyon designated Pre-Pottery Neolithic B (PPN.B). Their flint tool set was different from those of the PPNA, which had evolved from the Natufian, and included a higher percentage of heavy-duty core tools, such as polished axes for working the ground and cutting trees. Their houses were constructed with a type of brick different from and superior to that used by the PPNA people. Similarities with northern cultures suggest another early instance of migration from the north during a colder period as the northern highlands became less hospitable whereas the southern areas became greener and attractive.

The PPNB layers comprise over 20 building levels covering a time span of about a thousand years. In a niche of a particular building stood an elongated

---

[63] P. Smith, "People of the Holy Land from Prehistory to the Recent Past" in: T.E. Levy op. cit. pp. 58–74 (1998).

stone that might have been used as an idol. Similar stones, known as *stelae* or *bethyloi* (bethels) and referred to in the Bible as *matseba*, are found from now on in many archaeological sites and in different periods throughout the Near East. Other cult objects were found in these layers, including plastered and painted skulls and three life-size figures.

Horowitz studied the pollen spectrum in the sediments of the PPNB of Sde Divshon in the Negev. He suggests that this region, which today receives about 80 mm of rainfall per year, was humid during the PPNB and suitable for agriculture. He correlates this period with the Boreal of Europe.[64] Liphschitz and Waisel identified charred pieces of pistachio and olive trees in the same region, which also indicate a more humid climate.[65] Bar Yosef maintains that the PPNB economy in fertile areas was based on cultivation of legumes and cereal, together with hunting and herding. The economy in drier areas, where the inhabitants probably lived during winter, autumn and spring, was based on hunting and gathering.[66] The finding in Jericho of obsidian from Çatal Hüyük in the central Anatolian plateau established an exchange system between these two regions. One can assume that while obsidian traveled one way along the route, information and seeds may have traveled the other. In addition, a bone belt-hook found at Nahal Hemar (the "Bitumen Valley" in Hebrew) near the Dead Sea has identical counterparts in Çatal Höyük and other contemporary Anatolian sites, which could support the idea that the cold climate brought a migration of people from the high plateaus of Anatolia as well as trade. From Nahal Hemar we have a skull decorated with what looks like a bitumen hair net, and masks carved out of limestone, as well as some of the earliest textiles, displaying nearly all types of weaving known today.

Agricultural products supplied most of the needs of the people of Jericho. They domesticated the goat and no longer needed to depend solely on hunting. Goat meat became the main protein component of the diet. The desert plains of the area, which dry up during the summer but still provided sufficient dry pasture for the undemanding goat, could be used for grazing. In the future when the climate deteriorated the goat, well adapted to desert conditions, enabled Near East societies to survive during periods of desiccation by switching over to herding and pastoralism.

In the northern part of the Jordan Valley, on the shores of the shallow lake and swamps of Hula gazelle meat and wild pig remained the main protein component of the ancient dwellers' diet.[67]

Like Jericho, 'Ain Ghazzal in Wadi Zarqa in Jordan is situated near a perennial spring. In the Middle PPNB period (9250 to 8500 years ago), the houses were plastered and painted with ochre. The people harvested crops such as wheat, barley, peas, lentils, etc. Half of the animal bones found were domesticated goat, the other half, wild game. A variety of small clay figurines – animal as well as human – were found. Most of the latter were "decapitated," that is, the heads

---

[64] A. Horowitz, op. cit. (1980).

[65] Liphschitz and Waisel, op. cit. (1977).

[66] O. Bar Yosef, op. cit. (1986).

[67] J. Perrot, "La Préhistoire Palestinienne" *Dictionnaire de la Bible* 8 (supplément) pp. 416–438 (1968).

and bodies were found separately. Enigmatic finds are small geometric clay objects, interpreted by the excavators as counting tokens – perhaps forerunners of a counting or 'writing' system. The most imposing finds, however, were human statues of nearly a meter high, made of painted plaster over frames of reeds. There is scant evidence for some crudely made, low-fired pottery at the end of the period, a phase termed PPNC at 'Ain Ghazzal. Together with Syrian "White Ware," Rollefson and Simmons suggest these artifacts represent a "glide" of the Pre-Pottery into the Pottery Neolithic Period.[68]

Tell Ramad near Damascus and Beidha near Petra also date from this period. Bar Yosef and Avner suggested that during the PPNB, around 9700/9500 to 8000 years ago., the climate favored hunters and gatherers, attracting them to now drier desert-like areas such as the Negev, northern Sinai and the edge of the Syrian desert.[69]

The climate pattern of the Near East extends eastward to the Zagros Mountains and the Iranian plateau. One would expect that a more humid climate had enabled a connection between these two regions, as the Syrian Desert would have become more hospitable. Moreover, a colder and more humid climate would bring more precipitation to the Iranian mountains and enhance the springs flowing out from the foothills of the mountains.[70] Thus, with the improved climate during the PPNB extending to the foothills and intermountain valleys of the Zagros Mountains of Iran, sites such as Zawi Chemi, Shanidar and Karim Shahir developed parallel to those in the Levant. Outstanding architecture of *terre pise* (packed mud) and some of the earliest pottery of the Near East was found in western Iranian sites in Ganjdareh and Asiab in the Kermanshah province.[71]

The transition from Pre-Pottery Neolithic A to B can also be traced in the Syrian sites of Mureibet and el-Kowm near Palmyra. The smooth transition here supports our suggestion that the carriers of the PPNB culture had migrated south around 7000 B.C.E.

The best example of PPNB is the southeastern Anatolian site of Çayönü with its elaborate stone and clay brick buildings, baked clay figurines, attempts at pottery and, above all, use of native copper hammered into the shape of stone tools. The continuance of complex and monumental carved rock art at Gobekli and Nevali Cori into the mature phases of the Neolithic period opens entirely new aspects into the spiritual life of this time.

---

[68] G.O. Rollefson and A.H. Simmons, "The Early Neolithic Village of 'Ain Ghazal, Jordan: Preliminary Report of the 1983 Season" in: *Preliminary Reports of ASOR Sponsored Excavations 1981–83, BASOR Supplement 23:43–44* and Fig.7, W.E. Rast (ed.). Eisenbrauns for American Schools of Oriental Research, Winona Lake, Ind. (1985).

[69] O. Bar Yosef, "The Land of Israel during the Neolithic Period" in: *The History of the Land of Israel*, vol. 1, ed. Y. Ripel, Tel Aviv, Israeli Defense Ministry Publishing, pp. 27–46 (1986, Hebrew).
U. Avner, "Settlement, Agriculture and Palaeoclimate in Uvda Valley, Southern Negev Desert, 6th–3rd Millennia B.C." in Issar and Brown op. cit. pp. 147–202 (1998).

[70] A. Issar, "The Groundwater Provinces of Iran" *Bulletin of the International Association of Scientific Hydrology* 14/1:87–99 (1969).

[71] J. Mellaart, *The Neolithic of the Near East*. Thames and Hudson, London (1975).

An agricultural settlement was established at Çatal Hüyük in the Konya plain on the Anatolian plateau around 6200 B.C.E.,[72] contemporary with the late PPNB at Jericho. Whether the crops were rain-fed, irrigated, or both, has yet to be ascertained, but a greater variety of cereals was grown at Çatal Höyük than at Jericho. Wheat was the main cereal and wild cattle were the main source of animal protein. Obsidian from Hasan Dag near Çatal Höyük was found all over the Near East down to Egypt. Some time later the people at Çatal Höyük built shrines featuring bull heads plastered with clay and painted and a fertility goddess with leopard heads plastered too, which is similar to the plastered skulls found at Jericho and Tell Ramad. What material or spiritual exchange may have existed between the peoples of Çatal Höyük and those of Jericho is, for now, a matter of speculation.

It appears that seafaring also took great strides at this time, a fact largely neglected because of a lack of hard evidence. The first Mediterranean island to be discovered and settled was Cyprus. The ceramic Khirokitia culture is contemporary with the last phase of the Levantine PPNB, dated tentatively around the first half of the sixth millennium B.C.E. It is from here that the Neolithic revolution seems to have advanced throughout the Mediterranean world.[73]

Scientists at the German Nautical Institute at Kiel have for several years studied ancient petroglyphs of primitive boats from all over the world, some dating back to the epi-Palaeolithic/Mesolithic and Neolithic period. After experimenting with vessels of a variety of materials, they concluded that a large-sized reed boat was the most likely candidate for ocean travel during the Neolithic period. The vessel consisted of two enormous sausage-shaped structures made of reed and small pieces of wood, their ends tied together and bent upward to form the prow and stern. In the summer of 1999, Zohar witnessed the launching of such a contraption at the port of Alghero in Sardinia where the modern-day "Neolithic" sailors assured him that the boat is very flexible, adjusts itself to the shape of the waves, is practically unsinkable even in moderately strong storms, and can carry voluminous cargoes. The already mentioned Neolithic textiles found in Nahal Hemar near the Dead Sea support the possibility that sails could have been used. A short trip around the bay of Alghero demonstrated how Neolithic people, along with their animals and sufficient provisions of grain and other seeds supplemented by abundant fish, were able to cross with ease the relatively short distances between the Mediterranean headlands and islands.

The PPNB period represents the high watermark of Neolithic evolution and ends abruptly in the southern Levant around 6000 B.C.E. or a little thereafter. Again, the demise seems not to have been caused by earthquake or war. Rather, signs of erosion by water flow can be seen on the buildings of PPNB Jericho. The isotope data from Soreq Cave and Lake Van may corroborate a period of high precipitation and thus renewed floods. This cannot be stated definitely,

---

[72] J. Mellaart, *Çatal Höyük – A Neolithic Town in Anatolia*, Thames and Hudson, London (1967).

[73] J. Zilhão, "The Spread of Agro-pastoral Economies Across Mediterranean Europe: a View from the Far West" *Journal of Mediterranean Archaeology* 6, 5–63 (1993).

however, because of a problem with calibrating dates at Jericho during this period. Nevertheless, for whatever reason, the inhabitants abandoned their oasis as the more humid climate most probably caused the groundwater table below the town to rise and floods from the nearby valleys to cover the fields and destroy their houses.[74] It appears that the entire Jordan Valley had become a well watered and fertile land where farm could be built in a loose pattern without the need to congregate around springs.

In conclusion, since its earliest beginnings, the destiny of the human race was shaped by the changes of climate and their impact on the environment. These changes were also at the root of the exodus from the African Garden of Eden. The same changes opened the northeastern gates of the garden to enable the departure and closed them when the climate changed again. The primeval "flaming sword which turned every way" operated by turning the southern Levant into a parched desert which happened more than once. Aridization brought our ancestors to seek refuge in the mini-gardens of Eden, the oases dotting the desert. There, putting to work their intelligence which, according to the Biblical metaphor, was acquired by eating from the "Tree of Knowledge" they developed skills in farming and irrigated agriculture which enabled them to feed themselves, to multiply and fill up the land.

---

[74] J. Perrot op. cit. 1968 observed a reduction in the amount of tree pollen in core samples taken from Lake Huleh. He concluded that the climate became drier, which is probably mistaken, and the opposite may be more likely when the forest was cut down and grain fields took their place.

CHAPTER 4

The Great Transition –
From Farming Villages to Urban Centers

From the Pottery Neolithic to the Chalcolithic Period and the Beginning of the Early Bronze Age, ca. 6000 to 3000 B.C.E.

"And Zillah, she also bare Tubal-cain, an instructor of every artificer in brass and iron..."

(Genesis V:22)

The end of the Last Glacial Period and the increasingly sedentary way of life that began in the epi-Palaeolithic and Pre-Pottery Neolithic periods profoundly changed the human experience. Although many cultural elements coming to the fore in the new ages were undoubtedly noticed, and perhaps even occasionally used, during the Palaeolithic periods, they could flourish only when people began to live in fixed abodes. All aspects of human life – from technology to religious concepts and profound changes in social structure, such as the growing distinctions in social status and the deteriorating role of women – began to form an inextricable and apparently seamless net leading the majority of the human race towards complexity and eventually into civilization, leaving only a few marginal groups of hunters and gatherers.

Technology became the most important aspect of human existence, often at the expense of a moral or spiritual development. The fabrication and use of baked clay – the first manmade material, normally not occurring in nature – represents the first true technological invention after thousands, if not millions, of years. As with other sporadic early inventions, we can only speculate that Palaeolithic man had observed that wet clay hardened in a fireplace or that smearing clay over a basket and exposing it to fire had created a hard container, which, however, was too brittle and too heavy to be useful to the nomadic hunters and gatherers. Full exploitation of the "invention" awaited the increasingly sedentary lifestyle of Neolithic groups in various parts of the world. Experimenting with different raw materials and firing methods – first in an open fire, then in a kiln – they created ceramic vessels of various shapes for storing and consuming food, which revolutionized food preparation and changed dietary habits. Cooked food had a positive effect on child rearing in primitive societies, contributing in a considerable way to a steady population growth. Whereas women in hunter-gatherer societies were equal members of the group and actively participated in all activities, sedentary life and the availability of cooked food enabled women to give birth at shorter intervals, increasingly tying them to the "home" to care for their infants. By becoming "professional mothers" whose only legitimate interest was their offspring, they increasingly depended on men for their livelihood and led in the end to women's steadily deteriorating social standing. Earthen vessels soon assumed a central significance in the society, and eating and drinking habits became distinctive cultural elements, often with ceremonial and spiritual connotations. The large storage vessels for staples such as grains or for water became symbols of life-giving forces: they provided food between harvests, and in many cases they became the final resting place of the dead. They were soon embellished with impressed or painted decorations expressing spiritual and ceremonial ideas and, therefore, soon became symbols of the group's identity.[1]

Some of the oldest isolated evidence of pottery still within a Mesolithic/epi-Palaeolithic context is found in central and western Africa and in the Far East.[2] There is evidence of similar early attempts to produce vessels in the Near East, such as the "White Ware" on the Syrian coast – a calcareous ware made of finely crushed limestone that sets like cement without firing. However, the gradual appearance of fired clay pottery in the Near East had a more far-reaching effect on technology. Specialized kilns and skills necessary for earnest experimentation with metals, their alloys and other materials, such as glass were soon developed. Unlike the locally important but isolated innovations or cultural events in other parts of the globe, the repercussions of developments in the Near East were felt all over the ancient world, as this area was the crossroad linking Asia, Europe, and Africa.

4.1
The Progress of Climate

In the northern part of the Near East the level of Lake Van fell rapidly starting around 8000 B.C.E. and reached an extreme low around 6000 B.C.E. It remained low until around 4500 B.C.E. and then rose rapidly until 4000 B.C.E. The relative humidity curve calculated by Lemcke and Sturm shows low humidity starting around 6400 B.C.E. and continuing until ca. 5750 B.C.E. Immediately thereafter, the climate became more humid – reaching a peak between 4400 and 4300 B.C.E. – and then drier again, with dryness peaking around 3400 B.C.E. Accordingly, the lake sediments show a low percentage of tree pollen, such as oak and pistachio, and a high percentage of non-arboreal pollen, indicating dry conditions from 7600 to 4400 B.C.E. (Figs. 3 & 3a).[3]

---

[1] S.J. Shennan, ed., *Archaeological Approaches to Cultural Identity*, Routledge, London (1989).

[2] T.R. Hays, "Neolithic Chronology in the Sahara and Sudan" in *COWA*, 3d ed., vol. 1, ed. R.W. Ehrich Chicago, University of Chicago Press, pp. 309–329 (1992).
Song Nai Rhee "Japan and Korea" in *OCA*, pp. 359–362.

[3] S. Kempe and E. Degens. pp. 56–63 in Degens et al. op. cit. (1984).
Schoell, in Degens and Kurtman op. cit pp. 92–97 (1978).
Lemcke and Sturm, op. cit. pp. 654–678 (1997).

A study of the pollen assemblage from the bottom of Lake Zeribar in the Zagros Mountains and south of Lake Van shows that around 10,600 B.P. (calibrated to the middle of the tenth millennium B.C.E.), the percentage of Artemisia pollen decreased, while that of non-arboreal plants increased. Artemisia is a non-arboreal shrub, which usually indicates semi-desert conditions. Thus while climate became drier it was not extreme. A contemporaneous increase in the pollen count of oak and pistachio led the palynologist El-Moslimany to infer special climatic conditions enabling oak and pistachio to penetrate into the region: first, a decrease in snowfall on the northern Iranian highlands due to the general warming trend, and second, the prevalence of summer rains until ca. 4200 B.C.E.[4]

Proxy data for the ancient climate of the eastern regions of the Near East are scarce. Studies of core samples from sediments at the bottom of the Persian Gulf taken by the research ship *Meteor* provide clues about what happened in the Mesopotamian drainage basin. A low proportion of organic matter in the sediments indicates a humid period with much water, and vice versa. The data show a very dry climate from 6500 B.C.E. to 5500 B.C.E., followed by more humidity from 5500 to 3500 B.C.E. A reconstruction of flow of the Tigris and Euphrates, based on contemporary synoptic and hydrological data and other proxy data time series, such as pollen sedimentation rates, etc. shows a high flow rate in these rivers during most of the fourth millennium B.C.E.[5]

The isotopic composition of the stalagmites of the Soreq Cave show higher $^{13}C/^{12}C$ values, which is an evidence for a high percentage of grasses and legumes growing on the surface above the cave. The most likely explanation is a climate characterized by summer rains that kept the soil humid during most of the year, similar to that of the eighth millennium B.C.E. i.e. it had not yet become a distinct Mediterranean climate. Fluctuations in the composition of the oxygen isotopes indicate variations in the annual average rainfall. High values of both ^{18}O and ^{13}C are typical for the seventh millennium, most likely indicating mainly summer rains, whereas low ^{18}O and high ^{13}C are typical of the sixth millennium and indicate summer and winter rains. During such years one would expect both a rich arboreal and non-arboreal vegetation cover, which would lower the ^{13}C ratio, a trend that occurred during the sixth millennium. Excessive evapo-transpiration, due to the warm climate also would have negatively affected groundwater infiltration, reducing spring flow, water infiltration into the caves, and stalactite and stalagmite formation. Indeed, data from the Soreq Cave show a scarcity of speleothems and relatively slow deposition rates during most of these periods as well as enrichment in iron

W.W. Van Zeist and H. Woldring, "A Postglacial Pollen Diagram from Lake Van in East Anatolia" *Review of Palaeobotany and Palynology* 26:249–276 (1978).

[4] A.P. El-Moslimany, "Ecology and Late Quaternary History of the Kurdo-Zagrosian Oak Forest Near Lake Zeribar, Western Iran" *Vegetation* 68:55–63 (1986).

[5] W. Nutzel, "The Climatic Changes of Mesopotamia and Bordering Areas" *Sumer* 32:11–24 (1976). P.A. Kay and D.L. Johnson, "Estimation of Tigris-Euphrates Streamflow from Regional Palaeo-environmental Proxy Data" *Climatic Change* 3: 251–263 (1981).

oxides and detritus.[6] All these phenomena can be attributed to rainstorms of short duration and high intensity, causing erosion and flushing of the soils into the karst-solution channels through which they infiltrated into the cave.

The second half of the seventh millennium B.C.E. represents mainly a warm and dry period. Sometime before 6000 B.C.E. ^{13}C declined sharply, with an upward kick of ^{18}O, signifying a simultaneous decrease in summer and winter rains for a short period. This would indicate climatic and economic stress particularly severe near the desert border. The lower ratio of ^{18}O from about 5750 to 5250 B.C.E., with ^{13}C again becoming relatively high (but lower than in the preceding millennium) marks a climate of winter and summer rains that could have caused rather humid conditions. Later, it started to become dry, reaching a low peak of humidity around 4750 B.C.E. We can, therefore, conclude that during the fifth millennium the climate became much more Mediterranean in character, but mostly warm and dry. This dry spell was followed by a rainy interval with three distinct peaks occurring around 4500, 4200, and 3700 B.C.E., followed by another rapid deterioration with reaching an extreme low humidity rate around 3300 B.C.E.

Unfortunately, the records from the caves on the shores of the Dead Sea are not informative as the more recent, and therefore higher levels most probably abraded the lower older ones. On the other hand investigations based on pollen analyses, plant macrofossils and moluuscs from the pluvial deposits in the Fanyan area, in south Jordan revealed evidence that before 6K B.C.E. i.e. PPNB period this area enjoyed wetter conditions of about 200 mm/y thus being on the margins of the Mediterranean forest. It was followed by a drier phase between 6K to 5.4K B.C.E. probably the transition period between PPNB and PNA. Later the climate became more humid probably the PNA humid phase. Albeit the more humid conditions the arboreal vegetation did not return, which Hunt et al. attribute to increased grazing. [7]

There is also geological and archaeological evidence in the Jordan River Basin indicating that the Early-to-Middle Chalcolithic Period, i.e. the fourth millennium B.C.E., was a wet period.[8] Goldberg arrived at the same conclusions on the basis of substantial alluvations at Shiqmim in the northeastern Negev during the same time.[9] Further south in the Negev and Sinai deserts, evidence of a more humid climate during the sixth to third millennia B.C.E. can be deduced from the many remains of agricultural settlements in these regions.[10]

---

[6] Bar-Matthews et al. op. cit. (1998).

[7] A. Frumkin et al. op. cit. (1997).
C.O. Hunt, H.A. Elrishi, D.D. Gilbertson, J. Grattan, S. McLaren, F.B. Pyatt G. Rushworth, G.W. Barker, "Early Holocene environments in the Wadi Faynan, Jordan". *The Holocene*, 14, pp. 921–930 (2004).

[8] D. Neev and K.O. Emery op. cit., p. 90 (1995).

[9] P. Goldberg, "Geology and Stratigraphy of Shiqmim I – Prehistoric Investigations of Early Farming Societies of the Northern Negev, Israel" *British Archaeological Reports, International Series* 356: 35–46 (1987).

[10] U. Avner, I. Carmi and D. Segal, "Neolithic to Bronze Age Settlement of the Negev and Sinai in Light of Radiocarbon Dating: A View from the Southern Negev" in Bar-Yosef and Kra (eds.) op. cit. pp. 265–300 (1994).

In comparison with only three early Neolithic sites of hunters and gatherers dated to the eighth and seventh millennia B.C.E., 154 habitation sites dated to the Late Neolithic, Chalcolithic and Early Bronze Age indicate a population increase greater than 50-fold. The population density in the more southern Uvda Valley was more than 20 times higher than in the more humid and cooler high altitude Central Negev Mountains, peaking in the third millennium B.C.E.[11]

In Egypt, the trace element composition of the Nile River sediments yield information about climate changes and variations in the river's flow patterns. The composition of sediments originating from the river's tropical headwaters differs from that coming from the wadis of the western and eastern deserts draining into the river north of the cataracts. Since a warm climate strengthens the tropical and subtropical systems and weakens the westerlies Mediterranean system, and vice versa, this chemical indicator serves as an efficient tool to trace climate changes. A core taken from the Nile sediments near ancient Nekhen (Hierakonpolis) shows that from around 6000 to 5200 B.P. (calibrated to around 4800 to 4000 B.C.E.), the Nile brought and deposited headwaters sediments, indicating a strong tropical and monsoon regime. This confirms a warm and dry climate dominating the Near East in that period. It later changed to a humid Mediterranean climate and dryer tropical system from ca. 4000 to 3300 B.C.E. Other core samples taken from the Nile Delta show highest flooding from the Ethiopian plateau around 6400 B.P. (calibrated to around 5200 B.C.E.) whereas a millennium later, between 4400 and 4000 B.C.E., these floods were at a minimum.[12]

As already described, investigations in the Sahara Desert, particularly in the region influenced by the subtropical climate, contributed in great measure to the understanding of the impact of climate changes on pre- and proto-historic Egypt. The northward movement of the subtropical rain belt resulted in a wet interval, causing the inland sand dunes to become inactive and to be covered by vegetation. The last humid period lasted from the latter part of the seventh millennium to the second half of the fourth millennium B.C.E.[13] During the wet periods large parts of the Sahara were covered by vegetation on which thrived animals and groups of hunters and gatherers and then herders. The abundant rain falling on the porous sand dunes created a high groundwater level, leading to the formation of lakes, which were rich in fresh-water fauna. At the end of the humid period these lakes turned into *sabkha*-type depressions, increasing the salt concentration in both residual surface water and groundwater. The fauna and flora gradually perished or migrated, as did the human inhabitants.

---

[11] Avner op. cit. (1998).

[12] O.A. Allen, H. Hamroush and D.J. Stanley, "Impact of the Environment on Egyptian Civilization Before the Pharaohs" *Analytical Chemistry* 65/1:32–43 (1992).

[13] H. Faure and L. Faure Denard, "Sahara Environmental Changes During the Quaternary and Their Possible Effect on Carbon Storage" in: *Water, Environment and Society* pp. 319–322
N. Petit-Maire, "Climatic Evolution of the Sahara in Northern Mali During the Recent Quaternary" in: *International Symposium on Climate* pp. 135–137.
R. Kuper, and S. Kroepelin, "Climate-Controlled Holocene Occupation in the Sahara: The Motor of Africa's Evolution" *Science* 313:803–807 (2006).

M. Rossignol-Strick et al. observed a decrease in the $^{18}O/^{16}O$ ratio of epipelagic foraminifers between 6000 and 4000 B.C.E.[14] If this reflects a general change in the ocean composition, it would indicate the melting of glaciers, and therefore a warm period. Also B. Luz et al. found the ^{18}O minimum between 6700 B.C.E. and 4500 B.C.E.[15] They argue that this decrease is too massive to ascribe solely to influx of melt-water due to changes in ice, and is partly due to an inflow of fresh water resulting from heavy Nile floods.

More recently, Luz proposed an alternative explanation for the isotopic depletion: he claimed that a high influx of low-salinity water entered the Mediterranean Sea from the Black Sea when the rising sea surface reached the level of the Bosphorus.[16] Recent data show that, indeed, the influx of Nile water has a strong influence on the isotopic composition of the water of the eastern Mediterranean.[17]

Ancient sea levels are another source of paleo-climate data, because during glacial periods ocean water is captured by the glaciers, and vice versa. Tectonic activity must also be considered in seas – such as the Mediterranean – that are located in regions of active plate movements. Archaeological investigations by Raban and Galili along the Israeli shoreline show a mean annual rise in sea level of 5.2 mm between 6000 B.C.E. and 4000 B.C.E., with no evidence of tectonic movements in the area in the last 8000 years, while Y. Nir, according to the data from the PPN well of Atlit-Yam (See Appendix I) claims that sea level rise at ca. 6K B.C.E. was 20 mm/y at the beginning slowing to 6 – 8 mm/y at the abandonment of the site at ca. 5.5K B.C.E.[18]

Studying the Dead Sea sediments, D. Neev and J. Hall found extensive rock salt deposition between 5000 and 3000 B.C.E., indicating dry conditions in the northern part of Israel, part of the drainage basin of the lake.[19] This find corresponds with Dead Sea levels calculated from the elevation of outlets from the caves of Mount Sedom which show a very low level, starting around 4500 B.C.E. and continuing until 3200 B.C.E.[20]

---

[14] M. Rossignol-Strick, W. Nesteroff, P. Olive and C. Vergnaud-Grazzini, "After the Deluge: Mediterranean Stagnation and Sapropel Formation" *Nature* 295: 105–110 (1982).

[15] B. Luz and L. Perelis-Grossowicz, "Oxygen Isotopes, Biostratifigraphy and Recent Rates of Sedimentation in the Eastern Mediterranean Israel" *IJES* 29:140–146 (1980).
B. Luz, "Palaeo-oceanography of the Post-glacial Eastern Mediterranean" *Nature* 278:847–848 (1979).

[16] B. Luz, "Post-glacial Palaeo-oceanography of the Eastern Mediterranean." *International Workshop on Regional Implications of Future Climate Change, Jerusalem, Israel* (Jerusalem, Ministry of Environment, 1991).

[17] B. Schilman, *Palaeoceonography of the Eastern Mediterranean During the Late Holocene*, Ph. D. Dissertation, Hebrew University Jerusalem (2000).

[18] E. Galili, et al., *Quaternary Research*.
A. Raban, "Evidence of Sea-land Vertical Changes from Archaeological Sites" The Geological Society Conference, Acco, Guide 4, *The Israeli Geological Society*, 131–811 (In Hebrew) (1991).
A. Raban and E. Galili op. cit. (1985).
Y. Nir, "Middle and late Holocene sea-levels along the Israel Mediterranean coast – evidence from ancient water wells", *Journal of Quaternary Science* 12/2: 143–151 (1997).

[19] D. Neev and J.K. Hall, op. cit. (1977).

[20] A. Frumkin et al. op. cit. (1997).

4.2
The First Technological Revolution – The Pottery Neolithic Period (6000–4500/4000 B.C.E.) (Fig. 5)

On the canvas of the climatic changes described above, the archaeological record from the middle of the sixth to the fourth millennium will be investigated. This background might illuminate some of the reasons for the migrations and transferences of cultures and techniques extending from the Anatolian Plateau to Mesopotamia and to the headwaters of the Nile. As the southern Levant seems to offer the bulk of the relevant climatic and archaeological data, we shall discuss first Syria-Palestine and then Anatolia, ending with a bird's eye survey of the great centers of civilization – Mesopotamia and Egypt.

A general drying process beginning around the end of the seventh millennium B.C.E. marked a crisis that saw the disintegration of cultures along the desert margins, very pronounced in the south and less severe in the north. Until recently, it was assumed that the Levant was uninhabited between the end of the Pre-Pottery Neolithic B and the next culture, Pottery Neolithic.[21] There was no apparent hiatus of human settlement in other parts of the Near East at this time, and it now seems that this presumed gap resulted not so much from a lack of sufficient excavations but from differences in interpretation.[22] A perusal of the literature creates the first impression that each archaeologist has a distinct scheme, terminology and, often, chronology. For the Pottery Neolithic Period (8000 B.P. to 6200 B.P., calibrated to 6000 B.C.E. to 4200 B.C.E.), there is agreement on the early phase, called Yarmukian for M. Stekelis's excavations at Sha'ar haGolan on the Yarmuk River, and the late phase, called Wadi Rabah by J. Kaplan.[23] Y. Garfinkel's renewed excavations at Sha'ar haGolan changed the picture again: his analysis of the pottery suggests that the Pottery Neolithic consists of one phase only, which includes three facies, Yarmukian, Jericho IX (or Lodian) and Nizzanim wares. Wadi Rabah, the old Pottery Neolithic B, is now considered an early phase of the Chalcolithic period, followed by a Middle Chalcolithic Beit Shean and Qatif, and, finally, the mature Late Chalcolithic Ghassul-Beer Sheva.[24]

As we have seen above, climatic conditions improved around 5600 B.C.E. and in its wake the new Pottery Neolithic Yarmukian culture began to flourish in the central and northern Jordan Valley, the northern valleys of Israel and the central coastal plain. Notwithstanding its excellent pottery and sophisticated figurines, the Yarmukian was initially viewed as a cultural throwback. With the exception of a few sunken round floors surrounded by flimsy walls, the lack

---

[21] A. Moore, The Late Neolithic in Palestine, *Levant* 5:36–68 (1973).

[22] L.E. Stager, "The Periodization of Palestine from Neolithic through Early Bronze Age" *COWA* pp. 22–41.

[23] J. Kaplan, *The Neolithic and Chalcolithic Settlement in Tel Aviv and Neighbourhood* Ph. D. Thesis, The Hebrew University, Jerusalem (1958, Hebrew with English abstract).
J. Perrot op. cit. pp. 286–446 (1968).

[24] Y. Garfinkel, *The Yarmukians – Neolithic Art from Sha'ar Hagolan*, Bible Lands Museum, Jerusalem (1999).

of architectural remains led to the image of a primitive semi-nomadic people lacking any sophistication The new excavations at Sha'ar haGolan disproved this theory with the unearthing of extensive and even monumental buildings set in a planned settlement.[25] The most distinguishing characteristic of the Yarmukian is its pottery, sometimes quite coarse, with mat impressions on flat bases. The vessels are often covered by a reddish or brown slip and decorated with painted chevrons, cream-colored bands or chevron bands filled with incised herring-bone pattern between two lines or diagonal hatching. This pattern is sometimes found in northern Mesopotamia (Hassuna ware), Byblos and Merimde in Egypt. It most probably represents stylized palm fronds, underlining the importance of the date palm (*Phoenix dactylifera*) as a dietary staple not only for this culture but also for Egypt and southern Mesopotamia.[26]

Links of the Yarmukian culture with Byblos and other areas of the Near East are also found in pebble figurines and in the second most characteristi- artifact – anthropomorphic clay figurines with cone-shaped heads and 'cowrie- shell' or 'coffee-bean' shaped eyes. We suggest that these figurines personify dates, one of the most important food item in the diet. The cone-shaped heads resemble the fruit and the 'coffee-bean' eyes may represent the stones of the dates. The similarity with contemporary Mesopotamian figurine could be an indication of the spread of the date palm from the Rift Valley eastward as shall be discussed later on.

Very few sites from the Early Pottery Neolithic were found in the northern Negev and southern Sinai. On the other hand, six sites were discovered in the southern Negev. Most of the Pottery Neolithic sites in Israel are in the Jordan Valley and the coastal plain. U. Avner et al. found sites in the Uvda and Eilat area dated by ^{14}C to the Late Neolithic Period.[27]

From around 5500 B.C.E. on, a marked dichotomy arose between the cultures of the southern part of the Levant – Palestine, Lebanon and southern Syria, each with its distinct local characteristics – and the northern part centered in northern Syria, which fell under the influence of the Halafian cultural province centered in northern Mesopotamia. The differences probably have their roots in differing agricultural methods. In the north, the farmers could trust the more or less regular arrival of rain, whereas the south was much more dependent on artificial irrigation. This phenomenon had its parallel in Mesopotamia, where the north also depended largely on rainfall and the south on irrigation. Thus, although some cultural interrelations existed between north and south, in each case the two regions developed their own lifestyles and, therefore, cultures.

The Wadi Rabah culture (or Pottery Neolithic B) followed the Yarmukian culture (Pottery Neolithic A). It had a far more extensive geographic distribu- tion than its predecessor, and architecture, pottery and flint traditions were quite distinct from the preceding Yarmukian complex. The pottery is known in the archaeological literature as Dark-Faced Burnished Ware (DFBW) for its characteristic polished red or black surfaces, a part of the tradition originating

---

[25] Ibid.

[26] S. Lloyd, *The Archaeology of Mesopotamia*, Thames and Hudson, London, p. 19 (1978).

[27] U. Avner, et al. op. cit. (1994).

from the 'Amuq Plain, Cilicia and other parts of southeastern Anatolia. The repertoire of vessel shapes included hole-mouth and bow-rim jars, twin vessels, carinated bowls with splayed rims and other new shapes closely linked with the northern Mesopotamian Halafian ware. Mace-heads and bi-conical sling projectiles appear for the first time, and at Tell Tsaf walls were constructed of mud-bricks laid in a particular 'herring-bone' pattern typical of Syria and northern Mesopotamia but unknown in the Levant before this period.

The most important Wadi Rabah sites in Israel are 'Ein el-Jarba in the Jezreel Valley, Kfar Giladi and Kabri in the Galilee, Tell 'Ali, Munhatta and Jericho VIII in the Jordan Valley. Evidence of Wadi Rabah Neolithic is commonly found at the base of many major tell sites near a spring, such as Megiddo, Beit Shean, Shechem and many others. The reliance on perennial springs has already been discussed and can be traced back to the Pre-Pottery Neolithic period at Jericho.

The climate during this period was mainly warm with both summer and winter rains. As it gradually changed to a typically Mediterranean regime with less or no summer rains, the water recharge during the cooler winters into the subsurface water-bearing strata was sufficient to ensure a flow of the springs during summer, except in series of drought years. The inhabitants of these sites intensified their efforts to produce their food from the land through spring water irrigation and animal domestication, and relied less and less on hunting, a development evidenced by the declining number of arrowheads, compared with the growing number of agricultural tools.

In the Negev during the second half of the sixth millennium B.C.E. the winter and summer rains led to an abundance of settlements inhabited by farmer-herders who also hunted gazelles. Excavations revealed stone-built courtyards, or pens, containing layers of goat dung and agricultural flint tools and arrowheads. However, settlements declined sharply from the end of the sixth to the middle part of the fifth millennium B.C.E. This is confirmed by the Soreq Cave data showing a period of warming. Some summer rains probably continued, but not in amounts sufficient to sustain an agricultural way of life in the desert.

The Pottery Neolithic B period ended in the middle of the fifth millennium B.C.E., and was followed by the Chalcolithic period. The culture of the Wadi Rabah phase can be sometimes called Late Neolithic or Early Chalcolithic, although despite the absence of any copper objects, many scholars feel that it is, indeed, a prelude to the following Chalcolithic period.[28] A gap in settlement shorter than that between the Pre-Pottery and Pottery Neolithic periods seems to be established which might have resulted from another change in climate. A synthesis of the climatic and archaeological data led to the conclusion that an extreme deterioration in climate started somewhere around 4500 B.C.E.

Despite the remarkable achievements of the Pottery Neolithic cultures in the southern Levant, the cultural *floruit* of the period was in the northern parts of the Fertile Crescent, south of the Taurus in areas of northern Syria and

---

[28] Y. Garfinkel, "Neolithic and Chalcolithic Pottery of the Southern Levant" *Qedem* 39 Hebrew University of Jerusalem (1999).

northern Mesopotamia.[29] Syria's contribution is by far the most outstanding. The Dark-Faced Burnished Ware is the earliest pottery, found at sites such as Ras Shamra (Ugarit) V/A–B and the 'Amuq Plain (Tell Judeideh, Tell Dhahab), which was for a long time the ceramic yardstick for the archaeology of the entire Near East.[30] Dark-Faced Burnished Ware and characteristic lithic tools, such as the tanged 'Amuq point and denticulated sickle blades, are found also on the coast, e.g., Tell Sukas, where another peculiar ware, known as 'White Ware', is also found. It appears that these ware types are forebears of the polished pottery of the eastern Mediterranean. By ca. 5500 B.C.E. a new type of pottery – Halaf painted ware – appeared, linking northern Syria with northern Mesopotamia.

Going north, northwest Anatolia with links to the Balkans, the central Lake District is dotted with sites such as Çatal Höyük East, Can Hasan I (levels 4–7), Hacilar (levels VI–IX) Erbaba, Kurucay and Bademagaci. In the southeastern provinces particularly along the Euphrates rescue excavations (due to the construction of the Atatürk dam) revealed sites such as Tepecik and Kumartepe, which show strong local characteristics, but also links to Syria and Mesopotamia. In some sites we notice fortification walls for the first time – a clear indication of deteriorating social developments and rising aggressiveness.

Mesopotamia is divided into two clearly defined geographical entities by a line crossing the *Jezirah*, the "island" between the two rivers, from Hit on the Euphrates to Samarra on the Tigris. The north (later known in history as Assyria), once fertile farming land, is now a barren limestone plateau, with fixed river courses and rain-fed agriculture. The south (known in history first as Sumer and Akkad and then as Babylonia) is an alluvial plain with meandering rivers, canals constantly changing course, and irrigation-dependent agriculture. It is obvious that the fate and cultures of the peoples in the two parts of the country would differ significantly.

The Pre-Pottery cultural phase of the north, known from excavations at Maghzaliyah,[31] had ceramic material related to Çayönü and Bouqras in Syria. The earliest pottery in northern Mesopotamia, known as Proto-Hassuna, consisted of heavy straw or chaff-tempered medium-sized and large vessels, sometimes painted. It is found at Umm Dabaghiyah, Yarim Tepe, Tell Sotto, Kültepe, and the type-site, Tell Hassuna.

The Hassuna and Samarra periods are dated to the end of the seventh and the first half of the sixth millennium B.C.E. Settlements range from modest to large villages with substantial rectangular rooms arranged around courtyards, such as Hassuna level III. Hassuna pottery is incised and has a wide range of shapes, and a glossy red begins to appear in a later phase.

---

[29] M.J. Mellink, "Anatolia" *COWA* pp. 207–220.

[30] R.J. Braidwood and L.S. Braidwood, *Excavations in the Plain of Antioch I*, The Earlier Assemblages, Oriental Institute Publications no. 61, Chicago, Oriental Institute of the University of Chicago (1960).

[31] N. Merpert, R. Munchaev and N. O. Bader, "Investigations of the Soviet Expedition in Northern Iraq" *Sumer* 37:22–54 (1976). See also E. Porada et al. "The Chronology of Mesopotamia, ca. 7000–1600 B.C." *COWA* pp. 77–121.

Still later, a new painted ware was found at Samarra, and was named for its type-site. It is a fine pottery, with geometric and, sometimes, figurative motifs painted black and red. Samarra is in the south, near the beginning of the alluvial plain of Mesopotamia. The principal sites are Choga Mami and Tell es-Sawwan. A host of other, lesser known, sites are found in Upper Mesopotamia. A significant finding at Tell es-Sawwan was wooden molds about $80 \times 30 \times 38$ cm used to form standardized mud-bricks. This technological innovation was used exclusively thereafter and prepared the way for monumental architecture associated with the image of Mesopotamia. The excavators at Tell es-Sawwan found a very large, massively built structure with buttresses at the junctions of walls and at the corners. Other small finds include stone vessels, excellent figurines, and stone stamp seals.[32]

The following Halafian period, named for Tell Halaf on the Khabur, is dated to the second half of the sixth millennium B.C.E. Its pottery is the finest and most beautiful ever made in the ancient Near East and has been compared with Attic vases of the Classical Age.[33] Its highly burnished and lustrous multicolor paintings of static geometric designs are of an unprecedented artistic level. The principal site, Arpachiyah near Mossul, has several architectural phases characterized by large round structures, probably domed, called "*tholoi*" with rectangular vestibules or "*dromoi*". The structures were mostly made of mud-bricks on stone foundations. The abundance of figurines and some tombs in the immediate vicinity of these buildings suggest that they might have served a religious or ritual purpose.[34] Similar finds are reported from Yarim Tepe, Tepe Gawra, and Tell Hassan. Evidence for Halafian or Halfian-inspired pottery is found from the Zagros Mountains across the Mesopotamian plain to the foothills of Anatolia and northern Syria, ending around 5000 B.C.E.

The Halafian culture extended over the headwaters of the Euphrates and Tigris bordering the Taurus and Zagros Mountains. At present the area is situated between the isohyets of 500 mm in the north to 250 mm in the south, enabling a rain-fed agriculture. It is reasonable to assume that a cold and humid period would shift the isohyets southward, thus expanding the rain-fed agriculture into the more barren areas in the south.

The Halafian culture seems to have evolved from the local Neolithic cultures described above. At Tell Sabi Abyad in the Balikh River basin the entire development from Neolithic to Halaf is documented in an uninterrupted process.[35] Farmers who practice rain-fed agriculture – unlike those who depend on artificial irrigation based on harnessing rivers – show particularistic tendencies

---

[32] T.J. Wilkinson, B.H. Monahan and D.J. Tucker, "Khandijal East, A Small Ubaid Site in Northern Iraq" *Iraq* 68:17–50 (1997).

[33] Porada op. cit.

[34] M.E.L. Mallowan and J.C. Rose, "Excavations at Tel Arpachiya-1933" *Iraq* 2:1–178, I–XV (1935). H. Hijara, R.N.I.B. Hubbard and J.P.N. Watson, "Arpachiya, 1976" *Iraq* 42:131–154 (1980).

[35] P.A. Akkermans, "Tradition and Social Change in Northern Mesopotamia During the Fifth and Fourth Millennium B.C." in: *Upon this Foundation – The 'Ubaid Reconsidered*, Carsten Niebuhr Institute Publications 10, E.F. Henrickson and I. Thuesen (eds.), Museum Tusculanum Press, Copenhagen. pp. 339–364 (1989). Oates and J. Oates, *The Rise of Civilization*, Elsevier Phaidon, New York. pp. 98–109 (1976).

in their social structure. So, as expected, this culture seems to have spread not by conquest or annexation by some central power but by independent communities joining to form a horizontal network of settlements ranging in size from about one or two to more than eight hectares. The development of centrally located market places for exchanging goods or know-how and social gatherings enabled the growth of specialized crafts in the various settlements, as can be seen in the extraordinary pottery, figurines, seal impressions, and tokens/calculi found at Sabi Abyad. Interesting parallels can be cited from western Africa, in the Inner Delta of the Niger, where a similar horizontal economic system led to the beginning of urbanization. Also the Indus culture in northern India has been explained along similar lines but less convincingly.[36]

During the succession of these periods in the north, the southern marshes along the Euphrates were largely uninhabited or visited only occasionally by hunting or fishing parties from the surrounding foothills of the Zagros or the northern plains. The natural environmental conditions, as well as the archaeological evidence, suggest that the economy in the foothills was based not only on field crops, but also on an abundance of fish and dates.

By the middle of the sixth millennium B.C.E., some small permanent settlements were founded on natural or artificial elevations in the flood plain and close to the sea which at that time extended much further inland.[37] One such site was Tell el-'Oueili near Larsa. The origin of the simple painted pottery, now known as 'Ubaid 0, and its relation to that from level Ia at Hassuna and to Samarran ware is of great importance in terms of where the first settlers came from and whether they were, in fact, the first Sumerians. The first settlers were fisher-gatherers rather than hunter-gatherers. The temple of Enki, the God of Fresh Water, at the sacred town of Eridu shows seventeen successive re-built layers of identically planned temples, continuing without interruption into the historical period. In each layer a thick layer of fish bones was found on the floor. These remains suggest that an important component of the proto-historical and historical society was descended from the aboriginal group of people dwelling in the swamp areas at the outlets of the Two Rivers and profited from the abundance of fish in the area.

Additional groups arrived at various periods, creating a relatively heterogeneous population. Each immigrating group seems to have had its own pantheon with deities relevant to its subsistence economy. One cannot refrain from asking whether, and when, others might have come from the west, bringing with them the tradition of irrigation agriculture and, above all, the palm tree. The early Sumerian pantheon with the benign Enki as the original principal deity

---

[36] R.J. McIntosh and S.K McIntosh, "Excavations at Jenne-Jeno, Hambarketolo, and Kaniana (Inland Niger Delta, Mali), the 1981 Season" *Anthropology* 20 (1995).

R.J. McIntosh and S.K. McIntosh, *T le Peoples of the Middle Niger – T le Island of Gold*, Blackwell Publishers, Malden, Mass. (1988).

C.K. Maisels, *Early Civilizations of the Old World: The Formative Histories of Egypt, the Levant, Mesopotamia, India and China*, Routledge, London (1999).

Also J.M. Kenoyer, in his book *Ancient Cities of the Indus Valley Civilization* (American Institute of Pakistan Studies, Oxford and Karachi 1998) was the first to suggest a 'horizontal social system' for the Indus Valley civilization.

[37] Kennett and Kennet *Journal of Island Coastal Archaeology* 1:67; review in *Science* 312:1109 (2006).

and Innana as the primeval fertility goddess with the date kernel eyes might be a reflection of the merger of two or more waves of immigration to southern Mesopotamia. Their son "Dumuzi of the Dates" is seen on later 'Ubaidian cylinder seals pollinating female trees with a phallus-shaped instrument.

The date palm tree is of such fundamental importance to the Mesopotamian economy, particularly in the past, that it befits some elaboration. The tree requires specific climatic and hydrological conditions. First, it needs a warm and dry climate. High temperatures and low air humidity are particularly important for fruit setting and ripening. Date cultivation is therefore restricted to regions with high ground water table. To ripen in time, the fruits require mild winters, hot dry summers and, at the same time, a steady supply of abundant water, because the tree transpires large quantities of water, which may endanger the fertility of the tree, as it sucks up only fresh water, leaving a concentrated solution of salts in the water of the root zone. If not flushed by fresh water to maintain a delicate balance between fresh-brackish-saline water solutions in the subsurface, the salt level may impede the tree's fruit-bearing capacity, even if the tree itself could survive salt concentrations approaching seawater salinity.

The newly planted trees had to be irrigated at least for the first few years until the roots reach the groundwater table. Drainage is necessary to avoid the salination problem. Efficient cultivation of a fruit-giving date orchard requires propagation through offshoots, i.e. female suckers taken from a female tree and not by planting seed kernels, which would result in about 50 percent male trees without fruits – a waste of precious farmland and water. The farmers soon. learned to favour suckers from trees with pulpier and sweeter fruit and to gather pollen from the best male tree for artificially pollinating the female trees. This is reflected in the figure of a special deity, Dumuzi of the Dates, in southern Mesopotamia, just mentioned.

As long as there were summer rains in the Near East, the date palm was restricted to refuge areas protected from the atmospheric humidity caused by the rains. The rain shadow of the Syro-African Rift Valley may have been such a region, whereas the Persian Gulf region was then not optimal for this tree. Therefore, it is very possible that date palm horticulture and the trial and error process of finding the best variety with the highest yields of good fruits may have originated in the Jordan and Dead Sea valleys rather than in Mesopotamia, as is normally assumed.

Dumuzi, however, is also the shepherd and perhaps a representation of some of the earliest agro-pastoralists penetrating the drier stretches of land between the rivers and occupying the fringes of the irrigated lands.

Artifacts of the next culture were found at Eridu, an important port on the Gulf coast with beginnings dating to ca. 5500 B.C.E. or slightly earlier. The fine slipped, elaborately applied and delicately painted ware is now known as 'Ubaid 1. Its pottery represents a development of the earlier pottery style influenced by Halaf painted traditions. The first vernacular and monumental architecture – tripartite houses with a T-shaped central hall – is recorded from Tell Abada and represents the most characteristic house type of southern Mesopotamia. The people of the next phase, known as Hajji Muhammad, inhabited the entire southern part of Mesopotamia, and their pottery begins to appear along the shores of the Persian Gulf. This pottery has been shown to originate in Mesopotamia and was probably carried by fishers or traders opening up the sea-lanes towards the Indian Ocean.[38]

On the opposite end of the Fertile Crescent, in Egypt, the early sites of this period are too few and too scattered to supply a coherent picture. Egyptologists view ^{14}C dates with the greatest suspicion or even reject them out-of-hand, and no reliable and generally accepted chronology is available for the early periods. The Neolithic periods were the focus of much recent work, but the results are still very vague, at best, and are lumped together in the even vaguer term "Pre-Dynastic". Many local Palaeolithic and epi-Palaeolithic cultures can be discerned, some in the west and south of the Nile Valley already producing pottery several millennia before its re-introduction in the Neolithic period (see above). There seems to be a consensus that domesticated plants and animals, such as goat and sheep, were introduced from Syria-Palestine. We are sure, for the time being, that the only domesticated animal of African origin was the donkey. The question whether large cattle was independently domesticated in northern Africa is probable but not proven beyond doubt (see Appendix "Surviving the Desert" on pastoral nomadism below).[39]

Like Mesopotamia, Egypt consists of two quite separate geographical units – Upper Egypt (the Nile Valley proper) and Lower Egypt (the Nile Delta adjacent to the Mediterranean Sea). One of the traditional names the ancient Egyptians gave to their own country was 'The Two Lands'.

At the moment, there is no evidence in Egypt for a Pre-Pottery Neolithic culture, as we understand it in Near Eastern terms. In the last decade, however, Rudolf Kuper and Stefan Kroeplin of the Institute of Prehistoric Archaeology of the University of Cologne have shed light on the general environmental background of north-eastern Africa. During the Last Glacial Maximum around 20.000 B.C.E. and Allerod insterstadial (ca. 12,000–10,800 B.C.E.) the eastern Sahara was void of any signs of acquatic environments and, therefore, life. In the following early pre-Boreal some lakes began to form in the south, carbonate lakes in the Sudan and siliceous mud deposits on the Egyptian side.

By 8500 B.C.E. the tropical monsoon belt suddenly moved some 800 km northward (between the 16 and 24 degree N) resulting in summer rains and a semi-humid climate in the south and semi-arid in in the central part of the eastern, and probably also the western Sahara. The array of radiocarbon dates, 500 results from 150 excavation sites, shows a contemporary south to north progression of human settlements of hunter gatherers and fishers, perhaps already practicing some cattle husbandry (see below Appendix III), characterized by "wavy-line" decorated pottery, the oldest ceramics of the world. The subsis-

[38] J. Oates, T.E. Davidson, D. Kamili and D. Mckerrell, "Seafaring Merchants of Ur?" *Antiquity* 51:221–34 (1977).

[39] A.Muzzolini, "La préhistoire du bœuf dans le Nord de l'Afrique durant l'Holocène" Ethnozootechnie (1983).

M. Zohar, "Pastoralism and the Spread of the Semitic Languages" in Pastoralism in the Levant – Archaeological Materials in Anthropological Perspectives, Monographs in World Archaeology No. 10, (eds.) O. Bar-Yosef and A. Khazanov Madison, Wisc, Prehistory Press, 165–180 (1992).

tence was enriched by a variety of wild grasses and tubers. The near complete lack of settlements in the Nile Valley, with the exception of el-Kab in upper Egypt, suggests that the marshy and hazardous grounds densely occupied by crocodiles and hippopotami were avoided.

Around 7000 B.C.E. human settlement became well established throughout the Eastern Sahara. The arrival of Near Eastern bifacial knapping techniques for flint tools accompanied by sheep and goat herding around 6000 B.C.E. is an indication of contacts with south-west Asia, and perhaps migrations. Whereas the climatic and archaeological record of the Levant shows a period of crisis around that time, there is no sign of a rupture or change of the climate in north-eastern Africa until ca. 5300 B.C.E. when the regular summer rains became rarer and receded south. We can now see that living conditions deteriorated gradually, the population relied increasingly on their herds and had to concentrate around the ever smaller growing lakes and a few springs, e.g. the oasis of Kharga, Farafra, and Dakhla. Some found refuge in the hill country in the north, such as Abu Muhariq Plateau, or the south Gilf Kebir, wheras the majority simply followed the rains southward with mobility as the key factor for their survival and the spread of specialized pastoralism, their culture, and their languages.[40]

The exodus from the Sahara coincides with the first rise of sedentary villages in the Nile Valley. The earliest agricultural communities are found in the Fayum Depression, west of the Nile Valley. G. Caton-Thompson's excavations of the first two sites uncovered flimsy shelters, some hearths, storage pits and some debris, including simple buff or brownish pottery and an abundance of flint tools and weapons. She called this culture Fayum A. This inventory and the many animal bones indicate that hunting was still of great importance. The oldest flint assemblage, called Fayum B or Qarunian, is of epi-Palaeolithic ancestry and has probably no links with the agricultural hunters.[41] More recently, several campsites with Fayum B characteristics were discovered around Qasr es-Sagha.[42] There is growing evidence for a developed Fayum B pottery Neolithic in the Delta, particularly at Merimdeh Beni Salameh and a few other sites.[43] At Merimdeh bowls were found with the typical Yarmukian palm-frond (or herring-bone pattern) motif between two parallel lines on a reserved background, seeming to indicate Palestine as the origin of this culture, despite some important differences.

---

[40] R. Kuper and S. Kroepelin. "Climate Controlled Holocene Occupation in the Sahara: Motor of Africa's Evolution" *Science* 313:803–807 (2006).

[41] G. Caton-Thompson and E. W. Gardner, *The Desert Fayum*, Royal Anthropological Institute of Great Britain and Ireland, London (1934).

[42] B. Ginter, W. Heflik, J.K. Kozlowski and J. Silwa, "Excavations in the Region of Qasr el-Sagha, 1979: Contributions to the Holocene Geology, the Pre-Dynastic and Dynastic Settlement in the Northern Fayum Desert" *Mitteilungen des Deutschen Archäologischen Instituts* 39:105–170 (1980).

[43] J. Eiwanger, "Dritter Vorbericht über die Wiederaufnahme der Grabungen in der neolithischen Siedlung Merimde-Benisalame" *Mitteilungen des Deutschen Archäologischen Instituts in Kairo* (MDAIK) 36:61–76 (1980).
Eiwanger, Vierter Vorbericht ditto, MDAIK 38:67–82 (1982).
F. Debono, "Heliouan-El Omari: Fouilles du Service des Antiquités, 1934–1935" *Chronique d'Egypte* 21:50–54 (1945).

Excavations in the 1930s at el-Omari near Helwan, and Maadi, both now southern suburbs of Cairo, marked further developments.[44] Maadi pottery, also found in other sites in and around Cairo, is probably descended from Fayum A, but is of a higher quality and includes a larger range of vessel types. Most important are the new excavations of Maadi-related material at Buto/Tell el-Fara'in in the western Delta.

The information about Upper Egypt is not much better, in spite of the rich loot hauled from various cemeteries during the last two centuries. Hardly any living sites were found, however. Sir Flinders Petrie was the first to bring order into this mass of rather confusing material by arranging pots from Naqada and Diospolis Parva into a relative sequence ending with the historical First Dynasty. Every scholar of renown saw it as a challenge of scholarship to supply his or her own scheme – an extremely tiresome situation.[45] Around Thebes a culture of epi-Palaeolithic characteristics was identified, called Tarifian after the cemetery of et-Tarif, and tentatively dated to the first half of the fifth millennium B.C.E. This "culture" is similar to Fayum B, and like it, has no links to later developments.

In the 1920's G. Brunton excavated a cemetery at Deir Tasa south of Assiut on the east bank that is estimated to yield the oldest pottery of Upper Egypt – simple red and brown vessels with a dark rim – perhaps the earliest evidence for the well-known Black-Topped Pottery of the later periods. The first major phase of the pre-Dynastic age in Upper Egypt is the Badarian, characterized by its red polished Black-Topped Pottery, with no handles or only some ledge-handles and a profusion of hole-mouth jars. It is not known whether these and other "cultures" with similar types of pottery, i.e. Amratian and Gerzean, represent consecutive phases, expressions of ethnicities or regional styles and fashions.[46] In general, there was a rapid development of painted pottery, figurines and various art objects in the latter part of the fifth and the transition to the fourth millennium B.C.E., which belongs to the next cultural phase in the Development of the ancient Near East, the Chalcolithic Period.

Clear evidence can be found all over the ancient Near East of sedentary life, established agriculture, husbandry, a rich ceremonial life and the first signs of a complex ranked society by the end of the Pottery Neolithic period half-way through the fifth millennium B.C.E.

---

[44] O. Menghin and M. Amer, T he Excavations of the Egyptian University in the Neolithic Site at Maadi: First Preliminary Report, Misr-Sokar Press, Cairo (1932).

O. Menghin and M. Amer, T he Excavations of the Egyptian University in the Neolithic Site at Maadi: Second Preliminary Report. Government Press, Cairo (1936).

Menghin, O. "Die Grabung der Universitaet Kairo bei Maadi, Drittes Grabungsjahr" Mitteilungen des Deutschen Instituts für Ägyptische Altertumskunde 5, 111–118 (1934).

[45] H.J. Kantor, "Egypt" COWA pp. 7 (1992).

W. Kaiser, "Zur Südausdehnung der vorgeschichtlichen Delta-Kulturen und zur frühen Entwicklung Oberägyptens" Mitteilungen des Deutschen Archäologischen Instituts 41, 61–87 (1985).

D. Wildung, Ägypten vor den Pyramiden – Münchener Ausgrabungen in Ägypten. von Zabern, Mainz (1981).

[46] W. Kaiser, Ibid.

D. Wildung, Ibid.

4.3
The Metallurgical Revolution – The Chalcolithic Period (4500/4000–3500/3000 B.C.E.)

The division between the Pottery Neolithic and Chalcolithic periods is quite arbitrary and differs for each region of the ancient Near East. It hinges on the presence or absence of copper artifacts and whether an absence is real lack or the chance of not having found some. As we have seen, starting in eastern and southeastern Anatolia and northern Mesopotamia between the late seventh and fifth millennia B.C.E., one consequence of steadily improving kilns was an ever wider range of experimentation with metal-containing rocks, particularly copper (*chalcos* in Greek).

At first, copper nuggets found in riverbeds were used as ornaments, like colored stones or gold nuggets. It was discovered, however, that this material, also known as native copper, differed from other stones in that it was malleable and soft enough to be cut by a flint tool. It was also found that when heated in the fire, copper ore would liquefy, and when cooled, would take the shape of whatever container it was poured into. Whereas worn or broken stone tools had to be discarded, the new material could easily be re-smelted and re-used. These events could have taken place only in a region where native copper and copper ores with some trace elements, such as arsenic and antimony, were available. These two elements reduce the smelting temperature, and any stone containing these elements – mainly sulfide ores – thrown into the fire could produce metallic copper. Such ores were found in the north of the Fertile Crescent in a mountainous belt stretching from Anatolia to the Caucasus, the Elburz Mountain in northern Iran and into the Hindukush. Once the first step was taken, advanced production methods followed. Different methods of smelting were developed for copper ores that did not contain arsenic and that were in the form of carbonates, such as were found in the more southern parts of the Near East. These methods required raising the temperature by increasing the inflow of air, either by creating a draft, as is done in pottery kilns, or by using bellows made of hide. Considering the development of metallurgy as a logical evolutionary process from the simpler to the more complex, the various finds of metal objects can be placed in a space-time-information coordinate system. Accordingly, the first step occurred at a certain location (or a few locations), requiring the minimum amount of knowledge (i.e. ore requiring low temperature like copper-arsenide alloy). This step was followed by increased knowledge over both time and distance until, eventually, more sophisticated methods of production and use were developed and spread over other regions.

With the new techniques came new forms. Pottery took on new shapes and decorations, while the shape of copper artifacts embraced the decoration as well. New methods of chemical and mineralogical analysis enable scientists to trace the sources of the raw materials for the pottery and metal tools to where they were mined. It is now possible to analyze data on the source of the materials, for the artifacts and the traditions of their forms and styles, together with data about climate changes during this period.

After a warm and dry period from around 4800 to 4600 B.C.E., a rapid improvement of the climatic conditions took place in all areas of the Near East. The weather turned colder and more humid during what is known in Europe as the Atlantic period. The level of Lake Van rose rapidly and the relative humidity curve (Fig. 3a) reached a peak between 4300 and 4400 B.C.E. A similar picture can be seen from the elevations of the ancient levels of the Euphrates and the Tigris between 4500 and 4200 B.C.E.[47]

Such a climate most probably spelled harsher living conditions and a relative decline in agricultural productivity on the plateau and mountainous regions of Anatolia, the Caucasus and Iran. In contrast, the southerly plains of the Near East became more hospitable and attractive which was particularly pronounced in the lower stretches of the Two Rivers, where the inhabitants were well supplied by floodwater. Once the farmers learned to control the floods, they were able to profit from the abundance of water and fertilizing silt, transforming the hitherto marshy wastelands into the most fertile region of the ancient Near East.

During the first centuries of the fourth millennium B.C.E. we find scanty evidence of metal tools, ceremonial objects and jewelry throughout the Levant but with large lacunae in the distribution pattern. By now, metallurgy had reached a relatively high level of sophistication. The resources in the Arava Valley and Sinai were exploited during this period.[48] The magnificent finds from Nahal Mishmar[49] and Abu Matar, in the close vicinity of Beer Sheva, exhibit metal works in styles intimately connected to the northern highlands and Mesopotamia as well as Egypt, suggesting that the Levant was part of a metal-trading chain from eastern Anatolia to northeastern Africa.[50] Despite the absence of copper artifacts cultures once designated 'Neolithic' are increasingly being reclassified as 'Chalcolithic' because a growing number of scholars believe that the similarities in culture – particularly in the pottery – outweigh the differences.

By 4500 B.C.E. agricultural settlements were established in areas that today lie in the arid zone with an average annual precipitation (AAP) of less than 300 mm, and even in extreme areas , where the AAP is now less than 100 mm.

---

[47] Lemcke and Sturm op. cit. (1997).
Kay and Johnson, op. cit. (1981).

[48] B. Rothenberg, R.F. Tylecote and P.J. Boydell, Chalcolithic Copper Smelting. *Archaeo-metallurgy*, Institute of Archeo-metallurgical Studies, London (1978).
B. Rothenberg, "Timna" *Excavations and Surveys in Israel* 3:102–103 (1984).
A. Hauptmann and G. Weisberger, "Archaeometallurgical and Mining-Archaeological Investigations in the Area of Feinan, Wadi Arabah" *Annual of the Department of Antiquities of Jordan* 34:19–437 (1987).
O. Ilan and M. Sebanne, "Metallurgy, Trade and Urbanization of Southern Canaan in the Chalcolithic and Early Bronze Age" in: *L'Urbanisation en Palestine à l'âge du Bronze Ancien. Bilan et Perspectives des Recherches Actuelles, Actes du Colloque d'Emmaus 20–24 octobre 1986*, (ed.) P. de Miroschedji. BAR International Series, London 527(II), pp. 20–24 (1989).

[49] P. Bar Adon, *The Cave of the Treasure*. Israel Exploration Society, Jerusalem (1980).

[50] J. Perrot, "The Excavations of Tell Abu Matar, near Beersheba" *IEJ* 5:73–84 (1955).
J.B. Hennessy, *The Foreign Relations of Palestine during the Early Bronze Age*, Colt Archaeological Institute Publications, London. pp. 41–43 (1967).

These regions contain ample evidence – such as high river terraces, water vegetation, etc. – of higher precipitation rates, which is supported by paleo-climate proxy data time-series. We maintain that, in the face of this evidence, it is difficult to argue against climate change as an important trigger for cultural change. Moreover, we argue that as time progressed, the favorable climate fostered the economic and cultural prosperity of the entire Near East.

Not all archaeologists and historians share our conclusion and relegate the effect of paleo-climate to a marginal role, maintaining that the archaeological evidence in itself provides insufficient foundation for concluding changes of natural conditions.[51] T.E. Levy considers climate a factor in the collapse of the Chalcolithic culture around 3500 B.C.E., but not in the growth of more complex societies during this period. Rather, he attributes this growth to socio-economic development: *"These include a marked growth in the human population, the emergence of craft specialization and metallurgy, and the division of sites into spatial hierarchies with settlement centers which coordinated social, economic and religious activities"*. He also maintains that these changes, particularly in southern Palestine, reflect the emergence of chiefdoms where the rulers function within and thus supported by a religious organization. Therefore, such societies with a vertical hierarchical ruling system representing the divine powers and assuming responsibility for taking care of society evolved locally.[52]

The economic base of the Chalcolithic culture was rooted in agriculture and husbandry, not much different from the preceding or following cultures. However, this base was considerably broadened by the introduction of new plant species or diversification of known plants. For example, besides the customary cereals, pulses, dates, pistachio and other indigenous fruits and legumes, olives were used increasingly for their oil and the grape has arrived from the Caucasus.

The most economic advance was in animal husbandry. Until now, sheep and goats were used for their meat, skin, and perhaps, bones. A. Sherrat called certain advances in animal herding the 'Secondary Product Revolution' which had by now reached the Levant from the north.[53] Above all, it included the use of the milk of animals to produce dairy products such as yogurt, butter, and cheese as fresh milk was probably not consumed as only some northern European genetic groups had developed sufficient lactose tolerance. Sheep in cold climates had developed a fine and thick underhair and were now shorn for wool, opening up a whole new range of clothing and housing potentials. Last, but not least, some animals were used for carrying objects, pulling carts or riding. At Shiqmim, in southern Palestine, horse bones (*Equus caballus*) were found, the oldest remains of an animal originating in the steppes of Eurasia, an

[51] R. Gonen, "The Chalcolithic Period" in: *The Archaeology of Ancient Israel*, ed. A. Ben-Tor, Yale University Press and The Open University of Israel, Tel Aviv. pp. 40–80 (1992).

[52] T.E. Levy, op. cit. pp. 226–244 (1998).

[53] A. Sherrat, "Plough and Pastoralism: Aspects of the Secondary Products Revolution" in: *Patterns of the Past. Studies in Honour of D.L. Clarke*, I.R. Hodder, G. Isaac and N. Hammond (eds.), Cambridge University Press, Cambridge, pp. 261–305 (1981).

important indication of relationship with the north. Donkeys were probably Introduced from Africa somewhat later during the transitional period to the Early Bronze Age.

Settlements expanded considerably. From the Golan in the north of Israel to the Negev and Sinai in the south are found an abundance of settlements of all sizes. U. Avner has documented some hundred and eighty stone-built habitation sites, forty tent camps as well as threshing floors, granaries and terraces (Plate 8). Comparing the distribution of dated finds on the precipitation curve it becomes clear that settlements proliferated during the period of high precipitation, strengthening Avner's conclusion that the inhabitants of these settlements were farmers and herders. A medium-sized settlement was 1,600 square meters; three were triple that size. The finds all indicate farming activities – tabular scrapers for processing food, shearing or skin working, flint adzes, stone hoes, sickle blades and a large number of grinding stones. Granaries were paved with flagstones and threshing floors, some of which were dug into the rock surface, and their surroundings were littered with worked flint. The shape, size and micro-wear analysis of the flint suggest they were used as blades inserted in threshing sledges.[54]

The climatic conditions were optimal, ameliorating living conditions even in present desert areas the extent of which are yet unknown. Surveys in the Arabian Peninsula increasingly produce evidence for lakes and growing numbers of settled communities (unpublished except the *Abstracts* on the updated website of the *Seminar for Arabian Studies*). Winter rains created floods in wadi beds, which were used for irrigation by a method known today as water harvesting. The inhabitants built low soil and stone embankments crossing the water channels draining small water catchments areas. The floodwaters, carrying mud from the surrounding hills rose behind the embankment, soaked the soil, fertilized, and enlarged the fields.

Some regions remained relatively dry, particularly during the summer. To ensure sufficient drinking water for themselves and their livestock, the ancient inhabitants dug wells in the shape of shallow bells to reach groundwater in the saturated fractured limestone that overlaid impermeable chalk and marl layers. Digging wells was not a new idea as a well was found in a now submerged Pre-Pottery Neolithic village near Atlit.

The architecture of the Chalcolithic period also does not exhibit any innovations. Stone foundations are the rule, mostly topped by hand-made mud-bricks. The layout shows a variety of ground plans, from freestanding houses to rather amorphous combinations of rooms in an agglutinative fashion. The idea that the subterranean spaces cut into the loess near Beer Sheba were dwelling is now being abandoned; these rooms were probably used for storage.[55]

In contrast, metallurgy progressed enormously. The most extraordinary examples in the Levant were found in the 'Cave of the Treasure' in Nahal Mishmar, a canyon on the western shore of the Dead Sea. The cave was used for habitation for long periods and contained ash layers of more than two

---

[54] U. Avner, op. cit. (1998).

[55] J. Perrot, op. cit. (1955).

meters. Many house-hold objects and more than 400 cult objects were discovered, including hundreds of mace-heads (Fig. 7), axe blades, chisels, small decorated vessels, drinking horns and symbolic axes or 'standards' decorated with rams, crowns or stands decorated with house facades, birds and geometric incisions were created in copper rich in arsenic by a sophisticated 'lost wax method'.[56] The first metallurgical analyses of the copper showed that it was an arsenic alloy,[57] not known then in the Levant, which led the excavator, Bar-Adon, to assume that the ores originated in eastern Anatolia or Transcaucasia. Later such ores were found in southern Sinai, which could have led to the conclusion that the source of the treasures artifacts, or the ore from which they were produced, was in the Levant. Yet this conclusion is still questionable, as modern methods of analysis show that the composition of the treasure's items, as well as other copper artifacts found at Chalcolithic sites in Israel, are copper-arsenic-antimony alloys, whereas the ores found in

Plate 8. Aerial photograph of a threshing floor and dwelling sites of Chalcolithic Age in the Uvda Valley, Negev (Area of the floor about 20 by 120 m.) (Courtesy U. Avner)

---

[56] P. Bar Adon, op. cit. (1980).

[57] C.A. Key, "Ancient copper and copper-arsenic alloy artifact composition and metallurgical implications" *Science* 146:1578–1580 (1964).

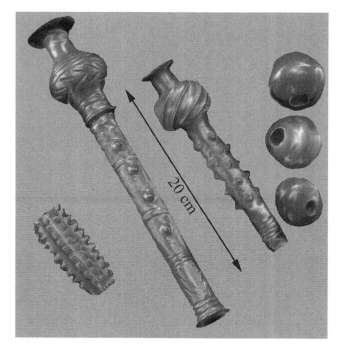

Fig. 7. Copper mace-heads and 'batons' from the Cave of Treasure

the Sinai contain only traces of antimony.[58] The suggestion that the source of the ores or even the tools was in the northern parts of the Near East or even beyond its border is, therefore, still valid. Moreover, stylistically similar contemporary objects are known from Anatolia and from Tepe Hissar in northern Iran, made from the same type of copper-arsenic alloy. Whether the objects were imported as finished products from these regions or produced locally from imported ores is still a matter of debate. Copper tools from other parts of the Levant suggest that in due time the artisans learned also to use the local non-arsenical copper ores, such as those from Feinan or the Sinai.

Clearly, the bulk of the Nahal Mishmar treasure consists of "ritual objects". The mace-heads were often hammered out of copper sheets and were too small and insubstantial to be actual weapons. Stone mace-heads were found in abundance at various sites ranging from Mesopotamia throughout the Fertile Crescent to Egypt, where it became the insignia of royalty. In the Levant, the copper mace-head may have designated rank and was probably used in ceremonies. At Can Hasan in Anatolia, a perforated mace-head of solid copper

---

[58] S. Ilani and A. Rosenfeld, "Ore source of arsenic copper tools from Israel during Chalcolithic and Early Bronze ages" *Terra Nova* 6:177–179 (1994).
S. Shalev and J.P. Northover, "The Metallurgy of the Nahal Mishmar hoard reconsidered" *Archaeometry* 35/1:35–47 (1993).

and a copper bracelet dated to 4900 B.C.E. were found near the body of a man trapped in a fire.[59]

A clearly defined and obviously hereditary ruling class appeared during the Chalcolithic period. At Nahal Qanah in central Palestine, a large quantity of gold artifacts was found at a burial cave containing graves. The archaeologists who excavated the cave suggested this represented evidence of a hierarchical social organization.[60] As mentioned, T.E. Levy although arguing for a cultural evolution from the local Neolithic, finds much evidence for the emergence of a group-oriented chiefdom, whereas Stager suggests a religious elite, perhaps shamans or an early priesthood.[61] The latter is supported by the earliest and near contemporary documents from Mesopotamia, which mention the *ensi*, a priest-king as the leader of a community.

The emergence of a ruling class is also reflected in burial customs of other regions. During the Neolithic period, the dead were often buried within the settlement and sometimes in the houses. We cannot be sure whether the rooms continued to be used or were abandoned after interment. Burial outside the settlement was common in Neolithic sites in eastern Anatolia and northern Iran. In Palestine, however, the first formal cemeteries outside of settlements were established during the Chalcolithic period, alongside the rather crude custom observed at some sites where human bones were found in rubbish heaps. In the hill country and the coast, natural or man-made caves were preferred. A peculiarity of these areas is the use of ossuaries made of terra cotta. They were shaped like granary houses or jars, often with architectural or anthropomorphic details and sized according to the longest bones, i.e. the upper leg bone, which indicates a custom of secondary burial after decomposition of the body. In the south, circular tumuli were erected over round graves or cist graves containing not only the bones of the deceased but what appears to be a standard set of utensils, such as a V-shaped basalt bowl and a cornet. Along the Jordan Valley cist graves were often built with large standing slabs of stone, a tradition leading to the dolmens and other megalithic tomb monuments of later periods. In the deserts further south, in the Sinai Peninsula, round charnel houses known as *nawamis*[62] were built of stone slabs and corbelled roofs, a custom found far away in Arabian Peninsula and southern Iran.

Burial sites are often in spatial and spiritual relation to certain cult sites. The most impressive temple stood alone, perched on a cliff overlooking the Dead Sea at ʻEin Gedi, a short distance north of Nahal Mishmar, which makes

[59] J. Mellaart, op. cit, p. 120 (1975).

[60] A. Gopher and T. Tsuk, *Ancient Gold – Rare Finds from the Nahal Qanah Cave*, Israel Museum, Jerusalem (1991).
A. Gopher, T. Tsuk, S. Shalev and R. Gophna, "Earliest Gold Artifacts in the Levant" *Current Anthropology* 31:436–443 (1990).

[61] Levy, op. cit. (1998).
L.E. Stager "The Periodization of Palestine from Neolithic through Early Bronze Times", in *COWA* vol. 1:22–41 (1992).

[62] *nawamis* in Arabic means "flies", as according to the local bedouin tradition, these structures were built for protection by the 'Children of Israel' when they crossed the Sinai desert and were attacked by flies/mosquitous.

it quite likely that the treasure discussed above had came from this temple.[63] A stone temenos wall linked the main sanctuary – a broad-room building with an altar or podium and benches along the walls – with another room, perhaps the treasury, and a gatehouse with benches. In the center was a large courtyard with a round libation altar. The main building was filled with the ashes of sacrifices; among the finds was a bull carrying two churns and an animal-churn figurine. Another impressive temple complex of similar layout but located inside the settlement and decorated with murals was found at Tulleilat Ghassul.[64] Rectangular buildings with round platforms and seven *massebot* (standing stones or *stelae*) were found at Gilat in the northern Negev. Also found there were figurines of a ram carrying three cornets and of a painted woman carrying a churn.[65] The motif of body paint and masks is also found in Hacilar in Anatolia, linking these figurines with the masks on the mural of Telleilat Ghassul. The attention to the sexual organs seems to symbolize fertility. In the southern Negev highlands we find open-air sanctuaries characterized by standing stones.

In the Golan, large settlements of broad-room houses linked in long chains dotted the countryside. Some houses contained large pillar-like figures with tops shaped like bowls presumably for offerings. Some were equipped with rams' horns and a goatee beard presenting according to the excavator C. Epstein a divinity responsible for the fertility of the flocks.[66]

Based on the cultural elements described above, as well as skeleton studies, a few scholars, such as J. Mellaart, A. Gophna and D. Ussishkin, suggested a northern origin for the mature Chalcolithic culture of Palestine. Most Israeli and other western scholars working in the Near East, however, insist that it evolved from existing local cultures; and both are probably right. Yet even the "northern origin school" does not relate the migrations with the climatic deterioration of the northern highlands. Once climate change is considered as a cause, other observation fall in place such as, for example, the extraordinary increase in population, rise in number and size of settlements, and increased productivity in nearly all economic endeavors on which all students of this period agree.

We claim that a cold spell did, indeed, negatively affect the living conditions in the steppes north of the Black Sea, the Caucasus and the mountainous regions of Anatolia and Iran during the first half of the fifth millennium B.C.E. Severely cold and long winters and short warm summers restricted the growing season, and the agricultural yield decreased. Growing social tension and warfare are indicated by the fact that settlements in Anatolia were heavily fortified for the first time during this period. Groups most affected by this trend began to move southward, more or less in the footsteps of their Ne-

---

[63] D. Ussishkin, "The Ghassulian Shrine at 'Ein Gedi" *Tel Aviv* 7:1–44 (1980).

[64] J.B. Hennessy, "Telleilat Ghasul. Its place in the Archaeology of Jordan" in: *Studies in the History and Archaeology of Jordan*. A. Hadidi (ed.), Department of Antiquities, Amman (1982).

[65] D. Alon, and T.E. Levy, "The Archaeology of Cult and the Chalcolithic Sanctuary at Gilat" *Journal of Mediterranean Archaeology* 2:163–221 (1989).

[66] C. Epstein, "Basalt Pillar Figures from the Golan and Huleh Region" *IEJ* 38:205–223 (1988).

olithic ancestors under similar conditions. By now their social organization had evolved into a tribal order under the leadership of a shaman, or chief, or both. A new warrior class emerged, equipped with new weapons – mainly copper or hematite maces heavier and less brittle than stone maces, axes and other metal weapons. They developed new fighting tactics, and their chieftains may well have been horse-mounted, enabling them to command more effectively. Comparisons with various peoples at roughly the same level of social and technological development show that religious personages, shamans, seers and similar 'spiritual leaders' accompanied these warriors in battle, inciting and urging them on. Examples from the Biblical stories of the war against the 'Amalekites and the conquest of the Land of Israel, the Homeric poems, Greek and Irish mythology, the battle scenes in the Bhagavadgita, and many others come to the mind. The copper items of Nahal Mishmar could easily be the 'baggage' of such hypothetical spiritual leader or leaders, which then were perhaps piously deposited first in the temple of 'Ein Gedi, and later, probably due to troubled times, in the cave where they were found.

This chain of events repeated itself throughout the entire known history of the Near East, which seems to consist of an endless register of roving peoples and invasions. The occupation by aggressive foreigners always resulted in a more complex social ranking and development of hierarchies, ranging from chiefdoms to incipient states, as will be discussed in the following chapters. It is true, however, that the archaeological record often does not reveal the influx of foreign ethnic elements. The local population continues to produce their pots and build their houses with materials close at hand. The newcomer is often, archaeologically speaking, 'invisible' except for a few tantalizing clues, such as tombs of warriors, a new art style, or the sudden appearance of exotic items, which, however, might also have been procured by trade. For example, if the correspondence of the Old Assyrian trading colonies in Anatolia had not been found, we would have never guessed their existence, as nothing in the material culture indicates it.

Once the climate became more humid again and what used to be the Syrian-Mesopotamian desert became fit for human habitation, the improved weather conditions enabled these peoples to maintain contact for a while. The eventual merger of the newcomers with the local population resulted in a variety of cultures, such as the Ghassulian-Beer Sheva and other Chalcolithic cultures in the Levant, the Upper 'Ubaidian in Mesopotamia and even the pre-Dynastic cultures of the Delta in Egypt. There certainly will be many surprises forthcoming in the discussion about the foundations of the Pharaonic civilization and whether it might as well be another outgrowth of this immigration from the north or from the increasingly arid desert in the west or most likely both.

We think that the similarities between the 'Ubaidian and Ghassulian culture have their roots in a common heritage as well as in distant but direct influences or trade contacts. Similarities can be seen in the band-painting of pottery, in metallurgy, in architecture, and in paintings such as the murals in Telleilat Ghassul, depicting, to our opinion, an introduction scene, and comparable with a common later Mesopotamian cylinder seal motif. This argument seems to apply also to linguistic features in the surviving pre-Semitic names of rivers and sites in Palestine, Syria and Mesopotamia. The obvious emphasis on rituals and ceremonies in many aspects of their material culture points to a similar spiritual life created out of the same need for one group of people to dominate another, comparable to the creation of the early Maya civilization by invaders from the east and north of Mexico and their eventual merger during the mature and late phases. The phenomenon is seen elsewhere in the world as well, such as early China, the culture of northwestern India, and many others.

The situation in Mesopotamia is, unfortunately, extremely unclear during this period. The end of the Halafian pottery style is also shrouded in mystery as the late 'Ubaid style spread throughout Mesopotamia and northern Syria. The next phases, now known as 'Ubaid 3 and 4 (formerly 'Ubaid I and II), correspond to levels XI to VI at the *ziggurat* or temple tower of Eridu. By now, all the important sites of southern Mesopotamia were settled, their temples and clerics were in place, and the first *ziggurats* were being constructed. Towards the end of the period, the painting was rather less skillful and the pottery declined, supplanting the magnificent Halaf ware in the north. S. Lloyd sees a spread from the head of the Persian Gulf in the south towards the north, gradually supplanting the older Halafian-related pottery.[67] The most important site is Tepe Gawra levels XX to XVI, but Tell Abada I and II and Tell Braq also yielded important finds from this period. Around 5000 B.C.E. the Halafian settlements in the north ended, and a new culture, characterized by southern Mesopotamian 'Ubaidian pottery, began its triumph throughout Mesopotamia and beyond. Not only does its pottery clearly distinguish the 'Ubaid culture from its predecessors, but also the use of bent clay nails for decorating mud walls, the beginnings of cylinder seal glyptic art, simplified female figurines, and a characteristic type of house *"consisting of a roofed central room, either rectangular or cruciform, running the length of the house and two rows of smaller rooms on each side"*.[68] The temples, however – which resemble older examples from Eridu and represent the first truly monumental architecture of the region – are the most distinguishing feature of the culture. The cultural unification of northern and southern Mesopotamia, ranging from the Plain of the 'Amuq phases D and E to the Persian Gulf made the mature 'Ubaid culture the unquestionable leader of the ancient Near East's march towards urbanization and higher civilization.

As already mentioned the basic difference in agricultural systems between the Halafian in the north and the 'Ubaidian in the south is of paramount importance. Rain fed farming was dominant in the north, while the south depended on artificial irrigation. At the beginning of the fifth millennium, the Halafian culture was restricted to the higher areas with more rains, whereas the 'Ubaidian 'people' descended to the low level plain to reach the river and the Persian Gulf. Trade, accompanied by cultural exchange, was apparently of considerable importance for their economic, social, and cultural development. The large quantity of 'Ubaidian pottery found in all sites around the Gulf led some researchers to speak of colonization When the climate improved,

---

[67] S. Lloyd, *The Archaeology of Mesopotamia*, Thames & Hudson, London. p. 19 (1978).

[68] M. Roaf, "Ubaid Houses and Temples" *Sumer* 43:80–90 (1984).

all cultures flourished and expanded. The rain fed agriculturists in the north maintained their dispersed and non-centralized social order. The 'Ubaidians, however, because of their dependence on a single source of water, developed pivotal nuclei for administering the water. These centers grew into conglomerates of settlements, eventually forming cities and city-states. A comparable general pattern of evolution can be observed elsewhere in the ancient Near East as well – societies born in mountainous regions and based on rain fed agriculture remained decentralized or relatively modest in size, whereas the singularity of the water supply, as for instance in southern Mesopotamia and the Nile Valley, brought about the rule of a central power during a rather early stage of civilization.

A new turn in the discussion was supplied by the excavations at Hamoukar in northern Mesopotamia, nearly half-way between Tell Brak and Tepe Gawra, northwest of Niniveh/Mossul. Here Clemens Reichel of the University of Chicago has found a large fortified proto-urban settlement with monumental architecture and claims that urban civilization in the north developed contemporarily with the south.[69] As it is common that most archaeologists attribute an extraordinary significance to their respective sites, such claims should be viewed with a grain of salt.

There are certain parallels in the Levant – Syria, in the north, was under the influence of Halaf-inspired Neolithic cultures, whereas Palestine, in the south, was more dependent on water-harvesting irrigation and, therefore, developed its special characteristics such a possibly more developed hierarchy. The present Syrian Desert becoming green and passable as a result of the humid climate may answer the question raised by R. Amiran on similarities between 'Ubaidian and Palestinian Chalcolithic pottery.[70] The same can be said about the relationship with pre-dynastic Egypt across the Sinai.

In this case, as in others, one has to free oneself from the today's impression of desolation of the present deserts. The already mentioned proliferation of agricultural settlements in the Negev during this period should convince every skeptic.[71] One should also keep in mind that the Sahara also was still green and that in the same period hippopotami and crocodiles inhabited lakes in what is now the Ruba' el Khali – the empty quarter of Arabia, today one of the driest areas of the world.

By the middle of the fifth millennium the two Mesopotamian cultures met. The better organized, and presumably more populous, southern 'Ubaidian culture was able to overcome and replace the northern Halafian – whether by conquest or simply by force of numbers is a matter of debate. Taken together, the evidence suggests violent conquest and expansion of a more aggressive people, used to defeating not only nature but other human societies as well. Unfortunately, the lack of historical documents leaves us largely in the dark about the details. However, the mentioned new evidence of brutal attacks, destruction, and massacre of the population at Hamoukar leaves little doubt

---

[69] A. Lawler, "North versus South, Mesopotamian Style" *Science* 312:1458–1463 (2006).

[70] R. Amiran, *Ancient pottery of the Holy Land*, Masada, Jerusalem, p. 23 (1969).

[71] U. Avner, op. cit. (1998).

about the direction civilization was heading. The fertility of the soil in southern Mesopotamia most certainly favored a steady, and presumably even steep, increase in population, leading eventually to increased competition for limited resources in a limited area, its people victims of their own success. The massive mud-brick walls surrounding the settlements points to the dangers of flooding as well as to growing internecine tensions and institutionalized war. It certainly did foreshadow the periods to come when, after the end of the 'Ubaid period around 4000 B.C.E., the Uruk phase witnessed the birth of urbanism with its devastating wars, human sacrifices, slavery, and vertically institutionalized religions and absolutism supported by the next most influential invention, writing. As the entire known history of the Middle East basically consists of not much more than incessant wars, conquests, and genocides, why should the prehistoric period, clearly being of the same substance and forboding this sad story, be different?

Southern Mesopotamia can now safely be called Sumer. The later 'Ubaid phase ('Ubaid 4) is characterized in the north by Tepe Gawra levels XVI to XIII. Monumental temples on high podiums dominate the upper city, filled with elaborate vessels in a variety of shapes and decorations, ranging from simple globular jars to large storage jar and incense burners, incised and, more rarely, painted. Its principal sites are still Eridu, 'Oueili, 'Uqair, and the type site 'Ubaid. The pottery continues to shows sign of gradual decline of its artistic value as mass production had to supply a constantly growing number of consumers. Figurines are now mostly of bulls or cows with a few anthropomorphic figures, mostly of slender women with small breasts and date-kernel ("coffee-bean") eyes. This trend towards fewer representations of slimmer and more elongated and schematic figurines can be observed from Anatolia to Palestine (compare the ivory figurines from Beer Sheba[72]), and from Egypt to Elam. The first stamp seals with simple patterns of symmetrical lines appear. Foreign relations flourished in this period, in particular with neighboring Elam/Khuzistan from where pottery was imported, and all around the Persian Gulf.

The obvious contradiction between immigration from the north and the spread of cultural elements from the south can easily be explained when considering that semi-civilized newcomers are always quick to adopt – and subsequently propagate as their own – the more sophisticated culture of the people they conquer. History is filled with similar examples as we shall see further on.

At the other end of the Fertile Crescent, another distinct dichotomy existed between Lower and Upper Egypt. The early cultural trajectory of the wide and flat Delta (Lower Egypt) differs considerably from the one in the narrow and constricted Nile Valley itself (Upper Egypt), until a growing cultural permeation took place toward the end of the Gerzean period, followed by political unification during the Early Dynastic period. These events show a certain similarity to the cultural but not political unification of Mesopotamia during the

---

[72] J. Perrot, "Statuettes en ivoire et autres objets en ivoire et en provenant des gisements préhistoriques de la région de Beersheba" *Syria* 36:8–19 (1959).

Late 'Ubaidian. It appears that in both cases, the unification of different cultures and subjugation of neighboring peoples would have been a prerequisite for the development of higher civilization and state formation. The question whether such a process was triggered or intensified by the immigration of people from the north during some cold and wet spell, or from the west in a warm and dry phase, can be debated but cannot be rejected out of hand.

A boost of cultural activity and widening of horizons can be observed in the transition from the various Neolithic cultures in both parts of the country, Merimdeh, Fayum A in the north, and the Badarian in the south, to the pre-Dynastic period around 4000 B.C.E. Large villages, and even larger cemeteries containing the first copper implements characterize the first phase called Amratian, after the site of el-Amra near Abydos in Upper Egypt (or Naqada I to some European scholars. Typical is a reddish-brown pottery with yellow-white paintings depicting hunting scenes that foreshadow the typical style of later, pharaonic artistic conventions. Famous are the 'dancers' – stylized clay figurines of women with arms raised above their heads as in a trance. The influence of new ethnic elements is controversial. In the north, at el-Omari near Cairo, remains of an ill-defined culture were found which could possibly point in that direction.

The transition to the Gerzean (or Naqada II) period around 3500 B.C.E. shows a light cream-colored pottery painted in a style that evolved out of the Upper Egyptian Amratian tradition with all the characteristics of later Dynastic ceramic production. New are representations of large many-oared boats, recalling Mesopotamian parallels. Other clearly Mesopotamian elements are found in the pottery, such as bent spouts on jars or lug handles on the shoulders of large storage jars. Native Egyptian pottery, from its beginning to its end, never had any handles or similar additions to the body of the vessel, due to the friability and brittleness of wares made of Nile mud, which dominated the tradition even when better clays were available. Lapis lazuli, a blue stone exclusively found in Afghanistan, was imported from Asia and traded west via Elam and Sumer.[73] Other items include stamp and cylinder seals, the latter more common than the former, at first directly imported, then local imitations with decorations closely following Mesopotamian prototypes. The cylinder seals had a strong influence on the development of early Egyptian hieroglyphic writing which, however, was probably of local Upper Egyptian origin.[74] Other typical Mesopotamian artistic motifs appear, such as hybrid animals, entwined serpents (the Two Rivers?) around a central motif such as rosettes, and the "Master of Lions," which appears on the famous knife handle from Gebel el-'Araq showing the long robe of a Mesopotamian ruler. Another handle was found at Gebel Tarif. The newest excavations in the Delta, at Buto and Tell el-Fara'in are overflow-

---

[73] M. A. Ohshiro, "Study of Lapis Lazuli In the Formative Period of Egyptian Culture: An Approach in Terms of Culture Contact" *Orient* 35:60 (2000).

[74] R.M. Boehmer, "Das Rollsiegel im Prädynastischen Ägypten" *Archäologischer Anzeiger* 495–514 (1974).
H.J. Kantor, "Relative Chronology of Egypt" *COWA* p. 15.

ing with characteristic Mesopotamian clay cones for mud-brick wall decorations.

Stone and pottery vessels imported from Palestine and relics uncovered at the recent German excavations at Minshat Abu 'Omar in the eastern Delta show a mixture of Lower and typical Upper Egyptian artifacts, such as slate palettes for the preparation of body paint, with Palestinian elements. The similarity, and thus the cultural connections, of many Palestinian Chalcolithic ivory figurines of slim stylized men and women with their Egyptian counterparts are still debated. An important item is the so-called wavy handled jar, an elongated storage jar with two ledge-handles to facilitate handling, which is generally believed a local copy of Palestinian prototypes. The site of Maadi near Cairo will play a crucial role in resolving the question of the relations between Palestine and Egypt, in particular for the last phase of the pre-Dynastic period.

The picture emerging from this admittedly rather meager evidence is that of a warlike and unstable period with various local principalities fighting for hegemony and echoes of countless struggles are found in early mythology and in some Old Kingdom texts. There is little doubt in our minds that an erratic climate pattern with some low, or a few excessively high Nile floods contributed in some measure to the increase of unrest and instability in Egypt at that time.

4.4
From Copper to Bronze – The Beginning of the Early Bronze Age

The relatively abrupt end of the Chalcolithic cultural complex of the Levant poses many questions. Did it die of its own accord as the old ideologies and social orders collapsed? Did the impact of external events lead to its demise and replacement by a different culture? Or was the ultimate cause, as usual, another change in the climate?

Our data (Figs. 3 & 3a) show an abrupt decline in precipitation around 3500 B.C.E., followed by a short revival for about a century, and then another extreme decline, reaching the lowest peak for the entire Holocene around 3200 B.C.E. This trend continued for about two centuries, turning, at around 3000 B.C.E. into a very humid climate. Thus one can conclude that at least the Levant went through two phases of extreme dryness between 3500 and 3000 B.C.E.

A survey of the archaeological record shows that these two dry phases coincided exactly with the record of fundamental changes in the material culture. The trouble is that the new period is characterized more by the absence than by the presence of definite and remarkable cultural features which contributed to the confusion in the archaeological terminology in Syria-Palestine for the half millennium between the end of the Chalcolithic period around 3500 B.C.E. and the establishment of the first full-size towns or cities around 3000 B.C.E. Following the American school of W.F. Albright and G.E. Wright, the term 'Early Bronze Age I' is widely accepted in Israel. The French school under de Vaux called everything before the establishment of the urban Early Bronze Age towns 'Late Chalcolithic', as de Vaux saw no connection between this phase and the Early Bronze Age. The later generation of French scholars, such as J. Perrot and P. de Miroschedji, called it 'Pre-Urbain'. The British under K. Kenyon, however, saw a great deal of similarities between the two, as did the Americans, and called this period 'Proto-Urban'. As already noted, and regardless of school, everything in archaeology must be divided into three parts, using Arabic or Roman numbers and capital or lower case letters denoting chronological periods, cultural entities, 'horizons' 'assemblages' or some combination thereof. In addition, every excavator felt free to expand and complicate the old system or invent his own, normally on the grounds of a particular excavation without any relationship to other sites or a larger framework. M. Zohar suggested to follow the neutral German term "Zwischenzeit" and called any ill-defined in-between period as 'Transitional' or 'Intermediate'.

The reason for this picture is that in the Levant the period following the end of the Ghassul-Beer Sheva culture is characterized more by the absence than by the presence of definite and remarkable cultural features. The archaeological evidence, culled from exposures at the base of excavated sites and cemeteries such as Megiddo XX/XIX, Beit Shean XVII to XVI, Beit Yerah/Khirbet Kerak, and Tell el-Far'ah (N) near Shechem in northern Israel, consists of a limited repertoire of simple pottery types and basic tools that had their roots in the preceding period and continued into the following one. Their descriptions may sound tedious but they represent the only substantial archaeological evidence of the possible ethnic and cultural heterogeneity of the population of this most crucial time just prior to the so-called Urban Revolution.

A type called 'Impressed' or 'Umm Hammad' Ware (Proto-Urban D) can be the earliest type of pottery and is found predominantly in the Jordan Valley and may represent a population with its roots in the preceding period. Another common type is called 'Red Burnished' and is found in the central and southern part of the country (in K. Kenyon's terminology 'Proto-Urban A'). It shows a clear break with earlier Chalcolithic fashions and the question is open whether this ware is a revival of ancient traditions of the Pottery Neolithic B, to which is has many obvious links, or an entirely new phenomenon. If this style is something new, the next open question is whether it developed locally or did it come from somewhere else?

A peculiar assemblage of small jars and juglets is called 'Line-Groups Painted' Ware, mostly found in tombs restricted to the east-central and eastern part of Palestine-Transjordan (Kenyon's 'Proto-Urban B'). Another restricted type consists of 'Grey-Burnished' bowls with a soft and soapy feel, often found together with red-on-white 'Grain-Wash' painted large storage vessels in the north of Palestine. These types of pottery appear late in our period and are a fashion common in Anatolia (Kenyon's Proto-Urban C). Our difficulty is the fact that the geographical distribution of these pottery types not necessarily coincide with the range of other types of artifacts.

The rare settlements consist of scattered villages or nomads' camps in an even distribution without larger concentrations of populations or any architectural aspirations. House plans range from simple round huts to rectangular houses of modest size. E. Braun found oval and apsidal houses in the Jezreel

Valley in northern Israel that are clearly linked to a northern tradition found at Dakerman near Byblos in Lebanon and Jawa or other sites in Transjordan.[75]

S.W. Helms report on his excavations at Jawa deserves a special comment. The relatively large settlement is located on the basalt plateau in the desert of northern Jordan. It consists of several layers of domestic architecture, retaining walls serving as defenses and a complex system of water collection and storage facilities. The finds include bones of domestic goats/sheep and cattle, the latter being equally important as the former. "The prevalence of sheep and cattle remains (relative to goats) implies irrigation to improve grazing" (p. 252). It has several occupation phases throughout the second half of the fourth millennium, with a minor phase of resettlement during the Middle Bronze Age. The principle phase of desertion occurred around 3000 B.C.E. Helms is at a loss to explain the growth, flourishing and decline of Jawa in such arid surroundings. He performs many hydrological calculations based on today's climatic conditions to explain its secret, which do not hold water, as he, like most of his colleagues, is blissfully oblivious of climatic changes. One glance at Fig. 3a. shows the extreme wet phase followed by an equally dry phase just at that time.[76]

In spite of the loose term "Bronze Age", flint tools knapped in a characteristic manner known as 'Canaanite blades' remained the most prolific and characteristic tool in the Levant.[77] There is hardly any art to speak of, whether painting on pots or walls, except a few crude figurines. Similarly, there is an absolute lack of evidence of an elaborate ritual or ceremonial life comparable to the Chalcolithic period, or of any 'higher' social organization, as appeared in the later phases of the urban Early Bronze Age. Each region has its own strong local flair in many types of artifacts, and a simple subsistence economy without evidence of substantial trade. The not very complex funerary cult mostly in natural caves and some artificial tombs indicates some form of ancestor cult characteristic of mobile and simple tribal social structure.

The most intriguing results are seen in southern and southeastern Palestine. The settlement and cemetery of Bab edh-Dhra' on the Lisan Peninsula in eastern Palestine, the caves of Lachish and Site H in Wadi Ghazzeh have close links with Egyptian sites such as Maadi and beyond.[78] The corbelled vaults of the "charnel houses" at Bab edh-Dhra are very similar to the *nawamis*, corbelled tombs in the Sinai and the deserts of Egypt and Arabian Peninsula and already mentioned. These similarities are hardly coincidental and an expression solely of trade relations, but could be signs of an increased mobility of people between northern Africa and Asia, as we shall see below.

All this clearly points to a time of subsequent great social and economic changes: the proliferation of pastoral nomads as a social phenomenon of the

---

[75] E. Braun, "The Problem of the 'Apsidal' House and Some Notes on the Architecture of the Early Bronze I Period in Israel, Jordan and Lebanon" *Palestine Exploration Quarterly* 121:1–43 (1989).

[76] S. W. Helms, *Jawa – Lost city of the black desert*, Methuen & Co., London (1981).

[77] S.A. Rosen, "The Canaaean Blade and the Early Bronze Age" *IEJ* 33:15–29 (1983).

[78] A.F. Roshwalb, *Protohistory in Wadi Ghazzeh: A Typological and Technological Study Based on the MacDonald Excavations*. Dissertation, University of London, p. 332 (1981).

greatest importance for the history of the Near East and for laying the foundations of the 'Urban Revolution'; man's first, and apparently pathetic, attempt
at 'civilization' – the building of crowded cities (*civitae*) inhabited by vertical
hierarchies with absolute power over their subjects through organized religion,
wars, and slavery, and expressed by an architecture which still haunts us, be it
the pyramids of Egypt, the ziggurats, temples and walls of Mesopotamia, or the
extravagantly massive defense walls of the Anatolian and Syrian-Palestinian
cities.

Civilization as we understand it has had its beginning in Sumer during
the Uruk period, ca. 4000 to 3100 B.C.E., and was contemporary with the
Chalcolithic period in most other parts of the ancient Middle East. Also in
Mesopotamia this problematic transitional period has led the archaeologists
and historians to give it various names: Pre-Dynastic; Protoliterate – or, following the archaeological practice of naming a culture after the site where it
was first recognized – the Uruk and then Jemdet-Nasr period in the south; and
the Tepe Gawra period in the north. The German excavations at the modern
village of Warka (ancient Uruk and Biblical Erech), and the home of Gilgamesh,
concentrated on the great temple complex of E-anna, dedicated to Inanna, and
the temple of Anu, the sky-god. The pottery is, quite significantly, simple and
undecorated, but with a fine gray slip and made for the first time on a fast wheel
and is found over a large part of the southern Mesopotamian plain. A peculiar
type of ware during the late Uruk period is the 'bevelled rim bowl', a mass-
produced mould-made open bowl with straight sides, found by the thousands
near construction sites of this period. It could have been a measuring device
for rations, or a bread form, or both. At the E-anna level VI we find a particular building method of burned bricks, decorated with stone or burnt clay
cones stuck into the plaster of the building. Most important was the finding
for the first time of clay balls and round tablets inscribed with pictographs and
numbers as accounting and mnemonic devices for economic activities of the
temple, sealed by cylinder seals. These early cylinder seals were undoubtedly
some of the finest ever made in Mesopotamia, and with them developed an art
style for stone and copper sculptures. The later Uruk phase also saw the expansion of the southern culture upriver along the Euphrates where the impressive
town sites of Habuba Kabira and Gebel Aruda are pivotal in understanding the
spread of the 'Urban Revolution' into other parts of the ancient Near East.

The following Jemdet Nasr period, from ca. 3100 to 2900 B.C.E., is a transition
to the Mesopotamian 'Early Dynastic Period' and the urban Early Bronze Age
in other parts of the Near East. Important finds are at the name site, Jemdet-
Nasr, at Uruk and Nippur, at Tell 'Uqair – the site of the 'Painted Temple,' and at
Khafajeh – the site of the temple to the moon-god Sin. In general, it was a period
of stagnation and even decline in some areas. And as so often, it nevertheless
did produce some of the most magnificent pieces of ancient art. It is possible,
however, that these unique finds were heirlooms of an earlier period. Cylinder
seals were manufactured using a drill, and pottery was mass-produced on
a wheel.

In Egypt, the end of the Gerzean period is even darker, corresponding to
an early phase of the Palestinian Early Bronze Age I. Some scholars, including

Kaiser and Wildung,[79] introduced a 'Late Gerzean' or 'Naqada III', based on the development of a particular storage jar with wavy ledge handles, and on some other artifacts such as particular flint tools. Another conundrum is the hypothetical 'Dynasty O' – a series of important rulers from Upper Egypt probably dominating most of the country before the established 'First Dynasty' of the official king lists recorded by ancient Egyptian scribes themselves. The commemorative mace-heads and stone palette incised with a primitive *serekh* (royal names in the shape of a temple facade topped by a heraldic animal, the forerunner of the later cartouches) of a king known as 'Scorpion' was found together with pottery imported from Palestine in an early temple at Hierakonpolis in Upper Egypt. The question of how many rulers there actually were and how long that period lasted is a hotly debated issue.

One of the conundrums of the origin of Egyptian civilization was the undeniable strong Mesopotamian influence during this period we have already mentioned – the relatively sudden appearance of writing, cylinder seals and mud-brick architecture with recesses, in conjunction with Mesopotamian art motifs such as serpent-necked animals, griffins, rows of animals, representations of high-hulled ships and people with Near Eastern or Mesopotamian dresses. The excavations at Tell el-Fara'in, ancient Buto, show the same decorative elements of cones and spikes stuck in mud-brick walls as those in the temple facades at Uruk, with a plethora of southern Mesopotamian trade goods dated to Uruk IV/V (Protoliterate B), and the periods immediately following, prove close contacts with Asia and rises the question about the ways these contacts were materialized. There is little doubt that the sea lanes linking Egypt with ports on the Syrian and Palestinian coast, indeed, flourished. But, they also ran parallel with much traveled overland caravan routes secured by Egyptian military and trading posts, which were established along the coast of the Mediterranean Sea and extended deep into the coastal plain of Palestine and beyond as they continued throughout the area now known as the Syrian and Arabian Desert. The largest Egyptian settlement in the southern Levant was at Tell 'Erani, once mistakenly called 'Gat' near Ashkelon,[80] and in Nahal Besor.[81]

---

[79] W. Kaiser, "Zur Südausdehnung der vorgeschichtlichen Delta-Kulturen und zur frühen Entwicklung Oberägyptens" *Mitteilungen des Deutschen Archäologischen Instituts* 41:61–87 (1985).
D. Wildung, *Ägypten vor den Pyramiden – Münchener Ausgrabungen in Ägypten.* von Zabern, Mainz (1981).

[80] B. Brandl, "Evidence for Egyptian Colonization in the Southern Coastal Plain and Lowlands of Canaan During the EB I Period" in: *The Nile Delta in Transition: 4th–3rd Millennium B.C.* (Proceedings of the Seminar Held in Cairo, 21–24 October, 1990, at the Nederlands Institute of Archaeology and Arabic Studies), E.C.M. van den Brink (ed.), and The Israel Exploration Society, Jerusalem (1992).
Ibid. "Observations on the Early Bronze Age Strata of Tel 'Erani," in *L'Urbanisation en Palestine à l'âge du Bronze Ancien. Bilan et Perspectives des Recherches Actuelles, Actes du Colloque d'Emmaus 20–24 octobre 1986*, ed. P. de Miroschedji,. BAR International Series, 527(II), London (1989).

[81] E.D. Oren, "Early Bronze Age Settlements in North Sinai: A Model for Egypto-Canaanite Interconnections" in *L'Urbanisation en Palestine à l'âge du Bronze Ancien. Bilan et Perspectives des Recherches Actuelles, Actes du Colloque d'Emmaus 20–24 octobre 1986*, ed. P. de Miroschedji., London, BAR International Series, 527(II), 389–406 (1989).

At Azor near Tel Aviv, A. Ben-Tor found a cemetery with allegedly "African" skeletons.[82]

It is astonishing that all these tantalizing hints of intense contact between Egypt and the Fertile Crescent has never aroused the suspicion of the archaeologists or historians of the region. The newest excavations in what is now the arid Sinai Peninsula and the Negev Desert show an increasingly dense pattern of interrelated village sites where "we can begin to understand how these cultures survived in the desert and, indeed, thrived" (according to S Rosen). No agricultural village, and much less dozens or perhaps hundreds of them, has ever "thrived" in the middle of a waterless desert. It must have escaped the attention of the excavator that the end of the fourth millennium B.C.E. witnessed the beginning of the most humid climatic phase in Near Eastern history. The now arid swaths of dry desert along the inner fringe of the Fertile Crescent were then covered with vegetation and dotted with springs and watering holes. The precipitation was clearly far above the minimum required for extensive agriculture as well as intensive local horticulture and pastures for herds of grazing animals.

The consolidation of the Egyptian state and the establishment of a state-controlled exchange system had a direct and far-reaching impact on her unfriendly neighbors to the south and to the northeast – there was apparently no threat of enmity in the west. The greatest importance of this expansion of Egyptian influence into Palestine is the penetration of groups of Semitic-speaking -groups of cattle herders riding on donkeys from northeastern Africa into the Near East.[83] We are quite confident that the preceding cultures of the Ghassul-Beer Sheva phase and their contemporaries in Syria and Mesopotamia were not Semitic. They were probably related to a general linguistic substratum of one or more Asiatic language families, one of which is represented by Sumerian, and identified with the help of toponyms already a time ago.[84] By the middle of the third millennium B.C.E. (the mature Early Bronze Age), we find the first Semitic names in the earliest readable documents from Tell Fara and Tell Abu Salabikh in Mesopotamia[85] and Ebla/Tell Mardikh in northern Syria (see Appendix III: Near Eastern Languages).[86]

The arrival of the pastoral nomads was an important turning point in the history of our region. They cannot live without the farmers from whom they obtain their cereals, fruits and other victuals in exchange for the cheeses, wool

---

[82] A. Ben-Tor, "Two Burial Caves of the Proto-Urban Period at Azor" *Qedem* 1. The Hebrew University Jerusalem (1975).

[83] M. Zohar, "Pastoralism and the Spread of the Semitic Languages" in: *The Levant: Archaeological Materials in Anthropological Perspectives*, Monographs in Old World.

[84] I.J. Gelb, "The Early History of the West Semitic Peoples" *Journal of Cuneiform Studies* 15:27–47 (1961).

[85] R.D. Biggs, *Inscriptions from Tell Abu Salabikh*. Oriental Institute Publications 94, University of Chicago Press, Chicago (1974).

[86] I.J. Gelb, "Ebla and the Kish Civilisation" in *La Lingua di Ebla: Atti del Convegno Internazionale, Napoli 21–23 Aprile 1980*, Seminario di Studi Asiatici, Series Minor XIV, L. Cagni, (ed). Istituto Universitario Orientale, 7–73, Naples (1981).

and skins of their animals.[87] And when they have nothing to offer, as it might have happened during drought years, they simply take. Nomadic people have always been warriors, 'lean and mean'; they not only fight continuously with other nomads over wells and pastures, but constantly hover, like birds of prey, on the edge of the fertile fields, ready to pounce on the stores of the farmers at the slightest sign of weakness. Their appearance and growing importance in the social fabric introduced an element of insecurity and turbulence into what appears to have been the hitherto relatively peaceful conditions of the Chalcolithic or the early E. B. I period.[88] Their presence undoubtedly encouraged the native population to build the first walls of defense around their settlements and organize themselves under warlords – very often nomads themselves – in order to withstand the danger – another important factor contributing to the evolution of urbanization and with it, civilization began.

[87] A.M. Khazanov, *Nomads and the Outside World*. Cambridge University Press, Cambridge (1984).

[88] P. Steinkeller, "Early Political Developments in Mesopotamia and the Origin of the Sargonic Empire" in: *Akkad, the First World Empire: Structure, Ideology, Traditions*, ed. M. Liverani, Padua, *History of the Ancient Near East Studies* 5, 107–113 (1993).

CHAPTER 5

The Urban Revolution and the Dawn of History

The Early Bronze Age, ca. 3100/3000 to 2400/2300 B.C.E.

"And they said one to another, Go to, let us make bricks, and burn them thoroughly. And they had brick for stone, and slime had they for mortar. And they said, Go to, let us built us a city and a tower, whose top may reach unto heaven..."

(Genesis, 11:3–4)

5.1
The First Cities and a New Order (Table 3)

The close of the Transitional Period at the end of the fourth millennium ushered in the first civilizations all over the ancient Near East. As the word "civilization" is derived from *civitas*, the Latin word for city, and despite the use of this word by some to include the Bushmen or Australian Aboriginals' cultures, we shall use this term strictly in its original sense, i.e. for urban cultures distinguished by the presence of stratified societies living in towns and cities. The first steps of incipient urbanization began in southern Mesopotamia as early as the Uruk period in the mid-fourth millennium B.C.E. and spread in a complex centrifugal movement, soon encompassing other great river basins, such as Egypt, the peripheral regions of Syria-Palestine and Anatolia and the Indus civilization. Sadly, we probably shall never understand the mechanism of the early phase of the spread of urbanization from its home in southern Mesopotamia, because modern dams on the Tigris and the Euphrates and their important tributaries are now drowning most of the archaeological evidence from this phase. The birth of the first cities in Mesopotamia proper is somewhat better understood due to the early onset of writing whereas in other parts of the Near East, information on incipient development is nonexistent, lost or scant.

Many parallel socioeconomic developments during the latter half of the fourth millennium B.C.E., laid the ground for the conditions for many people to live in ever-larger conglomerations. Security concerns heightened as steadily growing rival populations began to compete for limited agricultural land and water resources and in response to raiders from the outside. The anxiety of

Table 3. A historical-archaeological timetable of the Near East, from 3000 to 1500 B.C.E.

years B.C.E.	Egypt	Syria-Palestine	Mesopotamia	Anatolia
1500	2nd Intermediate Period, 13th–16th Dynasty	Middle Bronze Age IIC	Old Babylonian and Old Assyrian Period	Middle Bronze Age III (Old Hittite Period)
1600				
1700		Middle Bronze Age IIB		Middle Bronze Age II (Assyrian Trade)
1800	Middle Kingdom 11th–12th Dynasty	Middle Bronze Age IIA (Canaanite Period)		
1900				Middle Bronze Age I
2000			Isin-Larsa Period	
2100	1st Intermediate Period, 8th–11th Dynasty	Intermediate Bronze Age (M.B. I)	Ur III Period	Early Bronze Age IIIB
2200				
2300			Akkad Period	Early Bronze Age IIIA
2400		(Syria E.B. IV)		
2500	Old Kingdom 3rd–6th Dynasty	Early Bronze Age IIIB/C	Early Dynastic Period IIIB	
2600			Early Dynastic Period IIIA	Early Bronze Age II
2700		Early Bronze Age IIIA	Early Dynastic II	
2800	Proto-Dynastic	Early Bronze Age II	Early Dynastic I	
2900				Early Bronze Age IB
3000	1st–2nd Dynastic		Jemdet Nasr	

primitive city dwellers led to an unprecedented flowering of rituals favoring sacrifices of all sorts to pacify the wrath of imagined deities and fostered aversion and even aggression against everyone outside the pale of their own territory. We have already observed the inherent aggressiveness as an important element of the Mesopotamian urban culture in the 'Ubaid period, a trait which actuated the spread of civilized societies and was to be bequeathed to all its descendants. These and various other explanations for the origin of cities have been advanced and debated at length in the literature, ranging from the monumental work of L. Mumford[1] to the Chicago convention "City Invincible" and a host of most recent anthropological works on this inexhaustible subject.[2] Nearly no mention is made, however, of the importance of water and the impact of the vicissitudes of the climate on the birth, flourishing, and demise of those early civilizations.

---

[1] L. Mumford, *The City in History: Its Origins, Its Transformations and Its Prospects*, Harcourt, Brace & World Inc. New York, (1961).

[2] A. Raban, op. cit. (1991).
A. Raban and E. Galili, op. cit. (1985).
A.S. Issar, op. cit.(1998).
S. Kempe and E. Degens, in *Geologische Rundschau*.
M. Schoell, in *The Geology of Lake Van*; G. Lemcke and M. Sturm op. cit. (1997).

5.2
The Climate Background – When Cities Drowned and the Desert Bloomed

There is abundant evidence for a cold climate over the entire northern hemisphere for most of the third millennium B.C.E. Archaeological submarine investigations along Israel's shores have shown that the level of the Mediterranean Sea was lower during this period (Figs. 3 & 3a.).[3] This conforms with low sea levels on a global scale, and indicates a period of increased glaciation and capture of seawater by glaciers. The cool climate caused exceptionally high rates of precipitation in the Near East with high river flows and a high rate of sedimentation in Lake Van from around 3000 to about 2300 B.C.E. The depletion rates of ^{18}O isotope in the lake sediments also indicate a cold climate and, together with other data, show that in this region, humidity during most of the third millennium B.C.E. was at its highest in all the Holocene. Pollen analysis from Lake Zeribar in eastern Iran shows a maximum in the ratio of oak pollen during the same period, also indicating a more humid climate.[4] As these regions are the headwaters of the Euphrates and the Tigris, it can be expected that the regime of these rivers were influenced by this more humid climate. Indeed, the reconstruction of historical flood rates of these rivers shows a peak flow from ca. 2900 to ca. 2500 B.C.E.[5] Ample evidence for a humid climate is found in other parts of the Near East as well. The level of the Dead Sea rose to −310 meters below MSL, its highest level since the Last Glacial Period in the Pleistocene when it reached −200 meters below MSL.[6] Ephemeral and normally dry riverbeds in Palestine, locally known as *wadis*, became perennial streams.[7]

At the same time oxygen isotope composition in the deposits of the Sea of Galilee and the speleothemes of the caves of Galilee became lighter.[8] In addition, the isotope data from the stalagmites in the Soreq Cave showed that the precipitation during this period reached at least that of the Chalcolithic period and exceeded the peaks of precipitation of all the historical periods to follow. In detail the proxy-data from the Soreq Cave show that after a warm and dry period starting around 3500 B.C.E. and lasting a few decades to a century, a cold and extremely humid climate followed, continuing until 3300 B.C.E. Then came an extreme dry period, peaking around 3200 B.C.E., was followed again around 3000 B.C.E. by a cool and humid period that lasted, with some interruptions, for about seven centuries (Figs. 3 & 3a).[9]

---

[3] Ibid.

[4] A.P. El-Moslimany, op. cit. (1986).

[5] A. Kay and D.L. Johnson, op. cit.(1981).

[6] A. Frumkin et al., op. cit. (1997).
Migowski et al. op. cit. (2002).

[7] P. Goldberg and A.M. Rosen, "Early Holocene Palaeoenvironments of Israel" in T.E. Levy, op. cit. (1987).
A. M. Rosen, "Environment and Culture at Tel Lachish, Israel" *BASOR* 263, 55–60 (1986).

[8] A.S. Issar, Y. Govrin, M.A. Geyh, E. Wakshal and M. Wolff, "Climate Changes during the Upper Holocene in Israel" *IJES* 40:219–223 (1992).

[9] M. Bar-Matthews, et al., op. cit (1998).

With these records of the climate changes as a background, we will present an overview of the archaeological evidence of the first half of the third millennium B.C.E., starting with Syria-Palestine, followed by Anatolia, Mesopotamia and Egypt.

5.3
The Early Bronze Age in the Levant and Anatolia

In Palestine, the transition from the end of the Early Bronze Age I to the urban phase of Bronze Age II is extremely sketchy and can be followed in just a few sites such as Ha'Ai/ et-Tell, Tell el-Far'ah (N), Arad and Bab edh-Dhra', the latter two quite atypical and in now climatically marginal areas. It appears that scattered villages began to be organized into a local network of related settlements based on the exchange of tools and luxury items and gradually "imploded" into small or medium-sized "towns" around 3000 to 2700 B.C.E. These in turn developed into larger cities during the third phase of the Early Bronze Age around 2700 to 2400 B.C.E. The main motive for these "implosions" is still unclear. Until the 1970's the urban phenomenon was explained along the diffusionist line of argument whereas later it was attributed more to a process of local evolution.[10] In our opinion, the truth lies somewhere in the middle. As discussed in Chap. 2, we think that for most people, civilization – i.e. living in crowded towns under rather appalling hygienic conditions and under the thumb of a class of rapacious priests and megalomaniac rulers – was not an attractive option, but developed of necessity dictated by security problems.

The Early Bronze Age of Palestine is divided, as usual, into three parts:

- E.B. I (ca. 3300–3000 B.C.E), as discussed in the previous chapter is a transitional period inclusive of everything that is not Chalcolithic and not yet proper urban Early Bronze Age.

- E.B. II (3000–2700 B.C.E.) is the formative period of the first urban civilization of Palestine characterized by fortified townships.

- E.B. III (2700–2400 B.C.E.) shows in a few selected sites the mature phase of this culture, as well as the first signs of its disintegration.

The debate about whether the Palestinian Early Bronze Age II "towns" were real cities or just large fortified villages cannot be resolved from the information

---

[10] P. Lapp, "Palestine in the Earlly Bronze Age" in: *Near Eastern Archaeology in the Twentieth Century: Essays in Honor of Neslon Glueck*, J.A. Sanders (ed.) Garden City, New York. pp. 101–131 (1970).

R. de Vaux, "Palestine in the Early Brone Age" in: *Cambridge Ancient History*, I.E.S. Edwards, C.J. Gadd, and N.G.L. Hammonds (eds.), vol. pt. 2, Cambridge University Press, Cambridge. pp. 208–237 (1971).

K.M. Kenyon, *Archaeology in the Holy Land*, 4th ed, W.W. Norton, New York (1979).

R.T. Schaub, "The Origins of the Early Bronze Age Walled Town Culture of Jordan" in: *Studies in the History and Archaeology of Jordan I*, A. Hadidi (ed.). Department of Antiquities, Amman. pp. 67–75 (1982).

available on this period. The term "city" denotes a stratified society with a socioeconomic structure based on division of labor and, above all, of civility and ethos as, for instance observed in the Mesopotamian cities, the Greek *polis* or the early medieval cities of Europe, signs of which were not yet observed during this period outside of its original Mesopotamian home province. Evidence for a dense pattern of fortified "towns," or villages, is found in all parts of the country on both sides of the Jordan. The limited and sporadic excavations, however, provide scant information and, therefore, a quite un-representational picture. There is evidence of some spacious and well-constructed temples, which, however, lack the monumentality of those of the later ages, and some attempts at water harvesting. The only dominant and recurring feature is the presence of extravagantly massive fortification walls.

It has been claimed that settlements expanded to the mountainous areas, the Shefelah and the valleys of Jezreel, and with this expansion, a transition to a Mediterranean method of agriculture. A. Ben Tor, M. Broshi and R. Gophna claimed that the people of the Early Bronze Age mainly preferred areas within today's an annual rainfall of more than 300 mm.[11] However, more than 600 sites from the late Early Bronze Age I and II were discovered in the Negev Highlands and the Uvda Valley, as well as other areas that are now desert.[12] U. Avner's detailed study of the Negev's arid highlands shows that the number of agricultural settlements in this region reached a peak during the third millennium B.C.E. However, as most of these settlements were small farms, many of which occupied the same sites as the Chalcolithic farms, there is no sequence of layers indicating a settlement gap before or during this peak. On the other hand, one such gap could be observed between the Chalcolithic and the Early Bronze Age in most urban sites, and another between the Early Bronze Age I and II, starting around 3000 B.C.E. This lapse can be explained by the climate curve of the Soreq Cave, which shows a period of reduced precipitation ca. 3400 after a peak in 3500 B.C.E.

Not all the settlements of the Negev were walled, and most were located on hills near valleys with permanent water resources. Cohen claimed that the population growth of the Negev highlands represented the movement of settlers who could not adapt to the increasing urbanization of northern Israel. The economy of the Early Bronze settlements in the Negev was based on agriculture, pastoral economy and hunting.[13]

The best example of an Early Bronze Age township in the south of the country is Arad – a large settlement divided into quarters with an orderly street system and fortified by a modest stone wall and semi-circular towers (Plate 9). According to the excavator R. Amiran, Arad probably served as a way station on the routes linking the Dead Sea and the southern hills of Judaea with Egypt.

---

[11] A. Ben Tor, "The Early Bronze Age" in: *The Archaeology of Ancient Israel in the Biblical Period*, The Open University of Israel, Tel Aviv. pp. 11–73 (1989).
M. Broshi, and R. Gophna, "The Settlements and Population of Palestine during the Early Bronze Age II–III" *BASOR* 253:41–53 (1984).

[12] U. Avner, op. cit. (1998).

[13] R. Cohen, "Ein Ziq" *Excavations and Surveys in Israel* 7–8:57–58 (1989).

Plate 9. Aerial Photograph of partly reconstructed Early Bronze Age city of Arad. (Courtesy Albatros, Tel Aviv)

Amiran attributed the establishment of Arad to metal-hungry colonists from the "Canaanite" north.[14] In contrast, L.E. Stager, and later I. Finkelstein, suggested that a southern, basically nomadic population established the town. The layout of the dwellings, with a main building and some subsidiary storerooms surrounding a spacious courtyard betray the heritage of the earlier agro-pastoral village compounds all over the Negev and the Sinai, as do the cooking pots made of clay from southern Sinai.[15]

However, the abundance of olive pits and grape kernels found in all four layers of Arad attest to an economy based on a Mediterranean type of agriculture,

---

[14] R. Amiran, "Arad" in *NEAHL* vol. 1:75–82 (1992).
"Canaanite" is a term used for the Semitic-speaking population of the Middle/Late Bronze and Iron Age and is inappropriate for the Early Bronze Age third millennium B.C.E. whose ethnic affiliation is unknown and whose culture is profoundly different.

[15] L.E. Stager, "The Periodization of Palestine from Neolithic through Early Bronze Times" in *COWA*, pp. 22–41.
I. Finkelstein, "Living on the Fringe: The Archaeology and History of the Negev, Sinai and Neighbouring Regions in the Bronze and Iron Age" *Monographs in Mediterranean Archaeology* 6 (Sheffield, Sheffield Academic Press, (1995).
Finkelstein cites repetitions of the situation in the Late Bronze/Iron Age, tentatively identified with Edomites and "Amalekites", and the Nabataeans in the Roman period.

probably geared to exporting olive oil and wine to Egypt, with trade in other commodities a welcome extra. Although Arad is now surrounded by a desert landscape, its earlier thriving success can only be explained by a cooler and wetter climate; and that the transition to drier weather was responsible for its desertion?

The finds of this period consist mostly of a fine and elegant type of pottery also found in Archaic and Old Kingdom tombs in Egypt – particularly at Abydos where this ware, imported from the southern Levant, was found (and from which its name 'Abydos Ware' is derived) together with a metallic "combed" ware. Both pottery types probably served as containers for oil or wine exported from Palestine to Egypt.

The inhabitants of these fortified towns were already then called *a'amu* by the Egyptians, derived perhaps from the Semitic *'amu*, meaning 'people', or *'alamu*, meaning 'young man' or 'human being', terms by which these people referred to their kin.[16]

Unlike other parts of the ancient Near East, but similar to the preceding Early Bronze Age I, very little representational art is reported from this period in the Levant. Whether this lack is derived from religious conceptions or simply because relatively few luxury goods were produced in the Palestinian "cities" is still an open Question. To us it seems to point to a rather unpretentious or warlike ruling class with no need for elaborate artistic expressions.

The Early Bronze Age III, starting around 2700 B.C.E., is the mature phase of this culture and, in fact, seems to earn the term "civilization." The process of 'implosion' continued, emptying the existing towns and villages of their inhabitants, and concentrating them into some of the largest, truly monumental cities ever seen in this country before the modern era. The rectilinear architecture facilitated the agglutinative character of closely packed dwellings behind strong fortification walls. The temples, such as those at Ha'Ai and Megiddo, were also rebuilt on a monumental scale.

Two sites, Beit Yerah/Khirbet Kerak on the shores of the Sea of Galilee in the north, and Tell Yarmut in the south, had become veritable metropolises. In the latter, the French mission under P. de Miroschedji uncovered densely packed dwellings, small sanctuaries and an intricate palace complex. Built applying a standardized set of measurements, the palace was very similar to a contemporary structure at Byblos and prefigures the later palaces at Ebla and Mari. The city walls are the largest fortifications in Palestine consisting of combined systems of massive stonewalls up to 30 to 40 m thick and probably of corresponding height, with a monumental gate. Extravagant far beyond the exigencies of warfare of that period, the fortifications appear to be "prestige-oriented" or ritual military architecture similar to that seen in later periods (e.g. the walls of Hattusas, capital of the Hittite empire in the Late Bronze Age).[17] At the same time, they seem to point to a rise in militarism foreshadowing the events to follow and the collapse of the entire urban system.

---

[16] D.B. Redford, *Egypt, Canaan and Israel in Ancient Times*, Princeton University Press, Princeton. p. 32 (1993).

[17] P. de Miroschedji, unpublished.

When contemplating the contemporaneous and equally massive defense walls at Ha'Ai, a local workman remarked: "*These people must have been very much afraid!*"[18] We wonder whom they might have been afraid of and whether this simple question is the key to understanding the *raison d'être* of these cities.

The Early Bronze Age III pottery shows the influx of a peculiar type of pottery known as Kura-Araxes Ware in Transcaucasia and as Bet Yerah/Khirbet Kerak Ware in Palestine. It was probably carried by a small but culturally important group of immigrants from those distant parts and eventually copied locally. Evidence of this pottery is found continuously from eastern Anatolia to the 'Amuq Valley in Northern Syria, the Beqa' Valley in Lebanon and to the Jezreel Valley in northern Palestine. These immigrants might have been itinerant metal workers selling their arsenic copper products for weapons and other types of armory, such as the hoard of Kfar Monash, they might have been warlike invaders trying to subjugate the native population or simply immigrants trying to find a better place to live – we probably shall never know for sure. We tend to favor the third scenario, which conforms to the general theorem already observed in the preceding periods – i.e. that colder and harsher climates in the more northern regions of Eurasia and the mountainous part of the Near East may have pushed some people southward.

Whatever the case, the hypothesis cannot be firmly backed up, as this period is characterized by extreme climatic fluctuations and the accuracy of the paleo-climatic proxy data and the archaeological dating is insufficient. Moreover, contact and trade with Egypt – the life-blood of civilizatory developments and prosperity of the ruling class – seems to have stopped entirely, for reasons not clear for the time being but quite possibly due to the desertification and abandonment of the trade routes and declining crops in Palestine.

Notwithstanding its undoubted local importance and the impressive size of some of its towns and cities, the Early Bronze Age in Palestine never attained the level of organization and sophistication of its neighbors in the great river valleys. Nor did it develop a writing system or even attempted to borrow one from the higher civilizations. At Byblos an attempt was made, probably under Egyptian influence, to develop a local script known as Gublitic, used for a few inscriptions in an unknown, probably non-Semitic language. The fact that the entire culture disappeared without leaving any cultural heritage to its successors marks it as one of the great civilizatory failures in the ancient world.

The effects of serious droughts starting between 2300 to 2200 B.C.E. and lasting for several centuries caused the superstructure of this society to collapse when climate changed for the worst. Arad was deserted already by the onset of the Early Bronze Age III and, like other sites in the southern Levant, shows no evidence of destruction or other catastrophic disasters.

The picture dramatically changes when going northward: All the magnificent cities in northern Palestine and southern Syria were burned down in a great fury, not to be resettled until nearly half a millennium later. Remains indicative of similar waves of violent destruction and general upheaval are found in all the sites further north, extending deep into Syria and Anatolia,

---

[18] J.A. Callaway, *The Early Bronze Age Sanctaury at Ai (et-Tel)*, Bernard Quaritch, London (1972).

including Byblos on the coast of Lebanon, Hama, and the 'Amuq sites in Syria. In Anatolia, all the cities from Troy in the northwest to Beycesultan, Karatas-Semayük and dozens of other contemporary sites in the east and southeast were utterly devastated. But, unlike Palestine, where the cities lay deserted for several centuries, most of the Syrian, southeast Anatolian and northern Mesopotamian sites were immediately rebuilt and resettled, albeit within the framework of a quite different culture, known as the Syrian Early Bronze Age IV, the subject of our next chapter.

At the margins of the highlands of Anatolia in southeastern Turkey, excavations at Lidar, Kurban Hoyük, Arslan Tepe, and Tarsus show strong influences from Syria-Mesopotamia. The climate of central and eastern Anatolia with its strong continental conditions of hot summers and very cold winters was not overly conductive to a brilliant flowering of urban civilization. Yet, a network of initially modest city-states centered on fortified towns dominated by palaces and temples following the Mesopotamian example, developed and began to prosper during the second half of the third millennium B.C.E. when the climate showed signs of warming. The culture of Ikiztepe in the north is quite independent and compares well with the development in northwest Anatolia and northern Aegean sites such as Troy I and II and Poliochnoi on Lemnos. The economy of these city-states was similarly based on agriculture, trade and metalworking in arsenic copper. A characteristic simple painted pottery gave way to wheel-made and burnished ware.

During this time, eastern Anatolia was increasingly drawn into the cultural sphere of the vibrant Transcaucasian Kura-Araxes culture. Compared with its contemporary Near Eastern neighbors it appears quite unusual: a dense pattern of often large but unfortified settlements without any signs of a developed hierarchical structure, such as temples or palaces, nor monumental art of any kind or writing. It appears as steadily growing and expanding from the middle of the fourth to the end of the third millennium B.C.E. without visible evidence of crisis periods or other major disasters. During the earlier and colder periods, the settlements are found in the sheltered valleys whereas in the later and warmer conditions of the second half of the third millennium B.C.E. we find more and more sites in higher altitudes including the Trialeti Plateau, many of which were now fortified with megalithic walls. Its most characteristic pottery was handmade under rather low temperatures, with a greasy high-gloss black on the outside and buff on the inside. By around the 27th century B.C.E. this pottery was found throughout east-central Anatolia, northern Syria and the central valleys of Palestine. We have suggested that tribes practicing relatively advanced metallurgy immigrated into Palestine during this period due to demographic pressure.

Beginning around the middle of the third millennium B.C.E., during the peak of the cold period, pastoral groups from the steppes north of the Black Sea entered Anatolia and Transcaucasia in a constant influx, slowly forming a new elite. Their tombs under artificial mounds known as *kurgans* on the Trialeti Plateau in Georgia, at Alaca Höyük, Horoz Tepe, and Mahmutlar contained exquisite arsenic copper and gold artifacts, showing cultural connections with Eurasia and Iran. It is quite possible that these groups were the vanguard of some Proto-Indo-European speaking people who moved from their Eurasian homeland into the Balkans, Anatolia, Iran and India. As elsewhere around 2400 B.C.E., the first signs of a warming and drying climatic phase were accompanied by the decline, the destruction and the abandonment of some large highland towns.

5.4
The Great Civilizations

Despite their many differences, Mesopotamia and Egypt present a picture of a largely synchronous development throughout the third millennium B.C.E. The basic elements of their cultures – including their spirituality and economy, state ideology and politics, art and literature, religion and mythology, city and temple planning – were defined during this period and continued with only minor changes despite many changes until their disintegration with the Hellenization that followed the conquest of Alexander the Great.

Environmental influences led to significant differences between the histories and cultural development of these two civilizations which were of utmost importance not only for the course or Near Eastern history but for the world and deserve the following more detailed discussion. Egypt was an oasis along a river, isolated by sparsely inhabited grasslands or deserts and thus seemingly protected from its neighbors. As aridization of the deserts progressed over the millennia Egyptians could afford to develop a disdainful attitude to foreigners. In contrast, Mesopotamia was exposed to its neighboring countries and continuously had to confront and often absorb outsiders from the surrounding mountains or semi-deserts. Whereas the Nile floods with predictable consistency, the hydrological regime of the two rivers is capricious. Therefore, Egyptians developed a sense of security and stability extending even beyond the point of death, whereas the people of Mesopotamia felt a constant threat to their existence, finding spiritual refuge in rituals and oracles penetrating all social levels. The pharaohs of Egypt were divine and absolute, whereas the *en*, *ensi* or *lugal* of the Mesopotamian city states – and later territorial states – were originally chosen by an assembly, only later to become absolute and hereditary rulers.[19]

5.4.1
Mesopotamia During the Early Dynastic Period

Although it was settled and urbanized during the fourth millennium B.C.E., Mesopotamia's economic and cultural flowering was not to begin for centuries to come. Our data (Fig. 3) show at least three stages of aridization and three period of cold during this period. As has been suggested, these frequent and extreme changes brought influxes of different people from differing directions.

---

[19] T. Jacobsen, *Towards the Image of Tammuz and other Essays on Mesopotamian History and Culture.* W.L. Moran (ed.), Harvard University Press, Cambridge (1970).

In addition to the aboriginal fishermen dwelling in the swamps and related with the Halafian culture further north, the basic population were the farmers and horticulturists from the foothills of the Fertile Crescent probably speaking Sumerian. During the cold periods, people arrived from the northern highlands, bringing metallurgy skills, as well as relations with the people of the metalliferous provinces. The arid periods, on the other hand, saw the immigration of pastoralists speaking Semitic Proto-Akkadian from the east and west, with their proficiency in dairy and wool production.

Initially, because the population growth was limited and the people could remain near the rivers, the dry phases towards the end of the fourth millennium did not have a ruinous impact on the agriculture surrounding the urban centers. Crises developed with the most extreme climate changes: One occurring around 3600 B.C.E. which may have led to the end of the 'Ubaid period, and the other around 3200 B.C.E. which marked the end of the Chalcolithic period in Syria-Palestine and the Uruk period in Mesopotamia. As the Euphrates and Tigris draw their water from the mountainous regions where precipitation rates are generally rather high, their flow was probably reduced but was sufficient to maintain the local population. The dry phases posed the greatest threat to the economy and safety of the people on the periphery for whom reduced precipitation meant little or no pastures and thus famine on the parched plains bordering the green irrigated lands.

The paleo-climatic scenario based on the available proxy-data from the first half of the 3rd millennium B.C.E. is of auspicious precipitation and conforms well to the socio-economic scenario. The ample water supply in the rivers of the plain enabled the inhabitants to expand the irrigated areas into the desert. The low sea level, which resulted from a cold climate on a global scale during this period, enabled better drainage conditions in the most southern areas and an equally important precondition for a successful agriculture. Temperatures were moderate enough even during the winter to grow sub-tropical crops such as palm dates. In addition, the surrounding deserts became hospitable, so that the pastoral people prospered from their own resources, without the need of exaction from their farming neighbors. Thus, both communities could benefit from the exchange of food and goods, which brought prosperity to the people of Sumer. The wealth accrued by the ruling class was invested in temples, enlarging their cities, and surrounding them with walls.

Cities have been growing in a process of centrifugal consolidation of a ruling class headed by a king and supported by a priesthood that owned lands and extracted a mounting income from the peasantry. As suggested earlier, this might have begun with the steady infiltration of a pastoral warring society that quickly gained sovereignty over the farming society. This process may have begun during the Uruk period in the second half of the fourth millennium and gained momentum during the Jemdet-Nasr Period at the end of the fourth and beginning of the third millennium. After a reversal of climatic conditions ca. 2700 B.C.E. (Fig. 3a), the end of Early Dynastic I, it reached 'take off velocity' during the Early Dynastic II, during the mid-part of the third millennium. Not only the population in the urban centers increased but also the number and size of villages in their periphery. Needless to say, that various theories and models were advanced to explain this remarkable urbanization process[20] none taking the factor of an optimal climate regime into consideration which provides the only logical explanation for the "surplus of products" component and allowed the concentration of resources in the hand of the ruling class. As the rural population lost its independence and became subjects of the temples, the production system may have initially become more efficient once the ruling class had at its disposal the tools for administration – particularly writing, which was developed specifically for this purpose. The prosperity enabled a great part of the urban population to be employed in administration, in the temples and in the production of non-elementary products.

Local and foreign artisans and traders engaged in exchange of some utilitarian but mostly ostentatious luxury goods with neighboring cities and distant countries. In due time an army was needed to protect the city and even to try to enlarge its estate and sources of water. A class of managers, scribes and priests in charge of the temple warehouses was established, kept the records and, along with the gods and kings, enjoyed the fruits of the peasants' labor. They had leisure time to create literature and, eventually, to write down their hymns, myths, laws and the achievements of their royal patrons. Growing estates required more managers and written orders for subordinates, resulting in a mass of documents, archives and written directories. In many aspects, the invention of writing did for the Sumerian society what the invention of the computer and its binary script did to the international industrial and banking agglomerates toward the end of the twentieth century.

Each city had its own god, king, and army. The latter were needed to protect the inhabitants of one city-state from the others when disputes about borders, land parcels and canals could not be solved peacefully. There is quite explicit written and monumental evidence of these wars, despite the optimal economical conditions, indicated by the growth burst of cities, temples, irrigation networks and trade which fuelled the ambitions of local dynasties.

This historical fact contradicts the simple deterministic materialistic paradigm, which says that wars are a result of stress and the fight of survival in order to secure limited resources. In the case of Sumer this contradiction is sharpened by the fact that during this period Sumer became a kind of league of city-states, sharing the same language, pantheon, and belief in belonging to such a league, which they called *Kengir* which became Shumer or Sumer in later pronunciations.[21] In many ways this resembles the situation in archaic and classical Greece and Italy during the first half of the first millennium B.C.E. before Rome overpowered the various ethnic groups. Thus, although the written and monumental documents are in Sumerian, and give the impression of cultural unity, the ethnic substrata did not blend entirely. Each ethnic group

---

[20] R.McC. Adams, *Heartland of Cities: Surveys of Ancient Settlement and land Use on the Central Floodplain of the Euphrates*, University of Chicago Press, Chicago. pp. 88–89 (1981).
The reader may find an elaborate discussion of the processes of urbanization in Chaps. 5–8 of C.K. Maisels, *The Near East*, Routledge, London (1993).

[21] The earliest "lean" logogrammic inscriptions can actually be read in any language, in some ways similar to ancient Chinese. Only after grammatical endings had been added to the logogram one can be sure of the language involved.

maintained a certain part of the country where it continued to worship its ancestral gods. These sometimes were given a new title and were married to or proclaimed a descendant of one the conqueror's gods, but they remained the chief god of the city-state where the pre-invasion population predominated. Thus, ethnic-cultural differences as well as local patriotism may explain some of the tension and the wars between the city-states.

This situation could explain, for example, the story about the war between Agga, the king of Kish, and Gilgamesh, the king of Uruk. According to the Sumerian King List the city of Kish received "kingship" from heaven after the Flood. As we have seen, the proxy data show that after the two periods of extreme rains, either ca. 3100 or 2900 B.C.E., one or both of which could have been the period of the Flood or Floods, came periods of extreme drought, most probably causing the encroachment of the pastoral Semitic people into Mesopotamia. They settled in the region north of Sumer, calling it Akkad, and later, Babylon. Indeed, the Sumerian King List tells us that the First Dynasty of Kish on the border of Akkad includes among its kings some with Semitic names. This was not the case in Uruk/Erech, in spite the assumption that Semites largely inhabited this city. Although the principal goddess of both cities was Inanna, probably inherited from the 'Ubaidians, Uruk was also the city of An, the sky god, and later Enlil. An, most probably, was the chief god of the first Sumerians.

The story of the war between these cities is found on a tablet containing some hundred fifteen lines. It does not involve mythological figures or deified personalities, but is, rather, a historical document that has not been allegorized and adorned with mythological characters. It starts with the story of the envoys of Agga, the son of Enmebaraggesi, both kings of Kish, who arrived at Uruk/Erech and continues with Gilgamesh's words to the elders of Erech:

"To complete the wells, to complete all the wells of the land, to complete the wells, the small bowls of the land, to dig the wells, to complete the fastening ropes. Let us not submit to the house of Kish, let us smite it with weapons."[22]

Although these lines are obscure to most, to a hydrogeologist they are quite clear: Gilgamesh urges the elders of Sumer to prepare for siege, which will involve the cutting off the water supply to the city by damming and diverting its canals. He suggests digging wells and getting groundwater using ropes either to climb down the wells or to bring up the water in buckets. The elders are not enthusiastic about the plan and repeat Gilgamesh's words:

"To complete the wells, to complete all the wells of the land, to complete the wells, the small bowls of the land, to dig the wells, to complete the fastening ropes?" and conclude with, *"Let us submit to the house of Kish, let us not smite it with weapons."*[23]

---

[22] S.N. Kramer, *The Sumerians*, p. 187.

[23] Ibid.

Although there is no question mark in cuneiform writing, the elders' repetition most probably indicates a question, as though asking Gilgamesh, "Are you in your right mind to think that such a non-conventional water supply will be of any help?" Old and experienced, they preferred to submit rather than to rely on innovation. Gilgamesh then approaches the "men" of the city, and they vote for war, aware that a long siege is inevitable, as their army is no match for that of Kish. They remind Gilgamesh that Uruk is protected not only by Inanna residing in her "house of Heaven", the temple of Eanna, but also by the great sky god An, who built its walls. The war, after some exploits by the hero, has a "happy ending". According to the story, Agga raises the siege from Uruk and departs with his army for Kish after being awestruck by the sight of Gilgamesh. Objective historical reporting is not a strong point of these ancient stories, leaving the question as to whether king Agga's decision to depart, with all due honor to Gilgamesh's prowess, was not helped by a payment from the treasures of Uruk's temples.

Another possibility is that some citizens of Uruk, who were captured by Agga and tortured, were envoys sent to the kings of the neighboring Sumerian cities, such as Ur, to ask for help. Indeed, another tablet tells that the king of Ur replaced the king of Kish in the hegemony over Sumer. One can learn it from the history of Tummal, a district in Nippur, which had a temple to Enlil and a shrine consecrated to the Goddess Ninlil, his wife.[24] This complex of shrines, first constructed by Agga, fell into ruin and was rebuilt by a king of Ur named Mesannepada, who established the First Dynasty of Ur. Mesannepada later triumphed over Agga, putting an end to the Dynasty of Kish, and signed with the royal seal of Mesannepada, the King of Kish. The control of Enlil's and Ninlil's cult center at Nippur seems to have become a symbol for the control over Sumer. After falling into ruin again, it was reconstructed by Gilgamesh king of Uruk/Erech, only to be destroyed again and reconstructed a total of five times – the last by kings who are not mentioned in the canonical list of kings of Sumer.

Thus, after Kish, Ur became the dominant Sumerian city, followed by Uruk, with Gilgamesh its king. The great epic about this king, originally composed in Sumerian and expanded and translated into Akkadian, represents one of the earliest works of world literature. It probes basic philosophical questions of courage, friendship and the quest for eternal life, and, incidentally, contains the story of the Flood.

The positive influence of the colder and more humid climate on the economy of Mesopotamia had also a darker side: catastrophic floods. I In such years they may easily have destroyed one or more cities, bringing havoc on the rural population and their fields and palm date orchards – disasters to be remembered for millennia as the "Flood" or the "Deluge" mentioned in the Gilgamesh story, the Bible and other ancient folklores. In such times, the deep cyclonic troughs of the westerlies climate system arriving from the Atlantic

---

[24] Ninlil was affianced to Enlil by raping her, which convinced her to marry him despite the punishment afflicted on him by the other gods. She thus became the "mother goddess", a trick to "Sumerize" the Ubeidian mother goddess Inanna?

over Europe and the Mediterranean would be extremely beneficial for the desert areas of the Near East to the west and southwest of Mesopotamia, as well as the Iranian plateau.

Flood sediments devoid of artifacts were found in some layers of the excavated ancient tells of Sumer. These sterile layers were inter-bedded within layers containing potsherds and tools. One of the earliest findings was in the 1920s during an excavation of the 'tell' of the ancient city of Ur by Sir Leonard Woolley's team. Between layers containing 'Ubaidian artifacts of the mid-fourth millennium B.C.E., they discovered a layer of clay about 3–4 meters thick deposited by a river flood. It was found in deposits that for many centuries had been dumped outside the walls of this ancient city. Such layers were also found at other sites of Mesopotamia, such as at Kish, Uruk, and Lagash, dating to the Early Dynasty period, i.e. 2700–2400 B.C.E. whereas at other sites, for instance at Eridu, none were found.

The existence of these layers can be interpreted from the point of view of paleo-hydrology as evidence of periods of extremely high floods. We suggest that the flood stories serve as a kind of "proxy-proxy-data" which is in conformity with the paleo-climatic proxy-data. It is conceivable that extreme flood peaks of the two rivers brought disaster on some of the ancient cities, becoming part of the Mesopotamian mythology, whence reaching the Hebrew Bible and part of Christian traditions.

It can be surmised that during the peaks of the cold periods, Arabia, inner Syria and Iran down to Baluchistan in the south were considerably less arid than today, and evidence is piling up over the last years. At the same time, even though the northwestern part of the Indian Peninsula had weaker monsoon rainstorms in summer, it also enjoyed winter rains. This climate enabled a passable land bridge between Mesopotamia and the Indus Valley as well as the Hindukush and Himalaya mountains rich in mineral and precious stones documented by the finds of proto-Elamite commercial documents all over the Iranian highlands. At a later stage, the climate enabled the development of the Indus civilization, which benefited from a milder monsoon climate and a temperate winter precipitation, perhaps the legendary Magan and Meluhha. Thus the prosperity of southern Mesopotamia is founded not only in the fertility of its soil and water resources, but also on the prosperity of the surrounding countries whose wealth was largely due to the optimal climatic conditions and were, therefore, able to engage in profitable commerce. To the west were the pastoral communities of the Syrian-Transjordanian Plateau and beyond them the incipient urban centers and ports of the Levant. To the south and southwest were the pastoral communities of the Arabian Peninsula and the urban communities around the Persian Gulf. Finally, to the east were rural pastoral and urban centers on the Iranian Plateau extending to Afghanistan, the source of lapis lazuli and soapstone for sacred vessels, and the Indus Valley. In the same time, the rain-fed region of northern Mesopotamia benefited from the optimal climate conditions and from its location at the periphery of the richly irrigated countries to the south.

Such, the neighboring rain-fed agricultural and pastoral provinces shared the prosperity that Lower Mesopotamia enjoyed during the good four to five centuries of the early and mid-third millennium. Moving upriver and westward to the Levant, this socioeconomic prosperity peaked in northern Syria. Here the city-state of Ebla, situated between the Euphrates and the Orontes rivers, thrived between the 27th and 24th centuries B.C.E. and continued to prosper even into the following centuries. Documents from Ebla's royal archives show that the city's economy was based on the manufacture of wool and metals, whereas that of Lower Mesopotamia was based on agriculture. The raw materials for the metal products were imported, probably from the metalliferous mountain provinces to the north and northeast, while the manufacture of wool was based on a local surplus of sheep's wool. Gelb analyzed the environmental and economical differences between Lagash and Ebla. The great literate sheep-raising civilization blossoming at Ebla and other sites in the region during the middle of the third millennium B.C.E. led him to change his opinion that to flourish, civilizations of the Middle East had to be situated along a river. He suggests that higher forms of civilization were achieved in early historical times on the basis of leisure provided by the surplus of two kinds of commodities, a surplus of grain in the case of Lagash and Babylonia and a surplus of wool in the case of Ebla and northern Syria.[25] The present authors would like to add that one of the most important preconditions for agricultural surpluses in the Near East is abundant precipitation, and the precondition for this is a cold and humid climate, which characterized the period described by Gelb.

The pastoralists' pride in their dairy products becomes evident in a poem describing the courting of Inanna by the shepherd Dumuzi.[26] It begins with the sun-god Utu recommending Dumuzi to his sister Inanna. He assures her that the suitor is rich with precious stones, possesses date-wine and oil, and is a protector of the king, perhaps referring to the employment of some nomadic pastoral people as mercenaries. When Inanna refuses, and insists favoring the farmer Enkimdu, because his plants and grain grow abundantly, Dumuzi recites a list of the many wonderful products of his herds. Against the black and white garments of the farmer he will bring forth black and white ewes; against the farmer's date wine he will pour an assortment of milks; against the farmer's bread and beans he will give honey-cheese and small cheeses, more than anyone can eat. This abundance seems to overwhelm the farmer. He begs Dumuzi not to start a quarrel and proposes a compromise in which Dumuzi's sheep will be allowed to graze in the farmer's meadowland and field and drink water from his canal. Surprisingly, the shepherd does not agree to a friendly

---

[25] C. Bermant and M. Weitzman, *Ebla*, Weidenfeld and Nicholson, London (1979).

I.M. Diakonoff, "The importance of Ebla for History and Linguistics" *Eblaitica* 2:3–31 (1990).

I.J. Gelb, "Ebla and Lagash: Environmental Contrast" in: *The Origins of Cities in Dry Farming Syria and Mesopotamia in the Third Millennium B.C.*, ed. Weiss. 157–167.23. This scheme is still upheld in some recent publications, such as E. Porada, "The Chronology of Mesopotamia, ca. 7000–1600 B.C." in *COWA*.

Other scholars might have slightly different chronology, such as W.W. Hallo and W.K. Simpson, *The Ancient Near East, A History*, Harcourt Brace (1998) on page 37, suggest ED I 2900–2700, ED II 2700–2500, ED III 2500–2300.

[26] For a translation see S.N. Kramer, *The Sumerians: Their History, Culture, and Character*, Appendix E, University of Chicago Press, Chicago. pp. 328–331 (1963).

settlement, and the farmer surrenders, offering the shepherd his products, as well as the maid Inanna.

As can be learned from the poem, "Inanna and Dumuzi in the Nether World", Inanna submits and marries Dumuzi. They did not, however, "live happily ever after" as might be expected after so many adventures. As was suggested, decoding these poems could reveal the story of the encroachment and eventual integration of the pastoral societies from the surrounding desert into the irrigated lands during the warm and dry intervals, characterizing the transition from the Chalcolithic to the Early Bronze. Whether or not this interpretation is sound, still other sources of data show that during the same period Semites and later Elamites did gain more than a foothold in the fertile valleys of lower Mesopotamia.

The contribution of the semi-nomadic people from the west was manifold. First, from the "co-evolutionary" point of view, the influx of genes and cultural patterns resulted in a process of selection due to competition for survival in the nomadic and semi-nomadic society. This process was quite different from that of the sedentary farming society. The possible genetic traits diverged from lactose digestion faculties to perhaps adaptation to frequent migratory and transient ways of life. On the cultural level, it included a new pantheon of gods, rituals, myths, holidays, etc. Second, connections of kinship with the original population and intermarriage with the local population guaranteed a continuous inflow of genes, ideas and goods, ensuring genetic as well as cultural diversity.

As already mentioned, another contribution resulting from competition was the continuous rivalry between the cities, whose roots emerged from the different populations that settled the various parts of Mesopotamia. This rivalry led to wars, loss of life and capital, but at the same time, kept the leaders alert and demanded ingenuity and the development of systems of organization. In the end, chapters of history still to come, continuous warfare cancelled the positive gains of rivalry and competition, particularly when a foreign enemy or a climatic catastrophe emerged, to find a society destabilized due to self-attrition and continuous destruction of resources. Yet, during the formative centuries of the Mesopotamian civilization, the internal agitation may well have had a positive impact.

The infiltration of large numbers of pastoral people dwindled with the amelioration of the climate, which became considerably more humid after 2800 B.C.E. and remained more or less optimal for about five centuries. One has to live in the desert in order to grasp the change in the natural environment after just one year of above-average rainfall, much less decades and centuries. The reconstructed humidity and precipitation curves of Lake Van and Soreq Cave indicate an average 20 to 30 percent increase in precipitation. This increase was accompanied by a decrease in the annual temperature, resulting in cold winters and cooler summers with lower rates of evapo-transpiration and thus higher soil moisture rates. It meant fewer drought years and also more years in which the precipitation may have reached 50 percent above average. Evidence for this emerges from the paleo-levels of the Dead Sea and the Tigris-Euphrates. The higher rates but lower annual variability of precipitation and the reduced number of drought years enabled the farmers and pastoral people to benefit from post-harvest grazing in the semi-arid zones and to amass income without losing it during drought years.[27] In addition, stone- and soil-built diversion and barrier dams, like those found in the Negev and the Syrian Desert from the Chalcolithic period well into the Early Bronze raised the quantities of water available to the fields in the riverbed. This extended the reach of arable lands into the belt of marginal economic feasibility. At the same time, using similar methods, the semi-nomad pastoral people in this belt extended their area further into the desert. The higher rates of precipitation brought frequent floods, which recharged the groundwater in the rivers' gravel-beds The digging of wells in the riverbed guaranteed water for the flocks. Considering that well-digging was a known practice at this time, it can be concluded that during the Early Bronze Age the Fertile Crescent indeed extended southward far beyond the contemporary 200-mm line.

It is, however, too early to evaluate quantitatively the exact location of the paleo-precipitation lines or the quantities of harvest, as it is still not certain if the climate change was linear – that is, whether just the quantity of rain was higher – or whether it was also distributed over more months of the year. This lack of information on the quantity and variability of precipitation, coupled with inadequate knowledge of harvest methods, soil moisture conservation, crops and cropping system, manure application by post-harvest grazing etc. in the rain fed regions leaves us reluctant to give harvest estimates. Our reluctance to estimate harvests is not shared by Weiss in dealing with the origins of Tell Leilan.[28]

In the irrigated part of Mesopotamia, too, the high precipitation rates during the third millennium B.C.E. could have influenced the region's agricultural economy far beyond that resulting from simple augmentation of the water supply: The hydrological regime of the Tigris and Euphrates is influenced not only by the quantity of precipitation on their drainage basin but also by the percentage of rain to snow and its distribution over the months. Although the character of their flow is different, both rivers have peak flow during the spring months. At present, the precipitation rates on the Taurus-Zagros mountain ranges and Anatolian Plateau feeding the Euphrates and Tigris, extend from 800 mm in Anatolia and northwestern Iran to 400 mm in the area bordering southern Mesopotamia. The smaller flow is in the winter, fed by the rains on the mountains, creating floods in the riverbed. The greater flow in the spring is due to melting of the snow on the mountains and from the ebb of the many springs from the highly permeable limestone rocks from which much of these mountains are built. A colder climate thus would mean a higher rate of precipitation, largely due to an increased

---

[27] An experiment in growing a herd of sheep at the Institute of Desert Research of Ben-Gurion University's experimental farm at Wadi Mishash south of Beer Sheva in the Negev Desert showed that the loss during drought years due to having to purchase fodder drained the profits accumulated during good years (personal communication of Arie Rogel, former head of station).

[28] H. Weiss, "The Origins of Tel Leilan and the Conquest of Space in Third Millennium Mesopotamia" in: *The Origins of Cities in Dry Farming Syria and Mesopotamia in the Third Millennium B.C.* H. Weiss (ed). Four Quarters Publishing Co., Guilford, Conn. pp. 71–108 (1986).

snow cover on the mountains as the melting snow recharges the limestone mountains' aquifers more efficiently than rain. The higher storage capacity in the mountain regions extends the flow period well into the summer months.

The rivers' benefit to people living in the plain varies according to the nature of the rivers. With its higher gradient, higher and steeper banks and narrower bed, the Tigris is fast and torrential, befitting its Sumerian name, *idigna*, derived from the word "arrow". Thus, canals extending from the river for irrigation are few. However, during the third millennium B.C.E., this river had a separate delta, providing an additional body of fresh water to lower Mesopotamia and protecting it from the invasion of seawater, as will be explained immediately. At the same time, the increase in precipitation in the present more arid southern parts of the Zagros enlarged the quantities of water flowing in this river into the plain of southern Mesopotamia.

The flow of the Euphrates is more moderate and its banks are lower, enabling construction of diversion canals. At present it reaches its highest water levels in March and continues at a high level until the beginning of May. By then the fields have been sown and the crops are germinating. Thus, the fields have to be protected by dikes to prevent sudden flooding and erosion. Additional irrigation during the growing season can be achieved by opening a controlled breach in the dike, or by the use of a water elevation device.

Due to the low gradient of the river, drainage of the fields to ensure outflow of the irrigation water backflow was very important as protection against salt encroachment. This may have been achieved in a much easier way in the third millennium B.C.E. than at present because of the following:

The colder climate caused a retreat of the sea, which, together with higher rates of flow, probably ensured a regional lens of fresh water flow on the interface of saline water from the sea. Usually, the thickness of this lens, i.e. the depth from the surface to the saline water depends on the level of the fresh water. The level of the water in the rivers and flooded plain and the height of the water piled artificially on the fields due to irrigation decide this level. Considering the flow regime described earlier, it can be estimated that the lens of fresh water will be thickest during the spring and at its minimum from the end of the summer through the autumn. Additional influences on subsurface water salinity are the negative factors of evaporation from the soil and transpiration by plants. Evaporation is either direct from the water bodies spread on the land or through the soil from capillary movement depending on the depth to the saturated zone as well as the soil structure – the finer the capillaries, the higher the column of capillary water above the water table. Thus, one way to prevent reaching the capillary-saturated zone is to break the crust formed on the surface by desiccation. Indeed, one learns from the Mesopotamian "Farmers' Almanac" that the Sumerians were aware of the need to granulize the soil structure by plowing and breaking up the clods and irrigate their crops a few times during the growing period. Even after they adopted barley – a salt tolerant plant – as their main field crop, they continued to irrigate the fields four times to keep the water in the root zone fresh, thereby achieving a 10 percent higher yield.[29] Barley replaced wheat as the main field crop in Mesopotamia after 2300 B.C.E. when the climate became drier. The reason for this transition will be discussed later on.

Keeping a low salinity profile in the non-saturated and saturated parts of the soil is of primary importance in semi-arid regions with long dry seasons in order to ensure high yields. In general it can be said that even salt-tolerant plants will have higher production rates if irrigated with water of low rather than higher salinity, as the extraction through osmosis by the tree's roots demands an investment of energy by the plant. The higher the salinity, the higher is the investment of energy by the plant, or the smaller the amount of water it can acquire from the soil. In order to keep a low salinity profile in the subsurface, it is not enough to interfere with the capillary movement; soil and groundwater becoming saline by plant transpiration must also be controlled. Transpiration may affect depths even lower than the capillary zone because, due to the membrane osmotic filtration mechanism in their roots, most plants take up the fresh water, transpire it and leave the salts in the root zone. The salts accumulate in the groundwater, and if there is no flushing out – either horizontal or vertical – the subsurface water will become saline. In most cases the level of tolerance of the plants decides the peak level of salinity, since beyond a certain peak the plants will die and thus transpiration is stopped. With salt-tolerant plants such as barley or the date palm, salinity in the root zone may reach levels that will reduce the yield of the plant, yet the plant itself will survive, whereas salt intolerant plants will die.[30]

It can be assumed that the optimal freshwater conditions in the subsurface and the dry and relatively warm summers, especially in southern Mesopotamia, enabled a high yield of the date palms in Sumer even in the colder periods. The many references to this tree in the Sumerian literature, on memorial stelae and on cylinder seals, and the findings of date stones and wood in archaeological excavations indicate that the date palm was significant in the region's economy in the past, as it is in the present. Growing involved artificial fertilization of the female trees. From the early second millennium B.C.E., Sumerian lexicographical lists contain close to one hundred fifty words for the various kinds of palms and their different parts. The Sumerians no doubt ate the dates fresh and produced jam, which they called *lal*, often translated as honey, a term also Used in the Biblical text and both do not refer to bees' honey. As we learn from Dumuzi's argument with the farmer, they also made palm wine. All other parts of the tree were used in various ways, as in present times. Because the level of temperatures in ancient Mesopotamia are not known, it is difficult to say which regions produced dry varieties and which the moist variety of dates. The dry varieties need higher summer temperatures than the

[29] Kramer, op. cit. pp. 108 (1963).

[30] Except for plants such as the tamarisk, which takes up the salt with the water. After the salty water transpires on the leaves, the salt falls on the ground. This sterilizes the "*Lebensraum*" of the tree and prevents competition by other trees.
A.S. Issar, (ed.) "Brackish Water as a Factor in Development" *Proceedings of the International Symposium on Brackish Water, January 5–10, 1975 at Ben-Gurion University of the Negev,* The Jacob Blaustein Institute for Desert Research, Israel (1975).

moist one. From the abundance of the date palm and its importance to the Mesopotamian economy, it can be inferred that the summer months were dry, with temperatures over 18 °C from May to October and exceeding 10 °C in winter. When the climate became warmer and drier, the yield of dry varieties may have increased, but at the same time, the supply of fresh water in the riverbeds diminished, and with it the yield. In some regions production might have stopped entirely.[31]

According to the Sumerian king list,[32] the kingdoms in Sumer started with a line of kings from Eridu ruling all of Mesopotamia. Indeed, the most ancient 'Ubaidian layers were excavated at Eridu. Kingship then passed to rulers of other cities, each living and ruling for a few tens of years and continuing more than a hundred thousand years until "The Flood", after which the first dynasty began at Kish. This event seems to indicate an important historical transition to a new ruling class of Semitic origin, seen in the names of the kings, such as *Kalbum* (dog), *Qalumu* (lamb) and *Zuqaqipu* (scorpion). Yet, despite fundamental changes and various natural disasters, the culture of Early Dynastic Mesopotamia shows remarkable continuity. Some twenty city-states shared the "Land of Sumer and Akkad," as the ancients called their country, continuously forming alliances or fighting each other. The land and the city belonged to the city's tutelary god residing in a temple amid his people, a jealous deity often depicted in battle against neighboring gods. The rich soil enabled agricultural surpluses, allowing for tremendous population growth and creation of wealth for the gods and their stewards, the rulers. The earliest documents suggest that the temples were the focus of all economic activities, including trade links with neighboring countries from Syria-Palestine via Anatolia and Iran to the Indus Valley, which were established to satisfy the gods' and the rulers' appetite for exotic and ostentatious luxury goods.

The most important civilizatory development during the Late Uruk and Jemdet-Nasr periods was the standardization of writing. The number of pictographs was reduced, and instead of incising or drawing signs (as it is not simple to make curved lines on a flattened lump of clay), a cut reed was used to impress short impressions looking like small wedges (in Latin *cuneus*, hence cuneiform). This system soon developed into abstract signs, losing any resemblance to the original pictographs.

---

[31] S. Stoller, *The Growing of Date Palms in Eretz Israel*, Kibbutz Meouchad Publishing, (In Hebrew) (1977).
On the importance of the date palm tree to the economy of certain regions in the Near East see Chap. 5. The Date palm trees still bear fruit at salinities of up to 3000 mg/l Cl , Issar drilled into the root zone of a group of date palm trees in the Sinai Desert and reached water of 9000 mg/l Cl. The trees were alive but gave no fruit. During the late 1960s, while visiting the area along the Shatt El Arab lower delta, Issar was shown that many dates planted along the canals near the Persian Gulf stopped bearing fruit, and many even dried out. According to the local agronomist, it started after the Iraqis built dams on their rivers and the Iranians on the Karoon. During a visit to the Princedom of Oman in 1995, Issar was taken to a date palm plantation far from the sea, which stopped bearing fruit. Topographical considerations excluded the possibility of salination due to sea interface intrusion. Calculation of transpiration revealed the reason for the failure of the trees, namely salination of the water in the root zone by the trees themselves.

[32] For a translation see S.N. Kramer op. cit. Appendix E, pp. 328–331 (1963).

Following the rather shadowy Jemdet-Nasr Period, the tripartite division of the Early Dynastic period into E.D. I (2900–2750), E.D. II (2750–2650) and E.D. III a and b (2650–2400) is based on the stratigraphic groundwork laid by the American excavations of three sites in the Diyala region, northeast of Baghdad – Khafajeh, Tell Asmar and Tell Agrab.[33]

The Sin Temple at Khafajeh underwent several building phases from a single-room sanctuary, entered through a door in the sidewall in a characteristic "bent-axis approach", to a tripartite structure formed by adding rooms. More rooms were added in time and then surrounded by a massive double wall separating the temple from the city's residential quarters. The finds from a nearby cemetery laid the foundation for the sequence and development of Mesopotamian pottery.[34] At Tell Asmar, the capital of a state called Eshnunna was found the temple of Abu, the God of Vegetation. It contained a group of votive statues representing worshippers in constant adoration in a pose which was later to become standard: stiff frontal figures with clasped hands, long side-locks and square beards, huge eyes made of inlay of shell and lapis-lazuli, and dressed in traditional woolen skirts. Also at Tell Agrab, the *bothroi* (ritual deposits) and the "sacristy" of another temple yielded a large amount of refined artwork and ornaments used in the temple.[35]

Woolley's excavations at Tell Muqayyar, the Sumerian city of Ur, probably the Biblical "Ur-of-the-Chaldees", in the 1920s and 1930s astonished the world with its spectacular finds from what is known as the "Royal Cemetery", dated to E.D. III.[36] Ur was a large city, located in the extreme south of the country at a bend of the ancient Euphrates (now dry) not far from Eridu. Still visible massive walls surrounded the city. A sounding in the later "Great Temenos" led Woolley to sixteen corbelled tombs built of imported stones and mud bricks, approached by a sloping ramp or a shaft. Some of the tombs contained remains of sacrificed soldiers and attendants of both sexes, together with wagons and their draft animals. Some of the interred shared the burial chamber with the principal occupant. Sumptuous ornaments, weapons, musical instruments and other treasures, illustrating the high level of Sumerian craftsmanship, surrounded all. The best-known tomb belonged to Queen Pu-abi, according to a lapis-lazuli cylinder seal. In a nearby tomb, called the "King's Tomb", Woolley found some 59 bodies, including six soldiers leading two ox-drawn wagons and 19 court ladies in the dromos (approach ramp). Besides the golden ornaments and headdresses, the now-famous lyre with an inlaid sound box and a bull's

---

[33] P. Delougaz, *T he Temple Oval at Khafajah.* Oriental Institute Publications 53, University of Chicago, Chicago (1940).
P. Delougaz and S. Lloyd, *Pre-Sargonid Temples in the Diyala Region.* Oriental Institute Publications 58. University of Chicago (1942).

[34] P. Delougaz, *Pottery from the Diyala Region.* Oriental Institute Publications 63, University of Chicago (1954).

[35] H. Frankfort, *The Art and Architecture of the Ancient Orient.* The Pelican History of Art, Penguin Books, Harmondsworth (4th revised impression 1970).

[36] C.L. Woolley, *Ur Excavations. vol. 2.: The Royal Cemetery,* Publications of the Joint Expedition of the British Museum and the Museum of the University of Pennsylvania to Mesopotamia, Oxford University Press, London (1934).

head was found. Another tomb contained 74 human sacrifices – of these 68 were women in full golden regalia. This tomb also yielded the well-known ram rampant against a bronze tree, and the "Royal Standard", a magnificent box with ivory inlay depicting war scenes.

H.R. Hall uncovered additional outstanding finds at al-'Ubaid, including a large bronze lintel representing a lion-headed eagle between two stags, together with a pair of columns encrusted with colored inlay, animal friezes and ritual scenes.[37]

Kish was one of the most important capitals of the Dynastic Period mentioned in the Sumerian King List – a cluster of mounds a dozen kilometers east of Baghdad. The title "King of Kish" was employed by rulers claiming hegemony of the Land of Sumer and was perhaps an indication of some ancient political unification of the country. E. Mackay found a large palace with porticoes, buttressed walls and chambers with ceilings supported by columns,[38] reminiscent of the contemporary palace at Tell Yarmut in Palestine. The only *ziggurats* of the Early Dynastic Period were found at the top of the mounds and founded on a "flood stratum" covering the entire site. Below this layer of sediments, a residential area dated to E.D. I and II was exposed, again separated by other "flood strata".

In the center of the southern plain, another cluster of mounds forms the ancient state of Lagash. Tell el-Hiba, the ancient Uruk, was the capital, with Tello (ancient Girsu) and Nina (ancient Surghul) its principal religious centers. The French excavators found treasures from the time of the kings of the Sumerian King List, such as Ur-Nanshe, including the "Stela of the Vultures" celebrating the victory of Eannatum of Lagash over Umma, a neighboring rival city state. This "stela" – a stone relief carving on a wall plaque with a hole in the center for fixing on a wall – shows the king leading his troops, resembling a phalanx of Greek hoplites, riding a chariot. On the reverse side the god of war, Ningirsu, son of Enlil, gathers his enemies in a net, guarded by Imdugud, the lion-headed eagle. The excavations by H.V. Hilprecht and R.C. Haines[39] of Niffer (ancient Nippur), seat of the head of the Sumerian pantheon Enlil and religious capital of the land, confirmed the result of the other sites – particularly the correlation with the Diyala sequence – and yielded a series of these plaques. Level VII also showed a remarkable twin temple dedicated to Inanna and Dumuzi reminiscent of the twin temples at Arad and Megiddo in Palestine, with access through a doorway at the opposite end of the holy of holies, similar to the layout of other Near Eastern and Egyptian temples.

In the north, in the rain-fed zone beyond the alluvial plain of southern Mesopotamia, other sites represent outposts of the Sumerian culture. In Qal'at

---

[37] R.H. Hall and C.L. Woolley, *Ur Excavations. Vol. 1: Al-'Ubaid.* Publications of the Joint Expedition of the British Museum and of the Museum of Pennsylvania to Mesopotamia, Oxford University Press, Oxford (1927).

[38] E. Mackay, *A Sumerian Palace and the "A" Cemetery at Kish, Mesopotamia* pt. 2, Anthropology Memoirs, vol 1, no. 2 (1929).

[39] R.C. Haines, "Nippur" (*Illustrated London News*, 18 August 1956, 6 September 1958, 9 September 1961).

Sharqat (ancient Assur) on the Tigris, W. Andrae[40] found well-preserved remains of the oldest layers of the famous temple of Ishtar, dating to the Early Dynastic period. It is, in its design and its small finds, totally Sumerian in character. At Tell Hariri (ancient Mari) on the Euphrates, excavated by the A. Parrot, another temple of Ishtar and other shrines were discovered. Their contents included the seated statue of Ur-Nanshe, a singer at the court of King Iblulil. Most important, however, was the contemporary palace with its rich finds – particularly tablets – as, according to the Sumerian King List, one of the kings of Mari became overlord over Sumer. A cylinder seal inscribed with "Mesannipadda, King of Ur", which was probably presented to one of the kings of Mari confirmed the date, i.e. E.D. III, and the close connection between these northern city-states and the Land of Sumer.[41]

A rather peculiar site, excavated by A. Moortgat[42], is Tell Shuaira (or Chuera), between the Habur and the Balih in northern Syria. Undressed local stones topped by mud bricks were used to build its porticoed temples, which recall Anatolian prototypes whereas the figurines and pottery are clearly related to contemporary Early Dynastic Sumerian parallels.

Around 2600 B.C.E. – the transition from E.B. II to III in the Levant – a short period of warmer and drier climate occurred (Figs. 3 & 3a.). Although limited in both intensity and duration, it was felt in all parts of the Near East and negatively influenced the regions in the interior perimeter of the semi-arid belt that were dependent on rain-fed irrigation or that lacked a perennial water supply. Its effects were felt in the southern part of Palestine, where the city of Arad was deserted, and possibly by the people of Elam who settled along the Zagros foothills, as well as in the mountain ranges and intermountain valleys east of the Tigris. The reduced rainfall over these semi-arid valleys may have led the Elamites to venture westward to try to gain control of the river-fed plain. Indeed, the inclusion of kings bearing Elamite names in the Sumerian King List might indicate that, for a short period, these ventures were successful. The Elamite hegemony may even have contributed to the end of the Early Dynastic II period.[43]

The Early Dynastic III period, from ca. 2500 to 2300 B.C.E., saw the return of an optimal climate, and with it, continued economic and cultural prosperity for Sumer. Through purchase and conquest, real estate and capital slowly became increasingly concentrated in the hands of a small elite. The wealth of the king, who claimed direct divine patronage and who was supported by a wealthy ruling class, enabled mobilization of masses of citizens of the city-state for

[40] W. Andrae, *Die Archaischen Ischtar-Tempel in Assur*. Wissenschaftliche Veröffentlichung der Deutschen Orient-Gesellschaft 39, Hinrichs'sche Buchhandlung, Leipzig (1922).

[41] A. Parrot, *Mission archaéologique de Mari, vol 2: Le Palais.*
idem. vol. 4: Le "tr ésor" d'Ur. Institut Français d'Archaéologie de Beyrouth, Bibliothèque archaéo-logique et historique, P. Geuthner, Paris (1959, 1968).
See also S. Lloyd, *The Archaeology of Mesopotamia*, Thames & Hudson, London (1978).

[42] A. Moortgat, "Tel Chuera in Nord-Ost Syrien, Preliminary Reports" *Wissenschaftliche Abhandlugen für die Arbeitsgemeinschaft* (1959–1967).

[43] For the problems involved in interpreting the historical meaning of the canonical king list see W.W. Hallo and W.K. Simpson, op. cit. (1998) pp. 39–42 & p. 50.

military ventures to expand the city's dominion. The common people lost the last vestiges of economic independence and increasingly became tenant farmers, receiving only a portion of the wealth they produced.

With the formalization of writing, more and more historical information was recorded on tablets, stelae, wall plaques, or dedication vases, providing a clearer picture of the wars between the city-states themselves and with the neighboring countries. Other city-states, including Adab, Umma, Girsu and Lagash, vied for hegemony over Sumer. Located in the easternmost part of Sumer, these cities had become frontier states facing Elam. After the short interlude of Elamite hegemony during the transition period, the improved climate may have eased the pressure from the east, enabling the Sumerians to gather capital, strengthen their armies, and counterattack. As a result, they dominated their neighbors on the east for several centuries, superior both in combat and in culture during the Early Dynastic III period. The Sumerian culture spread throughout most of the Near East. Nearly all ancient Near Eastern peoples – from the Babylonians and Assyrians to the Syrians, Hurrians, Hittites, Elamites (and even the Persians in the first millennium B.C.E.) adopted the Sumerian cuneiform writing system and used it to record their own language, mixed with Sumero-grams and, therefore, had to study the Sumerian language. Sumerian literature, laws, mythology and worldview penetrated all levels of ancient Near Eastern civilization, a situation in some measure comparable to the much less important dominance of Latin in Europe until quite recently, or Chinese for Korea and Japan.

Yet, most of the economic and military effort now was spent on rivalries among the city-states. This continuous mutual attrition, together with an un-controlled overpopulation despite constant culling by wars accompanied by large scale massacres and accompanying deterioration of the socioeconomic system, as well as the concentration of power and property in the hands of a few proved to be a classic formula for disaster. Urukagina, the last king of Lagash, tried to remedy this demoralizing system with a series of land, finance and law reforms, but his efforts came too late; he was defeated by Lugal-Zagge-Si, the king of Umma, who destroyed Lagash and continued his successful march of conquest against the other sovereign cities. On a vase dedicated to Enlil in Nippur, the religious capital of Sumer, he boasts of extending his rule from the Lower Sea (Persian Gulf) to the Upper Sea (Mediterranean Sea). This might be more wishful thinking than historical fact, but most probably he was able to extend the control of Uruk over most of Mesopotamia, including Semitic Akkad and north to the city-state of Mari.[44]

The people of contemporary Palestine shared the affluence of Mesopotamia and Syria as seen in the archaeological record. These people *"clearly enjoyed a prosperity equaling that of its later halcyon days during the Roman Empire"*.[45] Redford's comparison is noteworthy in that, as will be shown later, the Roman period was also exceptionally cold and humid, although less so than during the Early Bronze period. Still the prosperity in the southwestern part of the Near

[44] G. Roux, *Ancient Iraq*. Pelican-Penguin Books, Harmondsworth p. 134 (reprint 1972).

[45] D.B. Redford, op. cit. (1993), pp. 30, 33.

East did not result in urban expansion commensurate with that of contemporary urban centers to the northeast in Mesopotamia and southwest in Egypt, or to the later cities of the Roman Period. The many settlements became walled cities into which the rural population could retreat in case of attack. Hostilities may have erupted between the city-states, as occurred in Mesopotamia or from military expeditions from Egypt. That the Egyptian army crossed the sand barrier and arrived in the Levant *"to smite the Asiatics"* is evidenced by more than one document.[46] The nomadic desert tribes were unlikely aggressors during this time, as the general agricultural prosperity had spread to the southern edge of the region, and during that time the desert bloomed.

However, signs of climatic changes and the pending disaster began to be felt sometime during the 24th century B.C.E. Periods of stability and prosperity invariably bring population growth that sooner or later exceeds the carrying capacity of a given territory. The effects of drier weather in the Anatolian highlands were increasingly felt in the irregular arrival of the floods of the two rivers and the consequent drying of previously green fields. The mushrooming outgrowth of superstructure and deprivation of the peasants of the fruits of their toil may have caused them to lose interest in investing energy and intellectual effort in their fields.[47] Needless to say their hardships went unrecorded; the illiterate peasants did not put their plight in writing, nor did the scribes, who served the rulers and had no interest in the lower class.

Excessive taxation widened the gap between the few rich and the many poor, and tensions arose within the ruling class as well – for example, between the priests, the aristocrats, and the king in Lagash – over the diminishing spoils of the peasants.[48] Coupled with the ruthless behavior characteristic of all tyrants, and exemplified by the enormous waste of human lives in the "Royal Cemetery" at Ur, the incessant internecine wars among the Sumerian city-states exhausted their physical resources. The end came with Lugalzagesi of Umma, who finally conquered Lagash and eventually all of Sumer, only to be overthrown himself within a few years by a newcomer. In 2370 B.C.E. (or 2334–2279 B.C.E., according to which chronology is followed)[49] Sargon of Akkad founded the first Semitic empire in Mesopotamia.

The warming of the Near East began to take a toll on farmers and pastoral people alike. Drought caused crop failures in the rain-fed areas. The irrigated

---

[46] Ibid., pp. 30, 33.

[47] Issar did hydrogeological research from 1961 to 1964 in regions dominated by feudal landowners in Iran before the last Shah instituted land reform. He discovered that the tenant peasants were reluctant to grow industrial crops, such as sugar beet or cotton, although they yielded better profits. This because the peasant's share was only one-fifth the value of the crops, while four-fifths went to the landowner as payments for his "contributions". First the landowner supplied the water (thus paid by fifth of the crop), as he owned the qanat or canal. Second the working animals and work tools (another fifth). Third the seeds or credit (another fifth) and fourth for the land. With wheat or barley, the peasant had a better chance to conceal a larger part of the crop and feed his family. On the other hand, cash crops could only be sold to the factory (which may have belonged to a group of landowners) only through the agent of the absentee landowner.

[48] W.F. Saggs, *The Greatness that was Babylon*, Mentor Books, New York. p. 65 (1962).

[49] J.A. Armstrong, "Mesopotamia: Sumer and Akkad" in *The Oxford Companion to Archaeology* B.M. Fagan (ed.), Oxford University Press, Oxford, pp. 456–457 (1996).

regions were affected as well because of the increased salt content in the non-saturated and saturated parts of the soil column. When prosperity of the pastoral life was ending, they responded with infiltration and increased pressure on the irrigated lands. With the entire socioeconomic system shaken by environmental disaster and battered by invasion of semi-nomads, the society finally collapsed. All these factors spelled a regional disaster for Mesopotamia and other parts of the ancient Near East, which will be discussed in the next chapter.

5.4.2
Egypt United Under One Crown

At the opposite end of the Fertile Crescent, Egypt's earliest history presented a similar picture of city-states and their immediate hinterland, later called *nome* in Greek. Manetho, a priest writing in Greek during the Ptolemaic period, relates in his history that the first king to rule a united Egypt during this period was "Menes", who founded his capital at Memphis, at the junction of Upper and Lower Egypt.[50] This statement oversimplifies a complicated and badly understood process. "Menes" was most likely a composite of several historical personalities, beginning with Narmer of Nekhen/Hierakonpolis in Upper Egypt halfway between Luxor and Aswan. Manetho then lists the rulers, divided into dynasties numbering from one to thirty-one. The system of 31 dynasties was adopted, with alterations, by modern Egyptologists and forms a broad chronological framework of ancient Egyptian history.

What is known in Syro-Palestinian archaeology as the Early Bronze Age II is paralleled in Egypt as the Early Dynastic period[51] and includes the first two dynasties. J. Baines and J. Malek dated these dynasties to 3050–2686 B.C.E., whereas W. Helck and others favor an earlier chronology.[52] The third, fourth, fifth and sixth dynasties are known as the Old Kingdom (2686–2181 B.C.E.), which was roughly contemporary with the E.B. III period and the beginning of the Intermediate Bronze Age in Palestine.

Our limited knowledge of the Early Dynastic culture is derived from the royal tombs of Abydos and Saqqara and finds from the temples of Elephantine/Aswan in the extreme south and Buto/Tell Fara'in in the western delta. The most characteristic artifacts are vessels of hard stone of the highest quality and cosmetic palettes for grinding copper or lead ores for eye paint, mostly of green schist but also other stones, sometimes decorated with representations of possible historical events. Art is by now already unmistakably Egyptian in proportion and harmony in form.

---

[50] In Hebrew and other northwest Semitic languages, Egypt is known as *Misraim, misru* meaning "country", and the suffix *-aim* denoting the dual, i.e. "The Two Countries".

[51] The term. "Early Dynastic" is used in both Mesopotamia and Egypt but in different ways: In Mesopotamia it includes most of the third millennium B.C.E. whereas in Egypt it includes only the beginning of the millennium.

[52] J. Baines and J. Malek, *Atlas of Ancient Egypt*, Andromeda/Facts on File, Oxford (1980).
W. Helck, *Geschichte des alten Ägyptens*, Brill, Leiden (1982).

The beginning of Egyptian history in the 31st century B.C.E. is based on annals, i.e. years' lists of rulers and their outstanding deeds – an administrative practice for some as yet unknown ritual purpose commemorating the original unification of Upper and Lower Egypt by the first pharaoh (as already mentioned, in Egyptian *per-'o*, meaning the Great House, and presumably a 2nd millennium B.C.E. term). This imaginary and cyclic unification is the expression of a dualistic concept of the physical world in balance, such as the opposition of *kemet* – the black soil and land of civilized people – against *deshret*, the red soil of the desert and wilderness, the abode of the "wretched barbarian"; or the two banks of the river, and sky and earth. We see similar dualistic conceptions repeated in many other spheres of life, such the division of people into *rekhyet*, the populace, and *pa'at*, the nobility.

Some pre-dynastic rulers had worn the white crown of Lower Egypt and the red crown of Upper Egypt. With the First Dynasty (probably Narmer) we see the crowns combined to denote a unified state with a defined ideology, a standardized writing system, the introduction of a calendar, and the establishment of an all-pervasive bureaucracy.

The populace, in exchange for safety and stability, became a workforce unimaginable in earlier times even though it numbered no more than about one million in the entire country. Their tools and agricultural techniques – in particular irrigation by the primitive *shadduf* or by hand – hardly changed since the Neolithic period and throughout history, as they were regarded to be sufficient. Their life remained static, as they produced all they needed, and any surplus was appropriated by taxation to create wealth for the ruling elite, a phenomenon already seen in Mesopotamia. The farmers were tied to the land because of the fact that land without their cultivators was quite useless. Naturally, they had no say in matters concerning land ownership. Whereas the warring Mesopotamian city-states could always employ superfluous manpower in their conflicts, internecine wars in Egypt ceased after unification and with no foreign countries in easy reach of conquest, the peasants could be employed in ever-larger building projects, to keep them occupied and out of mischief, particular during the three months of inundation.

According to the new ideology, the kings owned the land and could distribute it to loyal followers for services rendered. The newly acquired wealth was used to create luxury goods, either produced locally or imported. Artistry thrived as never before, laying the foundation for the unique art of the Old Kingdom and, in slightly shifting nuances of styles for all periods to follow.

Much of this wealth was invested in magnificent funerary monuments, some of which still are the glory of Egypt. Tombs of the ordinary people hardly changed from the prehistoric to the more recent periods, but a sudden great step can be observed in the case of royalty and nobility, with ever-larger tombs resembling underground palaces topped by *mastabas* (benches in Arabic) leading to ancient man's greatest monuments, the pyramids. The kings of the First and the early Second Dynasty were buried at Abydos, probably the original home of the dynasty, whereas the later Second Dynasty tombs are found in Saqqara. It was also customary to have a symbolic tomb in the north as well as in the south.

According to Manetho, the Old Kingdom started with the Third Dynasty, beginning with Nebka (Nekherophes) who was followed by the greatest king of this dynasty, Netjerikhet (Djoser). The Palermo Stone mentions the "erection of Men-netjeret in stone" in the reign of Nebka. Between the first and still standing stone architecture is the magnificent funerary complex of Djoser in Saqqara intended as the dwelling for the king's eternal spirit, or *ka*. The original *mastaba* was enlarged, topped by a series of slightly smaller *mastabas*, resulting in a stepped pyramid. It was surrounded by courtyards and shrines for ceremonies and by another, southern, cenotaph. The entire complex was surrounded with an impressive 'niched' stone-wall, imitating mud-brick construction.

The kings of the Fourth Dynasty, Snofru, Khufu (Cheops), Khafre (Chephren), Menkaure (Mycerinus), and others, magnified the idea of creating an "eternal" world in "eternal" material, stone. Snofru was the greatest builder and experimenter – he built three pyramids for himself. The most splendid of these structures are the pyramids of his successors in Gizeh southwest of Cairo.[53] An entrance structure, the valley temple, was built on the edge of the cultivated flood plain to receive the body of the deceased king for his final journey. He was brought through a covered passage towards the cult temple, behind which the tomb towered in the shape of a pointed, straight-sided pyramid of tremendous size, which was sealed after the body was laid to rest. In the vicinity of the tombs, dismantled boats were buried in boat-shaped pits. The principal and active part of the entire complex was not the pyramid, but the cult temple where daily ceremonies were conducted and offerings presented for several generations after the king's death.[54]

The desire to materialize the image of the *ka* – the immortal spirit and essence of the personality of the deceased – in eternal stone prompted the development of sculpture in the round. Some of the world's greatest masterpieces of antiquity, such as the statue of Khafre and the triads of Menkaure, date to this period. The figures were life-size or smaller. Standing figures had the left leg forward as if turning to the right, the preferred side of the body. Many cultures identify "right" with "life" or "good" and all the known writing systems were read from right to left, which remains true for present-day Semitic writing, including Hebrew and Arabic.[55]

Whereas the older royal tombs were in the desert, isolated and to be viewed from a distance, the Fourth Dynasty funerary complexes were more intimate.

---

[53] There exist numerous descriptions of the great pyramids. An instructive and very readable account is Jaromir Malek, *In the Shadow of the Pyramids: Egypt During the Old Kingdom* (photographs by W. Forman) Norman, University of Oklahoma Press (1986).

[54] The shape probably reflects the rising importance of the sun god *Ra'* of Iunu/On/Heliopolis, shown by the rays of the sun bursting through the clouds and found also on the top of the *obelisks*. The words used by the Greek historiographers have a humorous ring: *pyramis* were small folded cakes or cookies similar to Jewish "hamantaschen", while *obeliscos* was a meat skewer.
Khufu's pyramid contains 2.2 million stones ranging from 6 to 16 tons.

[55] Mesopotamian cuneiform also begins at the right, originally running from top right in vertical columns proceeding from right to left, like Chinese. In the mid-2nd millennium B.C.E., clay tablets were turned 90 degrees to the left, resulting in a left-to-right direction which by intermediaries such as Ugarit became standard in the first millennium and probably influenced the gradual change to dominance in western writing systems.

The *mastaba* tombs of the royal family, officials, priests and craftsmen huddled in the shade of the great pyramids as if to be assured of divine protection by the dead king throughout time. Similarly, all religious institutions were situated in or near the capital during this period. From the Fifth Dynasty on, the temples of local gods and their stewards in the provinces received increasing land donations and other privileges, which helped considerably towards establishing local independence in the First Intermediate Period that followed.

In addition to the funerary temples at Gizeh and the Fifth Dynasty sun temples at Abusir and Abu Ghurab, which were also crowned by small pyramids, temples to other gods of to the Old Kingdom are mentioned in the texts but remain largely unknown.

The divinity of kings was not an accident of birth, as claimed by pharaohs of later periods, but was a function of the office and became effective upon coronation. The king was the custodian of *ma'et*, best translated as "proper order", "justice", or "truth", and, therefore, was responsible for every event including disasters such as low Nile flooding or cold spells that proved to be fateful for the kings in the end. He could marry commoners and surround himself with an entourage from all social layers, indicating great social mobility as described in the "The Instruction of the Vizier Ptah-hotep".[56]

The monumental building projects of the Fourth Dynasty seem to have stretched the limits of exploitation of all resources, including the workforce that was increasingly supplemented by slave raids into Nubia. The campaign or campaigns of Pepi (Phiops) I into Palestine, described by the tomb inscription of his general Weni, were in reality glorified cattle-raids. Pepi II's 90-year rule showed all the signs of a weakened and collapsing economy and growing poverty – no more buildings are recorded after his death. The country split into several *de facto* independent principalities, their borders steadily under attack by equally famished foreign peoples from the desert where the crisis was even worst. The inundation level of the Nile is recorded as being too low to enable agriculture. Texts and vivid illustrations in the Sixth Dynasty tombs describe how famines ravaged the entire country: *"What never happened has now happened"* said Neferti in his Prophecies. Ankh-tifi, the lord of Edfu and Nekhen (Hierakonpolis) writes, rather boastfully, in his tomb: *"Upper Egypt in its entirety was dying of hunger, everybody eating his children, but I never allowed it to happen that anyone died of hunger in this district"*.

The calamities that befell Egypt at the end of the Old Kingdom and during the 1st Intermediate Period were the result of the major climate change described above: A very warm global climate regime has led to the melting of the world's major ice volume which, in its turn, cooled the surface of the oceans which then would weaken the souther oscillation's easterlies subtropical system, thereby causing reduction of the monsoon rainfall, increase aridity, subsequently lower the volume of rivers and levels of lakes particularly in Africa. Researches in Lake Shala in Ethiopia have showen that similar chains of events with varying severity have happened around

---

[56] J.A. Wilson, "The Instruction of the Vizier Ptah-hotep" in *ANET* pp. 412–414.

10,000 B.C.E., 8000 B.C.E., in the 5th millennium B.C.E., and between 2500 to 500 B.C.E.[57]

Presently still ongoing probing of sediments in Lake Tana, also in Ethiopia, by the University of St. Andrews in Scotland and Aberyswtwyth in Wales produced evidence that this lake (53 miles by 41 miles and only 50 feet deep) was nearly dried up by 2200 B.C.E. Unlike other east African lakes, the Lake Tana drainage basin represents the major direct source of the Blue Nile supplying around two-thirds of the water volume of the annual Nile inundation on which Egypt depends for her survival, its repeated and extensive failing, therefore, would be the principal cause for the collapse of her civilization around 2200 B.C.E.

As was to be expected, the pharaohs failed the trust in their all-powerful divinity, which proved to have been an illusion when the state vanished – but not entirely. The delay of its collapse, compared with the demise of the Early Bronze Age III cities in Palestine around 2400 B.C.E. and the peak of low Niles around 2200 B.C.E., was probably due to the socio-economic groundwork laid by the Old Kingdom administration, their mobile labor force and country-wide storage facilities and, perhaps, even a certain planning ahead.

---

[57] R. Gillepsie, A.F. Street-Perrott, and R. Switsur Post-glacial arid episodes in Ethiopia have implications for climate predictions. *Nature* 306:680–683 (1989).
A.F. and A.R. Perrot Abrupt climate fluctuations in the tropics: The influence of Atlantic Ocean circulation. *Nature* 343:607–612 (1990).

CHAPTER 6

Dark Age, Renaissance, and Decay

From the Intermediate Bronze to the end of the Late Bronze Age, ca. 2400 to 1200 B.C.E.

> *"Your river which had been made fit for the magur-boats – in its midst the... plant grows.*
> *On your road, which had been prepared for the chariots, the mountain thorn grows.*
> *O my queen, your city weeps before you as its mother;*
> *Ur, like the child of a street which has been destroyed, searches for you."*

(From the lamentation over the destruction of Ur, ca. 2000 B.C.E.)[1]

The second half of the 3rd millennium B.C.E. witnessed one of the most serious climatic events during the Holocene. Evidence gathered from all over the northern hemisphere of the world shows that the climate became warmer on a global scale towards the end of the millennium. The glaciers started melting causing the levels of the oceans to rise. The high sea level led to a general rise of the groundwater table. The immediate effect for the people in some of these northernmost regions was a considerable decrease of arable lands with famine and local migrations in its wake. The temperate zones, mountain ranges and their valleys saw a great improvement, its populations enjoying a benign climate with sufficient precipitation to maintain agriculture without being covered by snow for most months of the year. The northern highlands of the Near and Middle East, the Balkans and central Europe flourished.

In the Mediterranean countries and the Near East began a period of about three to four centuries of climatic setbacks starting around 2400 B.C.E. Notwithstanding that there always have been oscillations throughout the third millennium B.C.E., the peak of the dry period around 2200 B.C.E. is characterized by its extremity. The levels of the lakes dropped, perennial rivers turned into seasonally dry beds and the desert expanded. Mesopotamia dried up due to the reduction of the water volume of the two rivers and its soils became increasingly saline.

The desiccation of the semi-arid regions triggered the movement of its inhabitants toward areas, which remained green. Mesopotamian historical

---

[1] S.N. Kramer, *The Sumerians*, 143.

sources begin to mention tribes from the Syrian Desert, the semi-nomadic Amorites (derived from Sumerian MAR-TU, Akkadian *Amurru*) who invaded the river valleys of the Euphrates, the Tigris and their tributaries. Part of this migration pattern was also recorded by the Egyptians who mention an aggregate of warring folks, some coming from the west but most from the east and, particularly, from the northern parts of the Fertile Crescent. As they were moving south, they soon dominated Canaan and eventually invaded Egypt where they were called *Hyksos*, i.e. "Princes of Foreign Lands". Here, like their Amorite counterparts in Mesopotamia, they eventually managed to become rulers in some parts of the country without, however, being integrated into the local population.

Syria-Palestine and Anatolia show a wave of destruction and abandonment of its Early Bronze Age cities. As Karl W. Butzer recounts: *"Many, if not most, urban sites from the Balkans to Mesopotamia and Palestine were abandoned, destroyed or much reduced in size. Early states such as the Akkadian Empire and Old Kingdom Egypt collapsed ca. 2230 B.C.E., Troy II was destroyed and the Indus Valley civilization came to an end".*[2] This statement still holds in general but now we realize that a great variation in the ups and downs of the various civilizations can be observed and that each one had its own course of history. Once the nadir and the depression following these events was reached, each one of the human societies underwent profound changes and was forced to invent new methods to cope with the scarcity of water and food. In Mesopotamia as well as Egypt water projects on a vast regional basis were built. The inhabitants of the new Middle Bronze Age cities of Canaan began to develop their first subterranean water systems in order to irrigate land and to withstand siege in their fortified cities (see Appendix I). Presumably, the scarce arable land was augmented and preserved by building terraces even when solid archaeological evidence is still missing. Eventually, the climate ameliorated and a more humid climate followed after the Late Bronze and during the beginning of the Iron Age.

6.1
The Crisis Years – The Climate Evidence

The evidence for the deterioration of the climatic conditions in the Near East during the third millennium B.C.E. is the following: On the isotopic based time series of proxy-data, discussed in the previous chapters (Figs. 3 & 3a) we can see that a short phase of drier climate at ca. 2600 B.C.E. was followed by a more humid phase and again by abrupt ups and downs in the precipitation pattern of the region. Another dramatic regression occurred around 2400 B.C.E. when the climate became drier for a few decades. This event was again followed by a certain improvement until ca. 2200 B.C.E. when a major crisis occurred lasting for about a century until ca. 2100 B.C.E. followed by another century and a half of a slightly more humid climate. At ca. 2000 BCE the climate

---

[2] K.W. Butzer, "Sociopolitical Discontinuity in the Near East ca. 2200 BCE: Scenarios from Palestine and Egypt" in: *Third Millennium B.C. Climatic Change*, p. 245.

began to deteriorate again reaching a low from 1900 to 1800 B.C.E. Very soon afterwards the climate changed for the better facilitating the cultural and material renaissance of the Middle Bronze Age.

This occurrence of significant fluctuations during the last centuries of the third to the first quarter of the second millennium B.C.E. can be seen in all proxy-data series discussed in the previous chapters. The most spectacular effect can be seen on the level of the Dead Sea where from a relatively high level of −300 meters below sea level it fell to lower than −400 meters, which caused the entire southern basin of the lake to dry up. Parallel to this the level of the oceans rose by about 1.5 meters. The humidity curve of Lake Van reached a minimum around 1600 B.C.E. The real peak may actually have occurred about two or three centuries earlier, but appeared only later due to the retardation effect of the aquifers feeding the flow into the lake. This, however, is not the case with the Soreq Cave, which is fed by infiltrating rainwater and its curve thus represents climate changes with a retardation time of about three decades.[3]

The evidence for climate change in this period is much more than circumstantial. As early as 1966 the limnologist H.J. Wright reported that around 2000 B.C.E. lakes in the Zagros Mountains dried out.[4] In 1977 the palynologists W. Van Zeist and S. Bottema found that the percentage of arboreal pollen in cores from the bottom of Lake Zeribar in the Zagros Mountains was significantly reduced, possibly indicating aridization during the same period.[5] The palynologist A. Horowitz, in a series of articles published since 1971, concluded from pollen assemblage found in cores from the Lake Hula Basin in northern Israel that from 2500 to 500 B.C.E. the climate changed from cold and humid to warm and dry.[6] Findings in the same basin, supported by more ^{14}C dating than that of Horowitz, agrees that by 2000 B.C.E. the percentage of oak pollen was reduced, perhaps because of climate deterioration.[7] The sediments of the Sea of Galilee show a remarkable reduction of the oak and pistachio pollen with a parallel increase of olive pollen peaking around 2500 and followed by a reversal drop around 2000 B.C.E. Issar et al. interpreted these changes as an indication of the human response to climate change; i.e. during the humid phase of the Early Bronze Age, the inhabitants have cut the natural forest and

---

[3] A. Frumkin et al., op. cit. (1991).

   A. Raban and E. Galili op. cit. (1985).

   G. Lemcke and M. Sturm, op. cit. p. 669 (1997).

   M. Bar-Matthews et al., op. cit (1998).

   Personal communication from A. Kaufman of the Weizman Institute, Rehovot, Israel, who was in charge of the dating of the stalagmites.

[4] H.J. Wright Jr., "Stratigraphy of lake sediments and the precision of the palaeoclimatic record" *Symposium of the Royal Meteorological Society*, London. pp. 158–173 (1966).

[5] W. van Zeist and S. Bottema, "Palynological investigations in western Iran" *Palaeohistoria* 19:19–85 (1977).

[6] A. Horowitz, Climatic and vegetational developments in northeastern Israel during Upper Pleistocene-Holocene times *Pollen et Spores*; 13, 255–7, (1971).

[7] W. van Zeist and S. Bottema, "Vegetetional history of the Eastern Mediterranean and Near East during the last 20,000 years" in: *Palaeoclimates, Palaeoenvironments and Human Communities in Eastern Mediterranean Region in Later Prehistory* Oxford, British Archaeological Report, International Series 133: 277–321 (1982).

replaced it by olives, but once the climate deteriorated and the olives gave no profit, they were abandoned and replaced again by the natural vegetation.[8]

Analyzing Egyptian historical literary sources, A. Bell found evidence of two drought waves, one from 2180 to 2150 B.C.E. and the other from 2000 to 1900 B.C.E. He concluded that these severe droughts were responsible for the catastrophic famine especially in Upper Egypt as well as the disintegration of many Early Bronze Age cultures in the Mediterranean areas, the collapse of the Akkadian empire, and the fall of the Neo-Sumerian Ur III kingdom.[9] The geologist P. Rognon observed a process of rapid aridization affecting all of northern Africa between 2500 and 2000 B.C.E.[10] A.D. Crown in 1972 found ample historical documentation that the climate after 2500 B.C.E. was drier than the marked wet period extending from 3500 to 2500 B.C.E. and that at ca. 2300 B.C.E. the rise in temperature caused severe droughts and crop failures all over Mesopotamia.[11] From information derived from early Akkadian cuneiform tablets, the climatologist J. Neumann and his colleagues found indications of aridization around 2300 to 1900 B.C.E., marked by a lower water level of the Tigris-Euphrates and increased soil salinization.[12] Various proxy-data led A. Kay and D.L. Johnson to the same conclusion.[13] In 1967, D. Neev and K.O. Emery reported in their treatise on the sediments of the Dead Sea that salt tongues intertwined with the marl deposits around 2300 B.C.E. and interpreted this as an indication of a dry period. Later studies, already mentioned, showed that the ancient levels of the Dead Sea, which peaked during the Early Bronze Age, fell at the end of this period. After 2000 B.C.E. it dropped below the level of the bottom of the southern part of the lake, causing this part of the lake to dry up.[14]

The present authors do not suggest that each known historical event can be matched with every climate change seen on the Soreq Cave curve; they are well aware of the many other factors involved, including socio-economic and political, that may mitigate or amplify the impact of the climate changes. Nevertheless, one should expect that if the very humid climate during the third millennium brought prosperity to the Near East, then the very dry climate that

[8] M. Stilleret al op. cit. (1983–1984).

[9] B. Bell, "The Dark Ages in Ancient History. I. The first Dark Age in Egypt" *American Journal of Archaeology* 75:1–26 (1971).

[10] P. Rognon, "Aridification and abrupt climate events on the Saharan northern margins, 20,000 B.P. to present" in: *Abrupt Climatic Change*. W.H. Berger and L.D. Labeyerie (eds.), Kluwer, Dordrecht. pp. 209–220 (1987).

[11] A.D. Crown, "Toward a reconstruction of the climate of Palestine 8000 B.C.E. – 0 B.C.E." *Journal of Near Eastern Studies* 31:312–330 (1972).

[12] J. Neuman and S. Parpola, "Climatic change and the eleventh-tenth century eclipse of Assyria and Babylonia" *Journal of Near Eastern Studies* 46:161–182 (1987).

[13] A. Kay and D.L. Johnson, op. cit (1981).

[14] D. Neev and K.O. Emery, op. cit.(1967).
A revision of sediments dating, which was carried out a decade later (D. Neev and J.K. Hall, Climatic fluctuations during the Holocene as reflected by the Dead Sea levels), the non revised dating seems to fit much better with other data from the Dead Sea, especially that of Frumkin et al. (1991), herewith discussed, showing that the level of the Dead Sea was indeed lower between 4300 and 3500 B.P. Frumkin, et al., op. cit (2001).

ensued would have a negative impact on the region's socio-economic systems. Indeed, archaeological evidence throughout the region shows that most of the urban centers, which were established during the exceptionally cold and humid climate of the Early Bronze Age, were deserted towards the end of the third millennium B.C.E. and the surviving inhabitants turned to husbandry and became increasingly mobile and nomadic. They supplemented their food supply with seasonal planting of cereals and vegetables near steady water sources. Along with this, agricultural technology and international trade – the main source of urban elite wealth – dwindled and pastoralism became the dominant branch of economy spreading along the belt bordering the desert.

Until now, the main topic of debate among archaeologists was not about the reasons for this seemingly abrupt turn of events or the possible role of climate reversals on the history of the region but mainly concern nomenclature and how to subdivide this "Dark Age". Differences of opinions stem from the experts' preference for fixing chronological boundaries according to the emergence and disappearance of particular types of pottery and other objects believed to characterize the material cultures of the period.[15]

Ironically, there is a consensus that, indeed, a crucial phase of devastation afflicted the centers of populations throughout the Fertile Crescent during this period. This conclusion was based not only on physical evidence of destruction and desertion of the many city-states in the central part of the region, but also from the historiography of the kingdoms of Mesopotamia, which relate a continuing struggle to repel the marauding tribes from the west and east encroaching on their lands. Egyptian documents also describe a period of famine and strife marking the end of the Old Kingdom and the beginning of a period of disorder and incursions of nomads from the east and west. In mains stream archaeological literature, any mentioning of environmental and climatic changes was frowned upon and not taken seriously – for reasons already mentioed.

The agreement about climate changes, shared by scientists studying the natural environment, is not found among archaeologists and historians. Among the small minority of archaeologists who do concur is H. Ritter-Kaplan, who in the early 1980s carried out excavations in the Exhibition Gardens in Tel-Aviv located near the ancient sea outlet of the Yarkon River. She found that a hard black clayey layer, containing abundant pollen of oak overlying layers containing Early Bronze Age I and II type pottery and interpreted this as an indicating that a swamp formed under more humid conditions. Above this she found some Early Bronze IV pottery (which equals Middle Bronze I, and our Intermediate Bronze Age) in a layer of sand almost lacking any pollen. Ritter-

---

[15] W.F. Albright, "The Chronology of Middle Bronze I (Early Bronze-Middle Bronze)" *BASOR* 168:36–42 (1962).

R. Cohen, "The Mysterious MBI People" *BAR* 9/4:16–29 (1983).

W.G. Dever, "The EB IV – MB I Horizon in Trans-Jordan and Southern Palestine" *BASOR* 210:37–63 (1973).

K.M. Kenyon, J. Bottero, and G. Poserer, "Syria and Palestine ca. 2160–1780 B.C." in: *The Cambridge Ancient History*, 3rd edition, vol. 1, no. 2, Cambridge University Press, Cambridge. pp. 532–594 (1971).

Kaplan concluded that this represented a drastic change of climate, which caused the local settlements to be deserted and referred to this period as "*the crisis of the aridity*".[16] Observations of A. Raban and E. Galili, marine archaeologists who surveyed the submarine and coastal archaeology of the ancient settlements along the Mediterranean shoreline, supplied further support. They found a rise in sea level at the end of the Early Bronze Age that was synchronous with the invasion of sand dunes filling up the outlets of the streams serving as an anchorage site for the Early Bronze Age marine cities.[17]

Rosen, investigating the ancient environmental conditions at the vicinity of Tell Lachish in the western foothills of the Judaean Mountains, found in the *wadi* beds sands and sandy silts of Intermediate and Middle Bronze Age overlying the clayey layers deposited during the Chalcolithic and Early Bronze Age. She interpreted these as "*rapidly fluctuating rainfall patterns interspersed with drought leading to soil stripping from the hill slopes*".[18]

A detailed environmental study carried out in the plain of Harran in northwestern Mesopotamia showed that the Early Bronze Age network of large urban centers in this region collapsed and many sites were deserted at the end of the third millennium B.C.E.[19] Further east, in the northern part of Mesopotamia, the disintegration of the socio-economic system and the desertion of Tel Leilan were attributed to a climate change.[20] The archaeologists, however, attributed this climate change to the effects of a volcanic eruption, an explanation with which the present authors do not agree.

The simultaneous droughts in the Asiatic part of the Near East as well as Egypt may have resulted from a few extreme and abrupt fluctuations from around 2300 to 1900 B.C.E. During the colder spurs the Nile declined and rose during the warm ones around 1950 B.C.E. Evidence for such spurs could be found in the stalagmites and lake varves, but not in the morphological data from the Nile or the lake levels of eastern central Africa. The paleo-hydrogical sequence of events – in the case of ancient – river and lake levels – is blurred because higher and more recent levels usually obliterate the marks left by the earlier and lower levels. The anomaly of dryness in the low latitudes albeit a global warm phases may be explained by the fact that at this period due to abrupt melting of glaciers, there occurred a cooling of the ocean's surface, leading to a warm and yet weak monsoon regime.[21]

---

[16] H. Ritter-Kaplan, in *Eretz Israel, Braver Book* (1974).

[17] Raban and Galili op. cit. (1985).
A. Raban, "Alternated river courses during the Bronze Age along the Israeli coastline" in: *Colloques internationaux C.N.R.S. Déplacements des lignes de rivages en Méditerranée*, C.N.R.S. Paris. pp. 173–189 (1987).

[18] A. Miller-Rosen, "Environmental Changes and Settlement at Tel Lachish, Israel" *BASOR* 263:55–60 (1986).

[19] A. Miller Rosen, "Early to Mid-Holocene Environmental Changes and their Impact on Human Communities in Southeastern Anatolia" in: *Water, Environment and Society* pp. 215–240.

[20] H. Weiss et al. op. cit. (1993).

[21] A.F. Street and A.R. Perrott, "Abrupt climate fluctuations in the tropics: the influence of Atlantic Ocean circulation" *Nature* 343:607–612 (1990). A better answer to this question will be obtained in

It is quite surprising, therefore, that despite of all the evidence of a regional cataclysm and published reports of specific observations that unequivocally point to climate change as its cause, the majority of archaeologists and historians still seek other explanations for the ruin in each region without considering a wider connection. For example, the desertion of the Early Bronze Age cities of Syria-Palestine were explained by the "Amorite Hypothesis" which blamed the semi-nomadic people who overran these cities and causing their demise.[22] This hypothesis, accepted as fact by many archaeologists for a long time and was partially correct, indeed, did not even try to explain why the Amorites emigrated from their traditional territories, invaded the fortified cities of Syria-Palestine and fought the armies of the kingdoms of Mesopotamia? Moreover, archaeological remains show some continuation of the cultural material of the Intermediate Bronze Age from the preceding Early Bronze Age III tradition[23] indicating that there was no large-scale displacement of the original population by an alien invader but rather a pattern of system collapse accompanied by the destruction or abandonment of the towns and cities. Similar objections are relevant with regard to the perhaps equally valid "invasion theory" of tribes who supposedly arrived from Trans-Caucasus or Central Asia.[24]

The reasons for the soils of Mesopotamia to become saline towards the end of the third millennium B.C.E., is an example of an anthropogenic explanation to an environmental process. First suggested by T. Jacobsen and R.M. Adams,[25] it was subsequently adopted by most archaeologists and historians and became popular with environmentalists who adopted it as a classic example of the negative impact of human society on the environment. The onus placed on the ancient Mesopotamians was derived mainly from Sumerian economic texts. It demonstrated that between 2300 and 1800 B.C.E. the ratio of barley to wheat constantly became bigger in offerings and taxes delivered to the temples despite the fact that many long irrigation, and probably also drainage channels were dug. As barley is more tolerant to soil salinity than wheat, Jacobsen and Adams concluded that excessive irrigation had caused salinization. Although other Sumerologists debated this explanation by demonstrating from similar archive texts that the Sumerians were aware of the danger and had taken some

the near future once there is sufficient dated data from the cores of the eastern Mediterranean, now under investigation by different oceanographic groups.

[22] W.F. Albright, "From Jerusalem to Baghdad down the Euphrates" *BASOR* 21:1–21 (1926).
K.M. Kenyon, J. Bottero, and G. Posener, "Syria and Palestine ca. 2160–1780 B.C." in: *Cambridge Ancient History*, 3rd revised ed., vol. I (2), Cambridge University, Cambridge. pp. 532–594 (1971).

[23] S. Richard, "Toward a consensus of opinion on the end of the Early Bronze Age in Palestine-Trans-Jordan" *BASOR* 237:5–34 (1980).

[24] M. Kochavi, "The Middle Bronze Age I (The Intermediate Bronze Age) in Eretz-Israel" *Qadmoniot* 2:38–44 (1969).
P. Lapp, op. cit. (1970).
I. Finkelstein, op. cit. (1995).

[25] T. Jacobsen and R.M. Adams, "Salt and Silt in Ancient Mesopotamian Agriculture" *Science* 128, 1251–1259 (1958).
T. Jacobsen, *Salinity and Irrigation Agriculture in Antiquity, Diyala Basin Archaeological Project*, Bibliotheca Mesopotamica 14, Udenda Pub., Malibu (1957–1958).

preventive drainage measures,[26] the theory putting the blame on the people of Sumer prevailed.

Issar, after reaching the conclusion that there was a global rise in the temperature during this period has suggested a hydrological conceptual model to explain the Sumerian textual evidence. He argued that the climate change decreased the flow of the rivers and thus less water was available for irrigation and for flushing out the salts.[27] As southern Mesopotamia is very low lying, any small rise in the sea level would have been accompanied by a rise in the groundwater table bringing it close to the land surface, and thus evaporation and salt accumulation would increase due to capillary action. The results of field investigations show that soil salinity is not only a function of excessive irrigation but also of ill-drainage and that the latter could have been the effect of a rise in the groundwater table during a global warm period.[28]

As discussed at the beginning of this chapter, the time series of the paleo-climate of the Near East show that although there was a general trend of desiccation starting around 2200 B.C.E. there were periods of more humid spells lasting for a century or more, during which (albeit still relatively dry), most of the deserted urban centers in the more humid part of Syria-Palestine were resettled. In addition, all along the northern part of the coastal plain as well in the valleys of southern Lebanon and Palestine many new sites were established.

6.2
The Archaeological and Historical Evidence About the Intermediate Bronze Age

Similar to the earlier chapters let us have a bird's eye view of the evidence from Syria-Palestine followed by Anatolia, Mesopotamia and Egypt.

Evidence of desertion is particularly impressive in Palestine between 2350 and 2000 B.C.E., when all the urban centers were abandoned and the cities, which had existed for several hundreds of years, remained only as heaps of ruins. The social structure of the urban society was not able to withstand the effects of the environmental changes and collapsed. Arad was deserted already by the onset of the Early Bronze Age III and shows no signs of destruction or other catastrophic disasters. Other sites in the south of the Levant appeared to have been similarly abandoned by their inhabitants.

The picture dramatically changes when going northward: All the magnificent cities and towns in northern Palestine and southern Syria were burned down in a great fury, not to be resettled until nearly half a millennium later. The phenomenon of thick burnt layers ending all settled life at all the sites is also found further north, extending deep into Syria and Anatolia and evident at

---

[26] S. Pollock, *Ancient Mesopotamia: The Eden that never was.* Cambridge University Press, Cambridge, pp. 28–39 (1999).

[27] Issar, op. cit., pp. 69–70 (1990).

[28] K. Butz in: *State and Temple Economy in the Ancient Near East.* ed. E. Lipinski (Orientalia Lovainiensa Analecta 5, Lovain, p. 271 (1979).

Byblos and Ras Shamra/Ugarit on the coast of Lebanon and Syria, Homs, Hama in the Orontes Valley, Ebla, Aleppo and the 'Amuq sites in Syria, all bearing overwhelming witness to a wave of violent destruction and general upheaval ending all settled life. In Anatolia, this wave reached from Karatas-Semayuk and dozen of other contemporaneous settlements in the south-east to Tarsus and Mersin in Cilicia, Beycesultan in the west and Troy in the north-west – and even beyond.

Whereas the cities of Palestine and southern Syria lay deserted for several centuries, those of northern Syria, Anatolia and northern Mesopotamia were immediately rebuilt and resettled, albeit within the framework of a new and quite different culture, known as the Syrian Early Bronze Age IV. The new culture began simultaneously at sites such as Tell Nebi Mend (ancient Qadesh), Tell Mishrife (ancient Qatna), Hama, etc. The ongoing excavations of Ebla, with its archive, inform us about the economic background of this development, in particular about the trade network, which linked the Levantine coast, and eventually Egypt, with Anatolia and Mesopotamia. This new Syrian civilization continued without any major interruption into the Middle and Late Bronze Age and was largely responsible for the re-introduction of the urban system into Palestine. Many sites show a scattering of destructions and rebuilding, such as at Ebla, where the fire luckily (for us) baked the royal archive, as well as most other sites which appear to have been the result of local wars and invasions. These wars continued into the Middle and Late Bronze Age and can be contributed to the growing influence of Amorite warlords, as we shall see.

That a regional phase of desiccation may have been a prime factor can be deduced from the observation that not all settlements were abandoned in the same time, but that desertion moved slowly from the more arid parts to the more humid parts. It was ushered at the end of the Early Bronze II, with the destruction and desertion of the city of Arad located on the border of the desert, as well as many other small settlements in the Sinai and Negev Deserts. The towns and cities in the more humid parts diminished in their population, impoverished but staying alive, at least for a while.

As we have seen the collapse of the cities in Cis-Jordan was followed by the appearance of many small settlements in areas bordering the desert due to a shift from a sedentary agriculture to a pastoral based economy. From an ecological point of view this shift may well be due not only to reduced annual average precipitation, but also to a negative change in the statistical distribution of drought and normal years – i.e. the number of drought years became more frequent. Such a distribution does not enable food or capital produced in the agricultural areas surrounding the cities during more humid years to be stored for later use in years of drought. The only solution is to extend the area providing the food and commodities and vast areas that yielded a low and sporadic income were used most beneficially as grazing lands for animals adapted to semi-arid conditions, sheep and goats. Thus, instead of the granaries typical of the Early Bronze Age cities, livestock 'on the hoof' became the assets of food and capital on a seasonal and multi-annual basis throughout the Fertile Crescent.

We have seen already that transitional periods were, and still are, the subject of lively debates among archaeologists working in Syria-Palestine. Whereas there seems to exist a broad consensus about the chronology and general time frame with regard to the end of the urban phase of the final Early Bronze Age around 2400 B.C.E. and the beginning of the Middle Bronze Ages around 1800 B.C.E., the subdivision and meaningful nomenclature remained controversial. The differences in opinion concern the disappearance and emergence of this or that type of pottery and other forms of objects, which typify the material cultures of the various time intervals.[29] W.F. Albright hadsuggested the term 'Middle Bronze Age I' for the entire period, which implies that the material culture of this period foreshadowed the ensuing Middle Bronze Age. Other archaeologists thought that the links with the preceding period are much closer and called this period Early Bronze Age IV. This line of thought seem to have been confirmed by more modern excavations and was followed by a growing number of scholars. For K. Kenyon, the remains in the tombs of Jericho appeared so strikingly different from those of both, the preceding as well as the following period, that she regarded this culture as intrusive and labeled it with the rather cumbersome 'Early Bronze-Middle Bronze Transitional Period'. P. Lapp simplified this term to 'Transitional Bronze Age'. We follow a recent term, 'Intermediate Bronze Age' (IBA) which best describes the evidence at hand and has gained wide acceptance. Whatever its label, there is unanimity that Palestine went through a period of decline and revivals, extensive movements of people over nearly half a millennium.[30]

Thus, there exists a consensus among all scholars that, indeed, a crucial phase of decline has afflicted the centers of population all over the Fertile Crescent during this period. This is clearly seen not only in the physical evidence of desertion and devastation of the many cities in the central part of the region but also from the historical documents from Mesopotamia and Egypt. These texts depict a continuous exertion to push back the marauding tribes encroaching on the green lands from the desert as well as period of famine and strife demarcating the end of the Old Kingdom and ushering a period of disorder.[31]

Following the collapse of the cities in Palestine, many small settlements appeared in the areas bordering the desert, for instance, Transjordan, the Jordan Valley, the Negev and Sinai. This can be explained as a shift from an economy based on sedentary agriculture to a system primarily based on animal husbandry with less emphasis on agriculture. Such an economy has the advantage

---

[29] W.F. Albright, op. cit., pp. 36–42 (1926).

W.F. Albright, "Some remarks on the archaeological chronology of Palestine before about 1500 B.C." in: *Chronologies in Old World Archaeology*, R.W. Ehrich (ed.), University of Chicago Press, Chicago, pp. 47–60 (1965).

W.G. Dever, "The EB IV–MB I horizon in Trans-Jordan and southern Palestine" *BASOR* 210:37–63 (1973).

K.M. Kenyon, *Archaeology in the Holy Land*. Norton, New York (1979).

[30] P. Lapp, in *Near Eastern Archaeology*.

[31] The precise simultanity of droughts in Egypt and the Fertile Crescent is still a matter, which has to be verified. If it will be found out that they were simultaneous then it can only be explained by a global climatic anomaly, such as the sudden melt of the Laurentian glacier, an event, which had cooled the surface water of the ocean and negatively influenced the westerlies and monsoon systems at the same time.

of a higher degree of mobility and can adjust itself to seasonal availability of pasture and water for the animals that became the most important base of the economy.[32] Such a shift indicates that certain cultural, political and social interrelations had already existed earlier between the urban centers and the less sedentary population living around the centers, i.e. that both economies were not totally independent. The collapse of the urban system encouraged and amplified a shift from sedentary agricultural herding to more mobile nomadic pastoralism.

An interesting illustration of the socio-economic system of the Levant is found in the story of Sinuhe, an Egyptian courtier who fled to Canaan when the Pharaoh Amen-em-het I was murdered in a rebellion at around 1960 B.C.E.[33] During his exile he lived with the tribes in the land of the *Retenu*, i.e. in Canaan. These people, on one hand, enjoyed the products of typical Mediterranean type of fruit orchards, like figs, grapes and olives, whether owned by them or harvested by their neighbors, the sedentary farmers. On the other hand, they were organized as relatively mobile tribes thriving on their cattle and also on those of their neighbors'. Intertribal warfare over pasturage and wells and cattle raids always were the most favored past-time activity of a nomadic or semi-nomadic way of life.

The traditional archaeological evidence for this period in Palestine consisted nearly exclusively of tombs, occasionally grouped in large cemeteries and rarely adjoining settlements. This picture is changing as more and more sometimes large permanent settlements are discovered, particularly on the east bank of the Jordan. Abu en-Nias, Iktanu and Khirbet Iskander are some of the best-known quasi-urban sites of this otherwise non-urban period.

The proliferation of burial sites is a sign of the growing importance of a mobile life-style and changes in the social order. Whereas the Early Bronze Age populations buried their dead in caves often containing a large number of bodies, which had accumulated over the generations, the emphasis is now on the individual grave. This is a clear indication of nomadic and tribal way of life in many parts of the ancient Near East. It also shows the growing social importance of the warrior within society[34] as is confirmed by contemporaneous historical descriptions we shall discus below. The large variety in tomb typology, tumuli in the desert, megalithic dolmens along the Rift Valley and subterranean rock-cut or built shaft tombs, often in close proximity, was ex-

---

[32] W.G. Dever, op. cit. p. 237 (1973).
W.G. Dever, "From the End of the Early Bronze Age to the Beginning of the Middle Bronze" in: *Biblical Archaeology Today*, J. Aviram (ed.), Israel Exploration Society, Israel Academy of Sciences and Humanities, Jerusalem, and the American Schools of Oriental Research, pp. 113–135 (1985).

[33] J.A. Wilson, "The Story of Sinuhe" pp. 5–11 in: *The Ancient Near East,*. J.B. Pritchard (ed.) Princeton University Press (1973).
D.B. Redford op. cit. pp. 83–86 (1993).

[34] Interesting parallel and contemporaneous developments can be observed in Mediterranean and Western European societies where, at the same time, ancestor-worshipping and megalith-building agriculturist way of life was replaced by new societies dominated by warrior ethics, individual burial and the proliferation of metal weapons: "*The ancestors were pushed aside by the warrior*": J.P. Mohen, *The World of Megaliths*. Facts on File, New York, p. 276 (1990).

plained by K. Kenyon as "tribal characteristics".[35] Grave goods were simple and consist of some pots and a few personal items, a copper dagger was standard in most male burials and indicate the growing importance of the warrior within the context of aggressive nomadic societies.

The socioeconomic picture described seems to fit two important literary sources, the Egyptian story of Sinuhe's wandering, mentioned above, as well as the Biblical stories of the wandering of the patriarchs. In spite of the obvious shortcomings of early oral traditions being transformed, telescoped, simplified and fixed into written traditions centuries after the actual events, it is possible that the core of these stories could have been derived from historical events which might have taken place during this, or any other subsequent period.

The authors are divided in their opinion on this subject. Issar tends to adopt the conclusion arrived by many Israeli and American Biblical scholars and archaeologists based primarily on the theories of Albright that the origin of the patriarchal stories is in the socioeconomic situation of the Near East during the first half of the second millennium B.C.E.[36] Zohar, in contrast, accepts Thompson's conclusions, and those reached by most European scholars. They are convinced that, if there does exist a historic core of these stories, the majority of their *Sitz im Leben* (life-setting) should preferably be dated to the end of the second millennium B.C.E.[37] As a matter of fact, most situations described in the relevant portions of the Biblical narrative reflect a life-style and social conduct which could be dated to any time frame as the basic socio-economic situation of pastoral nomads and their environment has hardly changed over the last few millennia until the modern period (see Appendix III).

After a violent but short crisis, the north Syrian and southeast Anatolian cities began to flourish in a rather unprecedented development. Their rather northwesterly position in the Fertile Crescent procured enough precipitation in addition to the water, which descended as streams and springs from the Taurus Mountains into the plains to the south. This supply of water enabled to continue a flourishing economy based on agriculture as well as on husbandry. The Italian excavations at Ebla have yielded remains of a for its time prosperous and highly sophisticated civilization.[38] The archives of several thousand of clay tablets, written in Sumerian cuneiform but using an Old Semitic idiom closely related

---

[35] K.M. Kenyon, *Amorites and Canaanites*, Schweich Lectures, London (1966). Another important factor would be the local geology and the availability of building materials which, together with Kenyon's 'tribal characteristics', could be important indicators for ethnic groupings during this period.

[36] Summarized in the *Encyclopedia Hebraica*, Vol 1,. 111–124; and the *Encyclopaedia Biblica*, Vol. 1, 4–13 (Hebrew).

[37] T.L. Thompson, *The Historicity of the Patriarchal Narratives: The Quest for the Historical Abraham*. Beihefte zur Zeitschrift für Alttestamentliche Wissenschaft 123. De Gruyter, Berlin (1974). See also J.J. Scullion in *ABD* vol. 2, 949–956, includes recent bibliography (1992).

[38] The emphasis on 'Semitic' stresses the fact that the majority of the sedentary and urban Early Bronze Age population of Syria-Palestine, with its different material culture, was probably not Semitic and belonged to an unknown linguistic entity. Based on some, admittedly rather flimsy, evidence of river names and other toponyms, Zohar suggests a population speaking some form of "Asianic" which might have included Sumerian in its widest sense.

to Old Akkadian, give us profound insights into the working of a centralized economy and extensive trade relations. Artifacts imported from Mesopotamia, Anatolia, and Egypt indicate the integration of Ebla into a trade and exchange network with the rest of the known world at that time.

And Ebla was not unique – dozens of similar and as yet unexcavated sites bear testimony to the prosperity of northern Syria and northern Mesopotamia during the last centuries of the third and the early parts of the second millennium B.C.E. The pressure of a growing population and the menace of the by now ever present nomads made it sure that internecine war was always at hand. We hear not only about trade but also about war, for instance when the kings of Ebla conquered another great and competing city, Mari on the Euphrates, about which we will hear later. The wealth aroused the greed of the less fortunate kingdoms further south and Ebla was finally destroyed in the 24th century B.C.E., probably by Naram-Sin, king of Akkad. He burned the royal palace to the ground and, luckily for us, baked and preserved the archive.

Another treasure of archives coming to light in central Anatolia and dated several centuries later were the archives of Old Assyrian merchants at the *karum* of Kanes, now known as Kultepe. As we have seen, the warmer climate spelled prosperity for the inhabitants of the highlands of central and eastern Anatolia and Trans-Caucasia. City-states mushroomed and their newly affluent elite craved luxury goods and weapons of a new kind, true tin bronzes.[39] The archives consist of letters between the head of the company in the city of Assur on the Tigris and their representatives, mostly their own sons, in the 'colonies' over some four to six generations between the 21st and 19th century B.C.E. We learn not only about their business and how the trade was organized but, even more precious for us, their personal affairs and their fortunes and adventures.

The larger trading colony, known as *karum*, was outside the walls of the native city whereas smaller ones, called *wabartum*, were found inside the cities. The traders had married into local families and had adopted local customs. Without the letters, we would have had no inkling about this far-reaching overland trade network, as the material culture was entirely local. The caravans consisted of several hundreds of a special strong breed of black donkeys led by well-known and trusted personnel who knew the routes and were well acquainted with the territories of the various city states which – they had to cross – naturally paying healthy sums of tolls and other 'protection money' which made the merchandise quite expensive for the final customer. We hear about insurance and even scams to cheat, so what is new under the sun? The bulk of the merchandise imported to Anatolia consisted of a variety of finely woven and often embroidered cloth from Babylon, sometimes metal or some finished objects, but above all *annaku*, i.e. tin.[40] In return, silver, called *kaspum* (in ancient and modern Hebrew *kesef* means silver and money) and

---

[39] The situation has some similarities with the "barbarian" rulers of the Celts, Dacians or Thracians and their appetite for Mediterranean luxury goods during the Hellenistic and Roman period.

[40] See entry "annaku" in *CAD*, vol. A, part II, pp. 127–130.

a variety of produce ranging from honey and wax to precious stones, wood and finished metal artifacts were sent to Assur.[41]

We have seen that the new economy and the adoption of more mobile and often semi-nomadic ways of life by the local inhabitants swelled the numbers of the traditionally herders in the desert whose larger numbers allowed the penetration of these nomadic and semi-nomadic people into the fertile valleys of the big rivers. They often seized power after their settling down and became the ruling class. In Mesopotamia, these people were predominantly Semitic-speaking tribes originating from the arid and semi-arid plains and mountains of northeast Syria. They were called by the Sumerians MAR-TU, i.e. "westerners" and by the Akkadians *Amurru*, a name most probably derived from the western pronunciation of the Old Semitic word for pasture, *aburru*, which can be pronounced /awurru/ and /amurru/ and is probably related with the name *Habiru*.[42] The Biblical term for this group of people is Amorites, which can have also other connotations. Their traditional stronghold was in the mountainous area of the Basar Mountains known to the Sumerians as the "Mountain of the Amorites", today's Jebel Bishri, and spreading out over the entire semi-arid area west of the Euphrates. In documents from Ur III dating from ca. 2100 to 2000 B.C.E. they are described as warlike 'tent dwellers of the mountains' and a wall was built in order to keep them out, and, like all the other walls, to no avail.

The worsening climate and the continuous pressure by the Amorites from the west and north caused the weakening of the neo-Sumerian kingdom of Ur which exercised the hegemony over the southern part of this region. Other neighboring tribes such as the Guti joined the fray and descended from the semi-arid foothills of the Zagros on the eastern side of the valley in order to strike a devastating blow: the conquest and devastation of the capital Ur echoed over the ages as we have seen in those lines above from the "Lamentation over the Destruction of Ur".[43]

These Guti were later expelled and replaced by the Amorites who, despite of the wall built earlier by the Sumerians in defense of their land against their encroachment, assumed control over the northern part of Sumer and all of central and northern Mesopotamia. A century later, kings of Amorite origin also replaced the Sumerian dynasties ruling the city-states of the south.[44]

In order to illustrate the turbulence of this period, let us look at the history of three of these kingdoms in former Sumer. We shall begin with the kingdom of Larsa, which was established by an Amorite chieftain, Nablanum in 2025 B.C.E. After four generations, his descendant Gungunum overpowered Lipit-Ishtar, the Amorite king of Isin, in 1924 B.C.E. and annexed his city-state leading to the establishment of a new kingdom that extended over most of ancient Sumer.

During the reign of this dynasty new irrigation projects were built. This can be learnt from the archives of the irrigation bureau of Larsa, written in

---

[41] M.T. Larsen. *The Old Assyrian City State and its Colonies*, Akademisk Forlag, Copenhagen (1976).

[42] cf. *CAD* vol. 1/II.

[43] See footnote 1 in this chapter.

[44] W.W. Hallo and W.K. Simpson, op. cit. p. 71 (1998).

Old Akkadian mixed with many Sumerian terms. The names of the superior administration officers are Semitic, which shows their Amorite origin, while those of the field officers are Sumerian.[45]

In the first year of Sumuel, king of Larsa, in 1894 B.C.E., another Amorite chieftain by the name of Sumu-Abum, established a small kingdom in the former Sumerian town of KA-DINGIR-RA(K), perhaps pronounced Pabil or something similar. The Sumerian name was adopted into Akkadian and re-interpreted as *Bab-ilim*, i.e. the Gate of the Gods, or *Bab-El*, the Gate of God, in West Semitic and hence Biblical Babel or the Greek version Babylon.

Sumu-Abum's descendants slowly widened the domain, until in 1793 B.C.E. when Hammurabi ascended the throne and conquered all of Mesopotamia after campaigns lasting thirty-eight years. The conquest of Mari was one of the highlights in his career – his famous stela with an idealized code of laws is one of the most famous documents of this period. The death of Hammurabi at ca. 1750 B.C.E. marked the beginning of the decline of the Old Babylonian Empire. The culture of the Old Babylonian period is well known and its description would go far beyond the scope of this volume.

The climatic ups and downs may have caused the Kassites (Kashshu in Akkadian) to descend from the Iranian highlands and occupy the town of Hana. These invaders, or at least their leaders, this time were partially of Indo-European origin, the Kassites and Hittites. In the year 1595 B.C.E., Babylon was conquered and destroyed by the Hittite king Mursilis I after a brilliant campaign down the Euphrates. Mursilis could have undertaken this campaign only with the active assistance of the Kassites who were also the ones who reaped the fruits of this campaign, as they became the rulers of Babylonia for the next centuries.

Another vivid picture of the unrest, continuous displacement of people and constant warfare by tribal chiefs who resemble the *condottieri* of Renaissance Italy was revealed in the archives of the kingdom of Mari.[46] The location of brought it into a continuous conflict with the 'wild' Amorite tribes living on the edge of the Syrian Desert to its west as well as with the kingdoms of Mesopotamia ruled by kings of Amorite origin. Around 1800 B.C.E., the Amorite chieftain Yahdun-Lim overtook Mari and made it his capital. From there, he began a series of campaigns against the surrounding kingdoms and the chieftains of the Amorite tribes of the Hanu who inhabited the area between the Euphrates and the Tigris north of Mari. He also waged war against the tribes of Mar-Yamina (formerly, and erroneously, read Ben-Yaminites)[47] living in the area to the east of Mari. In the foundation inscription of the temple he erected to the sun god Shamash, he boasts also about a campaign, which had brought

[45] S.D. Walters, *Water for Larsa – an Old Babylonian Archive Dealing with Irrigation*, Yale University Press, New Haven (1970).

[46] A. Malamat, *Mari and the Bible*. Brill, Leiden (1998).

[47] Both pronunciations, however, the Akkadian *Mar-Yamina* or *Mar-Yamana* as well as the Canaanite-Hebrew *Ben-Yamin* mean "Sons of the South". A connection with the Israelite tribe of Benjamin is questionable as it is a common designation and found in other regions, for instance in South Arabian inscriptions.

him to the shores of the Mediterranean. He constructed irrigation channels in order to divert water from the Euphrates into the desert.

The rule of his successor, Shumu-Yam, ended when his servants murdered him, enabling the king of Assur, Shamshi-Adad, to take over Mari. Shamshi-Adad was an old antagonist of Yahdun-Lim, a descendant of another Amorite dynasty, which had established itself in the city on the banks of the Tigris. Shamshi-Adad ruled Mari through his younger son Ismah-Adad while his older son Ashma-Dagan ruled the eastern part. Numerous letters found in the archives of Mari tell about their intimate relationships and the persisting wars waged by the father and his two sons against the surrounding kingdoms and the tribes of Hanu and the Mar-Yamina and others. After the death of Shamshi-Adad, his eldest son inherited his throne yet the younger son remained the suzerain king of Mari until driven out by the king of Ashnunna.

The succeeding king of Mari was Zimri-Lim, who claimed to be the son of Yahdun-Lim.[48] Engaged in treaties and wars with the Hanu and the Mar-Yamina in an effort to restore Mari's supremacy in the region, Zimri-Lim extended his sphere of influence to become one of the most important kingdoms of Mesopotamia in alliance with Hammurabi, king of Babylon. Hammurabi, however, after annexing most of the other kingdoms of Mesopotamia to his empire, finally annulled his covenant with Zimri-Lim, and took over Mari, which surrendered without battle but revolted two years later. The second time, Hammurabi overwhelmed the city and destroyed its walls. It was the end of Mari as a capital city but the flames fired the clay tablets and thus preserved the archives, one of the most eloquent evidence of this turbulent period.[49]

In the end, two people with two different traditions and two empires had emerged: Assyria in the north, Babylonia in the south, and both headed by Amorite kings.

6.3
The 'Winning of the South' – the Migration Southward

The climate in the Near East had soon improved after 1800 B.C.E., although it did not reach the optimal conditions of the third millennium B.C.E. The amelioration brought a gradual revival of the urban centers in the Levant during the Middle Bronze Age (M.B.), i.e. ca. 1800–1500 B.C.E. The archaeological evidence suggests a division of this period into two phases known as M.B.IIA and B with the transition around 1650 B.C.E.

The continuation of very few older elements of the material culture, such as pottery typology and some architectural conceptions are evidence that the composition of the local population underwent a drastic but apparently not a total change. In the entire cultural history of the Levant we can witness two

---

[48] M. Anbar, *The Amorite Tribes in Mari and the Settlement of the Israelites in Canaan*. The Chaim Rosenberg School of Jewish Studies, Tel Aviv University, Tel Aviv (1985 Hebrew with English summary).

[49] The texts from Mari are still being published in over twenty volumes of *Archives Royales de Mari* (ARM) Editions Recherches sur les Civilizations, Paris.

near complete changes of culture: The first was the transition from the Pre-Pottery Neolithic B to the Pottery Neolithic A in the 5th millennium B.C.E., and the second took place now around the transition from the Intermediate to the Middle Bronze Age at the beginning of the 2nd millennium B.C.E. The racial composition of the population also seems to indicate a considerable but not complete replacement of the population.[50]

This renewed phase of settled and urban life can be seen above all in the more western humid parts of the region while in the more eastern and southern parts, the pastoral villages of the local population in the marginal zones of Transjordan, the Negev and Sinai were increasingly abandoned in favor of fewer and larger but mostly open conglomerations. We know little about their material remains, which appear to continue traditions of the preceding Intermediate period; and know nothing about their social order and ethnic or linguistic affiliation.

Like most archaeologists we support the view that the cultural renewal and the revival of the urban structure was not the result of autochthonous developments but of north Syrian and Mesopotamian origin with superficial characteristics of a developed urbanized society. In other words, it is thought that the people who migrated to Canaan were mostly Western Semites related to the Amorites who had first settled in Syria-Mesopotamia and later joined a wave of migration southward settling again in Canaan. The rather rapid revival of the urban centers and the social differentiation may have been, therefore, due to the fact that the Western Semitic newcomers were actually not any longer semi-nomadic barbarians, as were their ancestors who invaded Mesopotamia a few centuries before, but had gone through a phase of civilizatory acculturation.

After a period of domination by the newcomers followed a partial assimilation of the local population who seem to have been integrated into the material as well as the spiritual culture of their masters. There is some evidence of remaining sharp social differentiation between the upper and lower classes, a gap which grew over time and contributed to the demise of these cities.[51]

The main reason for the massive fortification surrounding the major cities appears to have been a very high level of hostility and aggressiveness between the "kings" of the various local city-states. Marauding tribesmen or disgruntled peasants from beyond the pale of civilization in a situation not very different

---

[50] P. Smith, "The Skeletal Biology and Palaeopathology of Early Bronze Age Populations in the Levant" in *L'Urbanisation en Palestine à l'âge du Bronze Ancien. Bilan et Perspectives des Recherches Actuelles, Actes du Colloque d'Emmaus 20–24 octobre 1986*, P. de Miroschedj (ed.), BAR, International Series 527, 297–313 (1989).

[51] W.F. Albright. "The Historical Framework of Palestinian Archaeology between 2100 and 1600 B.C.E." *BASOR* 209 12–18 (1973).
W.G. Dever. "The Beginning of the Middle Bronze Age in Syria-Palestine" in: *Magnalia Dei: The mighty Acts of Gods. Essays on the Bible and Archaeology in Memory of G. Ernest Wright*. F.M. Cross, W.E. Lemke, and P.D. Miller (eds.), (Doubleday, Garden City, N.Y., pp. 3–38 (1976).
N. Feig, in an unpublished M.A. thesis, commented on the distribution of modest huts and hovels compared to large and well-built mansions of the ruling class within the same city as expressions of pronounced social differences in the Middle Bronze Age cities.

from the one described in the archive of Mari[52] might have added to the general turbulence whereas a threat of outside invasions, for instance from Egypt, was less plausible.

The already mentioned 'Execration Texts' indicate that the political organization lacked the centralization or the hegemony of one city over a larger territory. The names of more than one ruler in the early group indicate that the tribal-patriarchal political system was still the dominant form of government during the earlier periods. The political regime based on one ruler and the establishment of a dynastic principle in each urban center crystallized only during the later phases of Middle Bronze Age.[53] The entire system was apparently still far from the politically complex cities, which characterized the city-states of Syria-Mesopotamia or even Anatolia, not to speak of the centralized state of Egypt.[54]

As before, the economy was based on agriculture and husbandry with some local crafts, domestic industry and trade in luxury items. An important advance was the introduction of the fast potter's wheel. It was a Mesopotamian invention and enabled the local pot-makers to create some of the finest and elegant vessels ever made in this region.

In spite of the fact that many Hurrians and Indo-Europeans had arrived in the Levant during this period, a Western Semitic dialect related with Amorite had become dominant. This language, despite some incertitude due to the lack of actual documentation, can be called Canaanite which over time evolved into Phoenician and Hebrew as well as some other dialects spoken in Transjordan during the followings periods (see Appendix IV).

The discovery of a large volcanic explosion, followed by some minor eruptions, on the island of Thera/Santorini in the Aegean Sea was taken as an important event in the history of Mediterranean. An ash layer attributed to this explosion was found in a lake near Sardis in western Anatolia and carbon-dated to 1627–1600 B.C.E.[55] The eruption covered a Minoan city on the island and its associated *tsunami* undoubtedly wrecked much destruction on the coasts of the Aegean and some parts of the eastern Mediterranean basin. It probably caused a few years of abnormally cool and rainy winters in the region but to attribute to it the decline of the Minoan, or any other civilization, such as the disappearance of mythical *Atlantis*, or causing migrations such as the Exodus of the 'Children of Israel', or even influencing the transition of the Levantine M.B.IIA to B period is stretching the evidence very thin and enters the realm of speculation.

As we have mentioned, the level of the precipitation during this time was not as optimal as during the Chalcolithic or Early Bronze Age. This in itself would suffice to keep the Western Semites in the desert plains on the move and

---

[52] W.G. Dever, "The Middle Bronze Age: The zenith of the urban Canaanite Era" *Biblical Archaeology* 50:149–177 (1987).

[53] B. Mazar, "The Middle Bronze Age in the Land of Israel" *Eretz Israel* 8:216–230 (Hebrew) (1967).

[54] W.G. Dever op. cit. (1987).

[55] W. L. Friedrich, B. Kromer, M. Friedrich, J. Heinemeier, T. Pfeiffer, and S. Talamo, "Santorini Eruption Radiocarbon Dated to 1627–1600 B.C." *Science* 312:548 (2006).

to continue with their attempts to settle down in the more fertile lands. The continuous pressure can be seen in references in Old Babylonian documents found in the archives of Mari, Babylon, Ur, Isin and other sites. The Amorites who now had ruled Sumer and Akkad for two or three generations, found themselves in need of fighting back new waves of invaders of the same stock from the west and north resulting in continuous warfare.

It can be envisaged that tribal groups of warriors of Amorite origin who were expelled from the kingdoms they had ruled for a while, were forced to move again into areas where they found less opposition. Albeit of a slightly later date, one of the most compelling stories illustrating the turbulence of the mid-second millennium B.C.E. was found on a statue of Idrimi, king of Alalakh, now Tell Achana in the 'Amuq Valley. He describes, how he and his family was driven from Aleppo, his ancestral home, to Emar on the Euphrates and then continued south into the land of Canaan, where he lived for seven years among the Habiru. With their and the help of the Sutians (presumably identical to the later Shasu or a northern sub-group) he fought with Barattarna, 'king of the Hurrian warriors', with whom he was eventually reconciled. He became king of Alalakh, where he gratefully built temples to the gods and a palace for himself, where this statue was found.[56]

The type of fortification consisting of a sloped *glacis* and *terre pisee* or mud-brick walls of the mature Middle Bronze Age symbolized the new era of walled cities in Canaan, the *"cities great and walled up to heaven"* (Deuteronomy 1:28). The construction of fortification reached its climax precisely during the Fifteenth or the Hyksos Dynasty in Egypt. Dever explained the need for a heavy defense system in the Canaanite city-states as a result of the growing strength of the Hyksos rulers in Egypt.[57] The connection with Egypt is clear, even though the local Canaanite rulers seem to have kept their independent status until the end of Middle Bronze Age.[58]

In order to guarantee a sustainable water supply, it appears that the inhabitants of the cities dug wells inside the cities' walls to reach the groundwater table. In cases that this was not achieved a gallery was extended from the bottom of the well to the direction of the nearby spring, in order to ensure supply in time of siege. It is also possible that galleries were dug from the outlets of the springs into the layers feeding the spring ensuring a longer period of supply from the spring for irrigation purposes.[59]

The problem of the chronology of the shaft and gallery system found in a remarkable number of urban centers in Palestine, such as Hazor, Gezer, Megiddo, Taanach, Gibeon, Jerusalem, and others, is still unsolved. In some

---

[56] A.L. Oppenheim, "Babylonian and Assyrian Historical Texts" in *ANET*, 557–558.

[57] W.G. Dever, op. cit. (1987).

[58] N. Neeman op. cit. (1982).

[59] Z. Ron, "Development and Management of Irrigation Systems in Mountain Regions of the Holy Land" *Transactions Inst. British Geographers* 10, 149–169 (1985).
Z. Ron "Geo-hydrological Breakthrough in the First Jewish Temple Period and Its Effect on the Development of Groundwater Resources" in: *Proceedings of the 12th Archaeological Conference in Israel Jerusalem*, Israel Exploration Society (1986).

cases, there is more and more evidence that, in all probability, they could originally be attributed to the Middle Bronze Age (see Appendix I).[60]

In Syria the flourishing Mature Bronze Age with its center now at Yamhad/ Aleppo continued unabated. The cooler and wetter conditions did not have such a positive effect on Anatolia and the other northern highlands of Caucasia and northern Iran where we hear about troubled times: trails of devastations show the path the Luwians took from the Balkans, with Hurrians and Hittites as well as other Indo-European speaking groups pressing in from the east, apparently trying to evade the grim cold weather further north and east.

Around 1780 B.C.E. Kanes and its *karum* were burned to the ground never to be rebuilt and the trade with Assyria came to a halt. In the destruction layer of Kanes, a bronze dagger with the cuneiform inscription "*belonging to Anitta, son of Pithana, king of Kussara*" was found and baffled the historians: Was Anitta the inhabitant of the palace which was burned down, or was he the attacker? This question is of great importance because in the later annals of the Hittites, Anitta is mentioned as one of their first kings and with him we can begin the story of the Old Kingdom of the Hittites, which lasted until 1600 B.C.E.

6.4
The Crisis Years in Egypt

With the collapse of the Old Kingdom around 2200 B.C.E. began the First Inter-mediate Period, conventionally known as the 7th and 8th Dynasties and which lasted until the re-establishment of the Middle Kingdom around 2040 B.C.E.

A slightly improving flooding system of the Nile was utilized by the rulers of the 9th and 10th Dynasties from Herakleopolis near the Fayum and the 11th Dynasty from Thebes to re-establish stability, only sporadically interrupted by local wars against some local dissident rul'rs. With Mentuhotep's II victory over the Herakleopolitan kings around 2040 B.C.E. Egypt was again reunited. The transition to the 12th Dynasty is unclear and the situation of the new capital somewhere between Memphis and Meydum is also unknown.

Large building projects were begun, trade routes to the Sinai as well as Nubia were re-opened. The new building projects, however, were not wasteful monuments to the dead kings but some lessons were learnt. For instance, the nearby Fayum depression in the Western Desert was originally filled with marshes and stagnant waters. A canal was dug in order to bring fresh water from the Nile and the area was developed into arable land with villages and towns springing up. The canal remains the lifeline of the Fayum to this day (locally known as the *Bahr Yusuf* after the Biblical patriarch Josef). Other monumental irrigation works undertaken in different parts of the country are still under investigation.

These works and their motives explain a dramatic change in Egyptian art: Whereas the kings of the Old Kingdom were semi-divine beings of perfect

---

[60] T. Tsuk. *Ancient Water Systems in Eretz Israel: From the Neolithic Period to the the end of the Iron Age.* Ph. D. Thesis, Tel Aviv University, Tel Aviv (2000, Hebrew with English summary).

beauty, their eyes gazing into eternity and existing only to be adored and admired, the kings of the Middle Kingdom are shown with pensive, severe and often grim expressions as if to impart the idea of rulers burdened by the responsibilities of taking care of the land and its people – the veritable 'Father of the Nation'. It was an undoubtedly effective propaganda and enabled the 12th Dynasty to stay in power for two centuries. Literature flourished and besides the already mentioned story of Sinuhe, the 'Prophecy of Neferti' and the 'Instructions of Amenemhet' are only some of the best-known works of ancient Egyptian literature.[61] The legacy of Middle Kingdom literature is still traceable in the Biblical story of Josef in Egypt, which represents a later version and elaboration of a well-known Middle Kingdom theme, the "Tale of the Two Brothers".

Important information about Nubia, Libya and Syria-Palestine can be gathered from the so-called Execration Texts generally dated to the end of the 3rd millennium B.C.E. These are short formulas and curses against foreign rulers written in hieratic (i.e. a cursive form of hieroglyphic writing) on small clay figurines, which were then broken in a magic ritual in order to transfer the curse on that particular foreign ruler. The enemies in Canaan and Syria have, for the most part, typical Amorite names and are generally referred to as 'wretched Asiatics'. The reference to chieftains and tribal areas in the older texts and to kings and cities speaks for a trend towards a sedentary way of life, which Canaan went through during the Intermediate and the beginning of the Middle Bronze Age.[62]

Alas, the curses did not help much. Around 1800 B.C.E. the Middle Kingdom began to crumble. Rival warloords, known as the 13th to 17th Dynasties of local and foreign provenance represent the 'Second Intermediate Period', a time which was called by the Egyptologist J.A. Wilson 'the Great Humiliation'.[63] For him, like for other scholars, the abrupt disintegration of a powerful structure was a mystery. A variety of reasons were advanced, ranging from social weakness rooted in a feudal system, internal dissent, and, naturally, foreign invasions. They were all true, with the last scourge arriving late, except that it was actually again the failing of the Nile, which was the underlying cause of what happened next.

The relationship between Palestine and Egypt, which were severed in the late part of the Early and Intermediate Bronze Age, had seen a revival during the Middle Kingdom, in particular during the 12th Dynasty. Archaeological finds and Egyptian annals prove trade connections and the Egyptian cultural and political influence grew all over the region. Towards the end of the period this strong relationship deteriorated and when Egypt began to decline around 1800 B.C.E. a vacuum was created which attracted the newly established warrior class in Syria-Palestine. They came with their horses and chariots and superior

---

[61] "The Tale of the Two Brothers" translated by John A. Wilson in J.B. Pritchard op. cit. pp. 12–16 (1973).

[62] "Hymn of Victory of Mer-ne-ptah. The "Israel" stela of Merneptah in the Cairo Museum" ANET pp. 376–378.

[63] J.A. Wilson, T he Burden of Egypt: An Interpretation of Ancient Egyptian Culture, The University of Chicago Press, Chicago (1951).

tin bronze weapons for which those of the Egyptian armies were no match, and around 1780 B.C.E. they poured into the Delta.

The invasion and the settling of Asiatic war lords with their mixed hordes of fighters was in fact similar to the earlier exploits described in the documents of Mari. The Egyptian perspective was described by Manetho and retold by Josephus Flavius in his work *"Against Apion"*, book 1, chapter 14, which is among our most vivid record of this invasion:

> *"There was a king of ours, whose name was Timaus. Under him it came to pass, I do not know how, that God was averse to us, and there came, after a surprising manner, men of ignoble birth out of the eastern parts, and had boldness enough to make an expedition into our country, and with ease subdued it by force, yet without our hazarding a battle with them. So when they had gotten those that governed us under their power, they afterwards burnt down our cities, and demolished the temples of the gods, and used all the inhabitants after a most barbarous manner: nay, some they slew, and let their children and their wives into slavery. At length they made one of themselves king, whose name was Salatis; he also lived at Memphis, and made both the upper and lower regions pay tribute, and left garrisons in places that were the most proper for them..."*
>
> *This whole nation was styled Hycsos, that is, Shepherd-kings.*[64]

The Egyptian archaeological evidence for the Hyksos period, which corresponds to the Syrian-Palestinian Middle Bronze Age 1800 to 1500 B.C.E. is mentioned only shortly. Its principal sites are found in the eastern Delta, such as Tell el-Dab'a, ancient Avaris and Pi-Ramses where a cemetery has yielded important results. The majority of personal names can be classified as West Semitic Amorite. The Egyptian name normally given to this people, *'3mw*, probably pronounced /'amu/, went back to the 3rd millennium B.C.E. and was an archaic survival. A sizable portion of the inhabitants of the Delta already were immigrants from Asia as the ongoing excavations in the western part of the Delta have shown. These people maintained a similar way of life to that of their brothers in the Levant and were often semi-nomadic. It is tempting to see a reflection of this situation in the Biblical story of the sojourning of Israel in the land of Goshen but, as usual, hard evidence is lacking.

It is questionable whether the Hyksos actually conquered the Nile valley itself but they certainly represented a menace demanding tribute from the native ruler further upstream. Small objects, such as rather carelessly executed scarabs with royal names of the 15th Dynasty were found all over the eastern Mediterranean and represent gifts or exotic souvenirs of traders. This was different in the Levant where Egyptian artifacts abound – in a sense one could claim that the eastern delta had become the southernmost province of Canaan during the Middle Bronze Age with a mixture of Canaanite and Egyptian cultural assemblage representing a mixed population. The high variety of tombs and burial customs, often accompanied by donkeys or other sacrificial animals,

---

[64] Jospheus Flavius, *Against Apion* book 1, chapter 14, translated by W. Whiston, Kregel Publications, Grand Rapids, Mich. (1960).

are an Asiatic and even northern Syrian-Mesopotamian hallmark, but without typical Levantine pottery. Instead, we find a large amount of imported Cypriote pottery. The early temples at Avaris were built according to a Levantine, not Egyptian ground plan and the gods initially venerated were probably closer to Astarte, El and Ba'al than to the Egyptian pantheon but soon identified with the Egyptian god Seth, the symbol of strength and the wilderness.[65] As time passes, however, the Egyptian gods were increasingly worshipped as many personal names with Ra', for instance, are attested. Towards the end of their rule, the Hyksos upper classes were largely "Egyptianized". Some important literary works such as the famous Rhind Mathematical Papyrus were composed during this time. Apophis, one of the last Hyksos rulers of the 15th Dynasty, calls himself *'a scribe of Ra'*.

6.5
The Late Bronze Age, ca. 1500 to 1200 B.C.E. (Table 4)

The mid part of the second millennium B.C.E. started with the warming up of the climate and thus may have initiated another turning point in the cultural and political development of the ancient Near East. As expected it brought a high Nile and prosperity for Egypt but dryness and declining fortunes in various degrees to the Levant, especially the southern border region where the cities of the Hyksos were situated, the central hills, and the Jordan Valley.[66]

In later historiography and on their victory stelae, a family of soldiers under Kamose and Ahmose serving in Thebes in upper Egypt and known as the 17th Dynasty claimed to have started a 'war of liberation' against the foreigners around 1550 B.C.E. They were followed by the illustrious 18th Dynasty, also from Thebes, who were able not only to evict the Asiatics from Egypt but also could carry the war into Asia and establish the Egyptian New Kingdom or the 'Empire', Egypt's most glorious period. The officer Ahmose, son of Abana, boasts in his tomb at el-Kab that he not only conquered and destroyed Avaris but many other fortified cities in southern Palestine, particularly S(h)aruhen/Tell el-Farah (South) and Tell el-'Ajjul. In any event, Canaan was reduced to a province under Egyptian rule and even when admitting that the basic local culture did not change, a new era has dawned over Syria-Palestine, the Late Bronze Age.

The campaigns of Thutmose I around 1500 B.C.E., which coincided with the peak time of the warm climate changed the political face of the ancient Near East. Pursuing the hated Hyksos into their own homeland, in one of his raids Thutmose I reached the Euphrates – the river that flows in reverse, an odd phenomenon and which appeared to him strange and despicable, like everything in Asia. When Thutmose was rebuffed by Mitanni, and thus barred from further expansion, he hunted elephants in the 'Amuq Plain in northern Syria in order to make up for his loss of face.

---

[65] M. Bietak, "Avaris and Pi-Ramesse" *Proceedings of the British Academy* 65:225–290 (1979).

[66] A. Mazar, *Archaeology of the Land of the Bible – 10,000–586 B.C.E.* Doubleday, New York. pp. 239–241 (1992).

Table 4. A historical-archaeological timetable of the Near East, from 1500 to 100 B.C.E.

years B.C.E.	Egypt	Syria-Palestine	Mesopotamia	Anatolia
100	Roman Period		Parthians	Roman Period
200	Ptolemies	Hasmonaeans	Seleucids	Kingdom of Pergamon
300	Hellenistic Period Near East Unified by Macedonian-Greek Civilization			
400		"Return to Zion"		Miletos Destroyed
500	Persian Period Near East Unified by Achaemenid Empire			
600	Late (Saitic) Period 26th–30th Dynasty	Neo-Babylonian Empire		Iron Age: Lydian, Archaic Greek, Phrygian and Urartian Periods
700	3rd Intermediate Period 21st–25th Dynasty (Libyan & Kushite Period)	Iron Age II	Assyrian Empire	
800		Israelite, Neo-Hittite and Aramean Period		
900			Neo-Hittite (or Luwian) and Aramean Kingdoms	Neo-Hittite (or Luwian) Kingdoms
1000	Ramesside Period 19th–20th Dynasty ("Sea Peoples")	Iron Age I		
1100				"Dark Age"
1200		Late Bronze III	Middle Babylonian (Kassite), Middle Assyrian and Mitannian Period	Late Bronze Age Hittite Empire
1300	New Kingdom 18th Dynasty	Late Bronze II		
1400		Late Bronze I (Canaanite Period)		
1500				Middle Hittite Period

After the interlude of the rule of Hatshepsut, the only female pharaoh, Thutmose's son Thutmose III (1490–139 B.C.E.) and his successor Amunhotep II (1438–1412 B.C.E.) ascended to become the greatest conquerors. During their reign the Egyptian empire became the supreme overlord of Syria-Palestine. Egypt had now reached its high water mark, its success reflected in the characteristic smiling faces of royal statues of this period, when in the Egyptian view *"the world was again in order"*. This artistic convention proved so powerful that it remained standard until the Greeks and neo-Hittites of Syria began copying ancient Egyptian statues complete with their well-known *"archaic smile"* during the Archaic Period of Greek art nearly a millennium later.

The effects of the incorporation of the Levant into the Egyptian empire, however, should not be compared with the Assyrian or Roman conquests of the region, with their systems of taxation and other extreme forms of exploitation. Life in the conquered territories continued very much as before, with no new order or basic changes in the social structure or the economy. The turbulence and infighting of city-states governed by independent rulers (in some cases supplemented by Egyptian "advisors") continued after the Egyptian army had departed.[67] The local arts, such as pottery and architecture, declined, and we see an increase of imported items and their local imitations.[68] Egyptian influence can be seen in the development of a local linear

---

[67] Among many publications, a classic and very readable account is G. Steindorff and K.C. Seele, *When Egypt Ruled the East*. The University of Chicago Press, Chicago (1971).

[68] R. Amiran, op. cit. (1969).

script called Proto-Canaanite – the ancestor of the later Hebreo-Phoenician alphabet – found on pottery at Shechem, Gezer and Lachish.[69] In northern Syria, a similar alphabet with the same order of letters representing the consonants, but based on cuneiform shaped letters for use on clay tablets was found at Ugarit/Ras Shamra and probably was current at other sites as well. Traditional Hurrian *migdal* temples (i.e. shrines in shape of tall towers) at Megiddo and Shechem co-existed with other types, such as the temple on Mount Gerizim near Shechem, the airport temple at 'Amman, or the fosse temple at Lachish. An increasing Egyptian influence is seen in the development of residential and sacral architecture, in particular the so-called residences or "governors' palaces" at some sites, and the classic tri-partite Canaanite temples found from Tell Judeideh in the 'Amuq Valley in the north to Hazor and Beit-Shean in the south – the prototype of Solomon's temple in Jerusalem.[70]

As to be expected, the level of prosperity increases further north, as seen in the excavations of tells in northern Israel, such as Hazor, and in Lebanon and Syria at Byblos, Ugarit/Ras Shamra, Alalakh/Tell Achana, Emar/Meskene and many other sites bear testimony to the high level of art and the sophisticated lifestyle of the urban centers of the Late Bronze Age in Syria.

The term Canaanites appeared during the Late Bronze period to denote the people of the Levant. Opinions differ regarding the etymology and meaning of the term. In the Bible it refers to one of several groups of inhabitants of the land, which was promised to the Hebrews and thus ought to be destroyed. Their worship of idols, sacrifice of children and temple prostitution were all *"abominations to the Lord"* and cited as additional arguments for their eradication. Another Biblical meaning for "Canaanite" is 'merchant,' perhaps because traders of the famous purple cloth, called *kinahhu* in Hurrian sources from Nuzi,[71] came from the land of Canaan. This cloth, dyed purple with the sap of the murex shell, was traded throughout the Mediterranean world. If *Kinahhu* is derived from a Semitic root (k.n.'.) it could also mean simply "Low Land" – that is, the coastal plain and its inhabitants.

Data from the archaeological excavations show that the region went through a socioeconomic crisis from the end of the Middle Bronze Age towards the end of 17th century B.C.E. This deterioration may have weakened the hold of the Hyksos on southern Palestine and their stronghold at sites such as Sharuhen (Tell el Far'ah-South), enabling the Egyptians to stage a rebellion and regain their independence. Many of the urban centers shrank in size dramatically, and the population of the rural hinterland diminished, if it did not disappear completely. At the same time, the same excavations revealed the existence of elaborate palaces and patrician houses, clear signs of an ever greater gap between the poor and the wealthy with the customary consequence: unrest and revolts. The local population suffered from the burden of

---

[69] J. Naveh, *Early History of the Alphabet*, Magnes Press, Jerusalem (1982).

[70] A. Biran, ed., *Temples and High Places in Biblical Times*, Nelson Glueck School of Biblical Archaeology, Hebrew Union College, Jerusalem (1981).

M. Ottoson, *Temples and Cult Places in Palestine*, Almqvist and Wicksel, Uppsala (1980).

[71] *CAD*, vol. 8, 379 (1970).

taxes and corvee labor[72] from which the Egyptianized elite was, of course, exempted.

The letters found at Tell el Amarna in Egypt vividly portray the political situation in the Levant in the 14th century B.C.E. – many small city-states headed by kings under the loose control of Egypt, all fighting each other, complaining and asking for help from the Egyptian suzerain in their quarrels. In the background is the threat of the 'Apiru/Habiru, pastoral nomads who were increasingly gaining a hold in the region their ranks swelled by some locals.[73] Whether these 'Apiru/Habiru are the forerunners of the Hebrew-Israelite settlers and included the later Biblical 'Children of Israel' is a question still under debate. What is obvious from the letters is that they infiltrated the sown land and, using a strategy of "divide and conquer" entering into agreements with some of the petty kings to fight their competitors, just as pastoral nomads always interfered in the affairs of their sedentary neighbors (see Appendix III). Between them and the invading Sea Peoples at the beginning of the 12th century B.C.E. the fate of the southern Canaanites was sealed: whether it was a *finale crescendo* or a whimpering out is an open question.

In Anatolia, after a period of turmoil and relentless wars with the Hurrians in the eastern and southern part of the country following the death of Mursilis I, the Hittite warrior aristocracy adopted the dynastic principle based on the "Edict of Telepinus" around 1500 B.C.E. They increasingly supported the so-called "Great Family" at Hattusas, whose leaders began to style themselves "My Sun", i.e. supreme emperor. Cities were rebuilt and fortified, their social order foreshadowing the constitutions of later Greek polis – one of the many enigmatic links with the later periods. The energetic rulers introduced new agricultural systems, most probably abetted by a warmer, more benign climate, when it comes to the Anatolian Plateau. Suppiluliumas I (1370–1330 B.C.E.) was the greatest of these rulers. He reorganized the army, defeated the Hurrians, occupied their southern kingdom of Kizzuwatna (the modern Cilicia) and married a Hurrian princess, the famous Tawananna Pudu-hepa, who became empress in her own right and introduced the Great Hurrian pantheon depicted on the rocks of Yazilikaya near Hattusas. He pressed into Syria, defeated the kingdom of Mitanni, and established a secondary capital at Carchemish on the Euphrates, but died of the bubonic plague, along with many of his countrymen. His successor, Mursilis II, conquered large areas of Syria and northern Mesopotamia and annexed most of the Hurrian principalities along the rain-fed valleys descending from the Anti-Taurus. This expansion, however, led to unavoidable clashes with the Egyptians, climaxing with the famous battle of Qadesh in 1296, where Ramses II claimed victory, even though he escaped just by the skin of his teeth. This battle is the main subject of the depictions on his temples at Luxor, Karnak, Abu Simbel, the Ramaseum on the West Bank of Thebes and other sites throughout Egypt, which still can be seen. In 1278,

---

[72] S. Bunimovitz, "On the Edge of Empires – Late Bronze Age (1500–1200 BCE)" in: T.E. Levy op. cit., pp. 320–331 (1998).

[73] J.B. Pritchard op. cit. pp. 262–277 (1973).

realizing the futility of the wars, the first known international peace treaty was signed by Muwatallis III and Ramses II, inaugurating a short time of peace known as *pax aegyptiaca*. Trade and the arts flourished in this time, and prosperity reigned throughout the eastern Mediterranean countries.

As in other areas of the ancient Near East, northern Mesopotamia experienced a general slowing of cultural development, which, in our opinion, was undoubtedly due to the warm and dry climate. Population growth declined and the level of prosperity and cultural development waned. The Hurrians had become the dominant ethnic element in northern Syria and Mesopotamia. Their customs and beliefs are detailed in rich cuneiform archival material from a variety of sites, primarily Nuzi/Yorgan Tepe, Urkish/Tell Mozan on the Habur, as well as Alalah/Tell Achana, Ugarit/Ras Shamra, and Hattusas/Boghazkoy. The kingdom of Mitanni can be regarded as representative of a Hurrian state, even though a ruling group closely related to the earliest Indo-European invaders of India had founded the state. Mitanni was the dominant power in the time under discussion, while Assyria was a vassal kingdom lacking the freedom to conduct a foreign policy. Assyriologists define this rather obscure phase as the "Middle Assyrian Period," characterized by harsh laws and strict discipline of the citizenry, as if preparing for hard times to come.

Further south, Babylonia had fallen under the rule of the Kassites (*Kashshu* in Akkadian), a people who came down from the Iranian highlands and rose to power in the aftermath of the campaigns of Mursilis I of Hatti. In order to rule the vast autochthonous population, the Kassites quickly slipped into the role of legal rulers by adopting the superior Babylonian culture and the Akkadian language, thus leaving no clues about their original language or ethnic affiliation. They carefully maintained all aspects of the Babylonian civilization in a most conservative manner for more than four centuries. Among the very few exceptional buildings erected during their rule were the Karaindash temple at Uruk and a new capital, Dur-Kurigalzu near Baghdad. Typical artifacts of Kassite art are the *kudurru*, elaborately decorated standing stones in traditional style, which served primarily as commemorative stelae or border markers between estates.

Under Mitannian domination for several centuries, Assyria was politically inactive until the 14th century B.C.E. when the Hittite defeat of Mitanni enabled the rulers of Assur to reclaim their position, expanded their territory, and built the empire so well known from later sources. In 1235 B.C.E. the Assyrian king Tukulti-Ninurta I defeated the last Kassite king, conquered Babylon for the first time and united Mesopotamia. He carried away the statue of Mardukh to Assur, a symbolic act: It illustrated the beginning of the political ascendancy of Assyria and its integration into the Babylonian "transnational" cultural sphere.

After a period of economic decline, religious reforms, and internal unrest during the Amarna Period, the Egyptian empire also began a slow but steady political decline. After the ignominious end of the 18th Dynasty, rulers such as Seti I and Ramses II tried to turn fate around, in vain. Ramses II's descriptions of the battle at Qadesh mentions the enlistment of a host of new people participating in the squirmishes. For instance, the Hittites were aided by their vassals of Naharina, i.e. Mitanni, the Luwians of Arzawa, the Dardany (the people of Troy in the northwest?), the Lukka (the Lycians of southwest Anatolia) and finally, their old enemies, the Kaska from northern Anatolia. With the Egyptians stood the Sardana (Shardana), allies of the Libyans and, like them, probably of North African origin. Some of them reappeared in later battles of Merneptah, son of Ramses II, against the the Meswes probably the Libyans and their allies, the Sherden or Sardana, the Ekwes (Achaians?), the Teres or Teswes (Etruscans?), the Tjekeres (Teucrans? other Etruscans?), the Sekeles (Shikilayu in cuneiform at Ungarit, Sicilians?) and the Lukka – all harbingers of tremendous upheavals and profound changes in the ancient Near East.

CHAPTER 7

Migrations and Settlings

From the end of the Late Bronze to the Early Iron Age, ca. 1200 to 900 B.C.E.

"The foreign countries made a conspiracy in their islands. All at once the l ands were removed and scattered in the fray. No land could stand before their arms, from Hatti, Kode, Carchemish, Arzawa, and Alashiya on, being cut off at [one time]. A camp [was set up] in one place in Amor. They desolated its people, and its land was like that which has never come into being. They were coming forward toward Egypt, while the flame was prepared before them. Their confederation was the Philistines (Peleset), Tjeker, Shekelesh, Denye(n), and Weshesh lands united. They laid their hands upon the lands as far as the circuit of the earth, their hearts confident and trusting: "Our plans will succeed!""

(Account of the war against the 'Sea Peoples', from Ramses III's temple of Medinet Habu at Thebes)

7.1
The End of the Age of Bronze and the Beginning of a New World (Table 4)

An abrupt change to a colder global climate took place towards the end of the second millennium B.C.E. (Figs. 3 & 3a). The cold began around 1200 and peaked by 1100 B.C.E. causing waves of people from Eurasia trying to settle down in new territories, either by force or by agreement, events we have seen already several times in the past.

A new metal – iron – was introduced and gradually replaced bronze; and an alphabetic script began to supplant the time-honored but cumbersome cuneiform script of old to become the writing method of the Levant and later of Europe and beyond.

Migrations were at the core of two of the greatest cycles of heroic exploits in world literature: the Greek Homeric epics of the Iliad and the Odyssey, whose idealized warrior ethics, honor, and valor foreshadowed European values; and the Hebrew Bible, whose mature spiritual and moral concepts seem as though tempered by time. Older literary works dated to the Bronze Age left only faint echoes in later history. Despite their undoubtedly great importance, the few surviving Egyptian, Babylonian and Hittite epics left no direct influence on later generations and their modern translations have little esthetic appeal for today's taste. The Greek and Hebrew masterpieces, on the other hand, are still widely read and continue to have an impact on the way we view the world. We can therefore, say that the events described in the Bible and by Homer had shaped our world as it began to unfold.

Many new peoples appeared on the stage of history, often quite suddenly. In many cases, we do not know where they came from, what language they spoke, or what was their material or spiritual culture. Some were seafarers roaming the Mediterranean Sea, others destitute nomads from the deserts, and still others uprooted and land hungry farming folks on the move, eventually turned into warriors.

One of the major avenues of migrations was the Balkan Peninsula where the Mycenaean culture had disintegrated. Some of the Mycenaean Greeks found refuge on the Aegean islands, Crete, and Cyprus and in their turn had created waves of refugees washing ashore on the coasts of the eastern Mediterranean Sea. Homer's Iliad in the saga of the 'Battle of Troy' immortalized the memory of many battles fought throughout this period. According to Egyptian sources, the turbulence caused the collapse of the Hittite empire, the destruction of Ugarit, and other Canaanite cities along the Levantine coast by what we today call the 'Peoples of the Sea'. The Egyptians allegedly fought them back and settled some of the tribes in Egypt, others were sent under Egyptian suzerainty to settle on the southern shores of the Levant.

Further disturbances were caused by migrations of peoples in the deserts in the south and east of the Fertile Crescent, among them the Aramaic-speaking pastoral nomads, the descendants of the Amorites. With the increasing cold, Egypt experienced a series of environmental calamities, most probably due to the weakening of the monsoon and penetration of most uncommon north-westerly rainstorms into Lower Egypt. This phenomenon had apparently caused considerable social unrest and perhaps the exodus of some immigrant Asiatic clans to the Sinai and Transjordan deserts and later into Canaan where they began to form a tribal alliance known as the 'Sons of Israel'.[1] They collided with the Philistines, who had formed a league of five cities, crushing and absorbing the remaining Canaanites towns between them. In time, the Israelite tribes prevailed, at least in the mountainous interior, and formed the kingdoms of Judah and Israel. The emergence of the people of Israel and their apparently revolutionary ideology is one of the most fascinating and controversial issues of ancient Near Eastern history. There exist as many theories as there are scholars, yet very few have questioned whether climate change was in any way related to establishment of the new nations.

---

[1] The proper translation of the Hebrew *B'nei Yisrael* is "Sons of Israel" however, as the term "Children of Israel" is entrenched in English we feel obliged to use it when there is a Biblical connotation.

7.2
Focussing on the Impact of Climate on the Events at the End of the Second Millennium B.C.E.

As we have already seen, a warm and dry phase started around 1600 B.C.E. and peaked between 1500 and 1400 B.C.E. A slightly more humid climate soon followed, reaching the average line at ca 1300 B.C.E. perhaps once more due to a series of volcanic eruptions, whose impact may be seen in the later phases of acidity in cores from Greenland.[2] By 1200 the warm spur switched to the cold – humid phase with its peak around 1100 B.C.E. (Figs. 3 & 3a). The paleo-climatic data from the cores of Lake Van show the same trends, although there is an approximate 200-year delay in the response, which may have been a result of retardation, as part of the lake's water comes from groundwater reservoirs. The ancient levels of the lakes and rivers in the Asian and European regions north of the Near East indicate variations of global temperature which had a different effect on the amounts of precipitation, and thus the hydrology of these regions. Thus, the rather high levels of the Caspian Sea during the 15th to the 13th century B.C.E. reveal a humid climate over the plains of Russia draining to the Volga, and an abrupt decline after 1300 B.C.E. The reverse is seen in the fluctuations of the Vistula levels,[3] which show high precipitation in the more central part of Europe from 1200 to 1000 B.C.E. In addition, the curve of movement of Scandinavian glaciers[4] shows retreat of the glaciers during a warmer period between around 1500 and 1200 B.C.E., followed by a rapid advance beginning around 1200 B.C.E. and reaching one of its highest peaks during the Holocene around 1000 B.C.E. Under a strict continental regime, the plains of western central Asia experienced a cold and dry phase phase bringing disaster to its inhabitants.

Pollen diagrams from the bottom of a few lakes on the Anatolian plateau give a more detailed picture. The pollen record from Lake Van shows a revival of pine and oak forests and a decrease in *Artemisia* from around 1800 to 1500 B.C.E., indicating a humid climate. A severe reversal occurred around 1500 B.C.E. and lasted two to three centuries. The forest began to rejuvenate, reaching its maximum around phase 1200 B.C.E. A slight decline was again followed by another expansion of the forests, this time peaking around 900 B.C.E. Although short-range fluctuations in the levels of Lake Van cannot be discerned by available methods, in general, a gradual rise is seen. The isotope record from

[2] C.U. Hammer, H.B. Clausen, W.L. Friedrich, and H. Tauber, "The Minoan Eruption of Santorini in Greece Dated to 1645 B.C." *Nature* 328: 517–519 (1987), but see further discussion for the latest date.

[3] A.L. Chepalyga, "Inland Sea Basins" in: *Late Quaternary Environments of the Soviet Union* (English-language edition), A.A. Velichkoe, H.E. Wright Jr. and C.W. Barnosky (eds.) Longman, London. pp. 229–247 (1984).
T. Kalicki, "The Evolution of the Vistula River Valley between Cracow and Niepolomice in Late Vistulian and Holocene Times" in: *Geographical Studies, Special issue no. 6*, Polish Academy of Science, Warszawa, pp. 11–38 (1991).

[4] W. Karlen, "Glacier Fluctuations in Scandinavia, in Temperate Palaeohydrology" in: *Fluvial Processes in the Temperate Zone during the Last 15,000 Years*, L. Starkel, K.J. Gregory, and J.B. Thornes (eds.), John Wiley and Sons, New York (1991).

the lake, however, portrays the same sequence, albeit with a retardation factor of about two centuries.

W. Van Zeist and H. Woldring noticed a zone showing the impact of agriculture on the natural vegetation in the pollen record of several sites in western Anatolia and Greece and date its beginning by ^{14}C to around 3200 B.P. (calibrated to about 1450 B.C.E.), i.e. the Late Bronze Age.[5] This change in economy was called the Beysehir Occupation Phase, as clear evidence for it was seen in cores drilled at the bottom of the Beysehir Gölü marsh in southwestern Anatolia. The palynologists also consider a climatic change involving an increase in precipitation. In Issar's opinion, there are too few dated samples, especially from the Beysehir core, to determine the exact timing of this event. Indeed, more recent work by N. Roberts and colleagues date this event as starting around 1250 B.C.E. and continuing to around 600 AD. They are aware, too, of the dating problems of this phase, which they suggest that it may be due to contamination by "older" hard water. The latter investigators made an interesting observation about a volcanic ash layer they attributed to the explosion of Mount Thera (Santorini Island). This ash layer was identified in a core drilled in Lake Gölhisar, near Sardis about 150 km southwest of Beysehir and dated to between 3300 to 3225 B.P. (calibrated to 1600–1450 B.C.E.). This falls within the dates speculated for the Thera volcanic eruption (i.e. 1627/8 or 1500 B.C.E.) At Gölhisar the Beysehir Phase is dated to some time between the volcanic ash layer (1600–1450 B.C.E.), which are close to the years 1627–1600 B.C.E. now confirmed for the Thera volcanic eruption and an overlying layer, which was dated to 2830 B.P. (calibrated to about 900 B.C.E.).[6] Closer investigation of the Gölhisar pollen curve shows a curious event just below the layer of the ash, characterized by an abrupt decrease in pine pollen and an increase in Artemisia, similar to that observed at Lake Van.

The impact of these climatic changes on the natural environments of the Near East can easily be appreciated. First, the warm and dry periods of the Late Bronze Age in the middle of the second millennium B.C.E. would have negatively affected the desert margins, which at present receive between 200 to 400 mm of rain per year. In the same time, it would affect positively the hydrological regime of the Nile. Later on the cold and humid period, which became more extreme towards at the end of the second millennium B.C.E., i.e. the beginning of the Iron Age, undoubtedly had a positive effect on most of the region, perhaps with the exception of the regime of the Nile due to the

---

[5] W. Van Zeist and H. Woldring op. cit. (1978).

E.T. Degens et al. op. cit. (1984).

Schoell, in Degens and Kurtman op. cit pp. 92–97 (1978).

Lemcke and Sturm, op. cit. pp. 654–678 (1997).

S. Bottema and H. Woldring, "Anthropogenic Indicators in the Pollen Record of the Eastern Mediterranean" in: *Man's Role in the shaping of the Eastern Mediterranean Landscape.* S. Bottema, E. Nieborg and W. Van Zeist (eds.), Balkema, Rotterdam. pp. 231–264 (1990).

[6] W.L. Friedrich et al., op. cit. (2006).

N. Roberts, W.J. Eastwood, H.F. Lamb, and J.C. Tibby, "The Age and Causes of Mid-late Holocene Environmental Change in Southwest Turkey" in *Third Millennium B.C. Climatic Change*, pp. 409–427.

weakening of the monsoon system. Accordingly, the very cold phase might negatively impact the highlands of Anatolia and the plains of Eurasia north to the Black Sea and Caucasus, causing relatively harsh and unfavorable for farming and herding, although the forest thrived.

7.3
The Wave Pattern of Migrations and the "Sea Peoples"

Seeking the trigger for the chain of events connected with the appearance of the "Sea Peoples" in the 13th and 12th centuries B.C.E., the cooling trend of the climate first comes to mind. If this proves to have been true, then the regions in the northern parts of continental Eastern Europe and adjoining parts of western Asia would have been negatively affected leading the people in these regions to exert pressure on their more southern and western neighbors.

Indeed, there are observations about disturbances of the agrarian and pastoral communities in the Trans-Pontian and Eurasians steppes beginning in the 13th century B.C.E. Soviet scholars excavating in the southern parts of Turkestan in central Asia found the initiation of impressive irrigation works in, for example, ancient Dakhistan, out of which grew the great oasis cultures of Bactria and other sites along the Silk Road.[7] In the more northern and eastern parts, on the other hand, they found that nearly all the sedentary settlements had been deserted, with a general turning to a pastoral nomad way of life, called 'the nomadic alternative' in Russian anthropological literature.[8] The rationale is that pastoral societies need much larger territories for their livestock than sedentary farmers would have required in order to obtain the same amount of food. Expansion is, therefore, an integral part of the socioeconomic conditions of any pastoral people because free access to pastures without political or any other boundaries is seen as essential to their survival (see Appendix III).

It may turn out to be that not only the people living in the Russian steppes and along the northern and eastern coasts of the Black Sea, but much further east took the 'nomadic alternative'. It is said that Mongolia is a lush island between the ice in the north and the desert in the south. In good years, it can provide for a large number of animals, and therefore people; in bad years it drives them out into all directions.

Contrary to the situation in Central Asia or the Near East, temperate Europe of the 12th century B.C.E. enjoyed a period of prosperity albeit of profound culture change.[9] The subsistence economy and food production of the Late

---

[7] P.L. Kohl, "Central Asia (Western Turkestan): Neolithic to the Early Iron Age" in *COWA* p. 194.

[8] A. Khazanov. "Nomads and the Outside World". *Cambridge University Press*, Cambridge (1984).

[9] S.P. Wells, "Crisis Years? The 12th Century B.C. in Central and Southeastern Europe" in *Crisis Years*, pp. 31–40.
In Spring 2001 M. Zohar inquired about recent works in these countries at the various archaeological institutions at Bucharest and Sofia and could confirm these assumptions. See also Mervyn Popham, "The Collapse of Aegean Civilization at the End of the Late Bronze Age" and Anthony Harding "Reformation in Barbarian Europe, 1300–600 B.C." in: *The Oxford Illustrated Prehistory of Europe*, Barry Cunliffe (ed), Oxford University Press, Oxford. pp. 277–335 (1994).

Bronze Age societies did not change in essence, it became more extensive than intensive. Agriculture expanded into different and hitherto unexploited ecological niches and new crops were introduced. Cattle raising had become extremely important which could confirm colder conditions. The volume of bronze industry and trade had considerably grown, producing a wide range of bronze tools, ornaments and weapons. The grave goods found in some of the tombs of the elite provide clear evidence of far-reaching alterations in social structure and increasing status differentiation in certain regions.

An important change is seen in rituals, particularly burial customs. Most Eurasian peoples had practiced various forms of interment, but mostly inhumations under burials mounds. From around 1300 B.C.E., cremation, which had been practiced centuries earlier in the Plain of Pannonia in Hungary, began to spread in all directions and in ever-increasing numbers – the Urnfield phenomenon.[10] It was a new social, cultural and ideological order of a different class of warriors, which from modest beginnings soon developed into socialy a wider and highly mobile elite with increased commerce and contact with people in other regions, including the eastern Mediterranean. Their hunger for land was insatiable and gradually they began pushing south. Equipped with novel types of weapons and new military tactics they were able to expand by incorporating the neighboring people into their society once the native elite accepted their ideology as the new standard.

One would expect that pressure from nomadic or semi-nomadic peoples from the east on those living along the coasts and rivers of the Black Sea would force them to initiate inroads for the purpose of looting and colonizing. The turmoil might well have caused other groups to migrate to the Balkan Peninsula who subsequently merged with the autochthonous population and became the forefathers of the Thracians and Illyrians known from later classical sources. Were these the same people who destroyed Mycenae, Troy and the other Greek Late Bronze Age sites? Supporting this assumption is the fact that with the destruction of the Mycenaeans' palaces and strongholds, the typical tholos-tombs also disappeared and were replaced by cist-graves,[11] characteristic of contemporaneous populations living in the north-Pontic region.[12]

In support of this hypothesis, Troy layer VIIb and the layer overlying the destruction, i.e. VIIa, contain foreign and crude handmade wares with Balkan affinities which is also found scattered at some other places in Anatolia. The mapping of these locations indicates that the local population was augmented by a new ethnic element from southeastern Europe. Many scholars consider this a confirmation of Herodotus' story that the ancestors of the Phrygians

---

[10] Cremation was never a uniform practice. In most cases, the deceased were cremated, their ashes collected in containers (urns), often cooking vessels, and buried with no further votives or grave goods in defined cemeteries, i.e, the 'urn-fields'.

[11] M.I. Finley, *Early Greece, The Bronze and Archaic Ages*, Chatto and Windus, London, p. 58 (1981). Herodotus tells about the *"earlier Cimmerians attack on Ionia being a mere plundering raid and in no sense a conquest"*. (*The Histories*, Book I).

[12] C. Graham and P. Piggott, *Prehistoric Societies*. Penguin Books, Harmondsworth, p. 269 (1970). The still best summary of literature on this problem in English is found in *The Oxford Illustrated Prehistory of Europe*, B. Cunliffe (ed.), Oxford University Press, Oxford. pp. 492–493 (1994).

and Armenians came from Europe. However, there is no evidence that the newcomers were the actual sackers of Troy or the other Anatolian Late Bronze Age civilization. It is more likely that the "*arriving Europeans may have entered a territory whose local inhabitants were by then surviving at a subsistence level*".[13] The same can be said about the Dorians in Greece who may have entered an exhausted land. Another possibility is that the destroyers were "proto-Vikings" coming down the rivers of Europe from Scandinavia and the Baltic countries, which, according to Finley,[14] went through a period of severe crisis.

The evidence of destruction by an unidentified enemy, extending from Thessaly in the north to Laconia and Messenia in the south, including the fortress of Mycenae, is dated to about 1200 B.C.E. This is approximately the time of the collapse of the Hittite empire, destruction layers on Cyprus and of Ugarit, and of disturbances not only in Mediterranean Malta,[15] Italy, the Lipari Islands and Sicily, but perhaps even in France and all the way north to the Baltic Sea. A wave of destruction with large-scale movements of people can be seen sweeping over all the regions in a 'domino-topple-chain' fashion – namely in a 'wave action' in which each collapse propels another. Thus the Mycenaeans had probably become a "domino piece" forced from its position and consequently exerting force on the next "domino pieces" – in this case, the urban centers bordering the Eastern Mediterranean. This was not an organized onslaught. M.I. Finley maintains that "*It appears rather to have been broken in rhythm, pushing in different directions at different times, as in the case of Egypt, attacked once from the west and a second time, a generation or so later, from the northeast. There was little stability in the inter-relationships among the migrants, uncertainty in the ultimate objectives. All that is analogous with the later Germanic movements, as is the fact that trading and cultural interchanges and influences, at least with Greece, had been going on for centuries before the raids began*".[16] If we accept the Dorian invasion theory, then were the Dorians the domino that pushed the Mycenaeans to become one of the 'Sea Peoples'? Carpenter opposes this suggestion, maintaining that "*even if there really was some sort of invasion from the outside world into southern Greece during the opening decades of the thirteenth century, the invaders could not possibly have been the Dorians, for the simple and sufficient reason that when the Dorians did come they came to stay – as their dialect, widespread over most of the Peloponnese, coupled with the testimony of the ancient historians, unambiguously testifies*".[17] Carpenter continues to argue that it was not an invasion from the outside but disintegration from within, probably due to a climate change during the thirteenth and twelfth centuries B.C.E., involving a warmer climate and less rains over the Aegean zone. As much as the present

---

[13] G.K. Sams, "Observations on Western Anatolia" in *The Crisis Years*, pp. 56–60 (1992).

[14] M.I. Finley, op. cit. (1981).

[15] M. Zohar, "Malta" in *The Oxford Companion to Archaeology*, B.M. Fagan (ed.), Oxford University Press, Oxford, pp. 87–99 (1996).

[16] I. Finley, op. cit. p. 58 (1981).

[17] H. Carpenter, op. cit. p. 36–41 (1966).

authors are inclined to see climate change as an important historical factor in transition periods, it is difficult to accept Carpenter's arguments. In contrast, all time-series of proxy data, which were not available to Carpenter, actually show that the long-term changes during these centuries were toward a colder climate, causing the Mediterranean region, including the Aegean, to become more humid. This scenario is confirmed by investigations of the sedimentation rate in the Marathon plain, demonstrated in cores drilled in this area.[18] The soil profiles obtained from these cores show that, starting with the 13th century, deposition rates diminished. Issar interpreted this to indicate decreased soil erosion because of increased forest coverage due to increased precipitation.[19] Yet, even though the proxy data time series show no long dry phase, one cannot rule out short-range changes of a few decades. Such a dry spell may have brought famine to the semi-arid non-irrigated parts, just on the eve of the fall of the Hittite Empire, around 1190 B.C.E. Letters from the last Hittite emperor to the last king of Ugarit, urging him to send grain to relieve his hungry people[20] suggest there was such a famine, which is confirmed by our data.

But it was not from the north where the first people from beyond the horizon of the Near East came from: They were the Libyans, a name given by the Egyptians to a variety of tribes from northern Africa. Under the leadership of a certain Meryry, a group had left their parched homeland on the edge of the Sahara after a period of drughts and came with their families, cattle and all their other possessions, clearly not to raid but to find refuge and to settle in the fertile fields of Egypt. Merneptah, the son and successor of Ramses II, could boast that in mid-April 1220 he defeated and captured the entire invading army, disarmed the warriors and settled them on state land in the frontiers of the country, thus actually doing their bidding.

The next, and, in this context, most important event is the battle of Ramses III, depicted and described on the walls of his mortuary temple on the west bank of Thebes and known today as Medinat Habu. As usual, the story started with the Libyan wars in 1189, the fifth year of Ramses' rule. The Libyans, according to the wall engravings, were slaughtered in large numbers. The question remains whether these bombastic descriptions are a true reflection of actual events or merely empty propaganda inherited from his predecessors?[21]

The "Great Land and Sea Raids" took place in his eighth year, i.e. 1177, 1186 or 1190 B.C.E.[22] A description of the event excerpted from the account on the temple walls appears at the beginning of this chapter. It continues with another

---

[18] R. Paepe, I. Marolakos, S. Nassopoulou, E. van Overloop and N. Vouloumanos "Quaternary Periodicities of Drought in Greece" in Angelakis and Issar eds. op. cit., pp. 77–110, (1996).

[19] A.S. Issar, op. cit. p. 32 (2003).

[20] A. Malamat, "The Egyptian Decline in Canaan and the Sea Peoples" in: *The World History of the Jewish People*, vol. 3, B. Mazar (ed.). Massada Publishing Co., Tel-Aviv p. 30 (1971).

[21] We should remember that these "wretched" Libyans constituted the 22nd Dynasty from 935 to 715 B.C.E. and were undoubtedly the real masters of Egypt even before that.

[22] The exact dating of the events depends on which chronological "school" is applied (High, Middle or Low). The range of accuracy is ±15 years.

boastful description of the defeat of the invaders,[23] which is of less interest except for its detailed depictions of the combatants, their dress and arms. It is unclear whether the land battle depicted on the same panel was the same event as the battle at sea or a different event; in any case we should not place too much trust in their veracity. Here we see the Sardana in their characteristic horned helmets actually fighting on both sides, and, most strikingly, ox-carts with women and children, clearly showing a people on the move in search of a new home. It has been suggested that these carts resemble contemporary vehicles with solid wheels from Anatolia or the Caucasus, or even Iran. Again, it seems to be certain that some northern people were moving south, indicating to us a worsening of living conditions in their home countries. (Plates 11a & b)

The Ramses' III inscription tells us that before the invaders reached Egypt, they had destroyed Hatti, the Hittite empire, Kode, the coast of Cilicia, Carchemish, the city on the Euphrates, Arzawa, a kingdom somewhere in southwestern Anatolia, and Alashia, probably Cyprus.[24] The Hittite archives had fallen silent during the reign of the last emperor, Suppiluliuma II, and we know nothing about the end of the city and the empire. Its capital Hattusas was destroyed, yet the question remains by whom? As archaeology has not yet been able to provide the answer, the story told in Ramses' III inscriptions is the principal historical source. We are left to wonder how much the Egyptians knew about the outer world and their enemies. They had trouble distinguishing between foreign sounds such as 'l', and 'r', or 's', 'sh' and 'tj' – the readings of which could have been as confusing for them as they are for us. Moreover, since there are no native words in Egyptian for sea or island, the hieroglyphs can be interpreted in different ways. The greatest problem for modern scholars, therefore, is to ascertain how much of this 'historical documentation' is in fact, merely pious or propagandistic oratory woven around a few relatively unimportant or even non-existent skirmishes, whose intend was to prove the king "master of all lands" – even if it was restricted to the walls of his mortuary temple. The question of the homeland of these various peoples has entertained scholars for a long time. The European origin of some of them – particularly the Aegean origin of the Philistines – was believed to have in Crete. This is based mainly on Biblical references and on a few correlations between Caphtorian-Cretan-Philistine linguistic similarities and certain words that infiltrated the Hebrew language.[25] Recent excavations at sites such as Ashdod, Ashqelon, and 'Eqron and the discovery of a characteristic monochrome painted pottery at these sites in the coastal and some inland regions of Canaan have left little doubt that they actually did come from the Aegean-Mycenaean cultural sphere, which included the islands of Crete and Cyprus. Provenience tests of the clay confirmed that this pottery was manufactured locally in the fashion of the Late Helladic/Mycenaean IIIC and close in style found in the last stage

---

[23] For a vivid description of how these monumental inscriptions were deciphered, see T. Dothan and M. Dothan, *People of the Sea, The Search for the Philistines*, Macmillan, New York, pp. 13–28 (1992).

[24] J.A. Wilson in J.B. Pritchard, (ed.), op. cit. pp. 185–187 (1973).

[25] C.H. Gordon, "Cultural and Religious Life" in: *The World History of the Jewish People*, vol. 3, ed. B. Mazar, Massada Publishing Co., Tel-Aviv. pp. 52–65 (1971).

of the Mycenaean culture on the Greek mainland, the Aegean islands, and the coast of Anatolia. These wares were found in Greece at sites destroyed in a wave of destruction of at least a dozen fortresses and palace complexes, and even abandoned cemeteries.[26] The devastation was connected with the dispersion of population within the Aegean region and affected the islands and peninsulas of the entire central and western Mediterranean Sea.[27]

Some scholars have maintained the romantic notion that the first cause of the upheavals in the Near East was the fall of Troy as told by Quintus of Smyrna.[28] It is more likely that this episode in itself may be not more than a distant memory of battles in the past and not necessarily a specific historical event. The destruction of Hissarlik/Troy is traditionally dated between 1200 and 1250 B.C.E. and roughly concurred with the fall of the major Mycenaean cities on the Greek mainland and might represent another aspect of the same phenomenon. The stratum of Troy VIIa shows clear signs of destruction by warfare and contains imported Mycenaean IIIC pottery quite similar to but distinct from that of the Philistine settlements in Palestine.[29]

There is no, and probably never will be, a consensus on the question of the original home of the different "Sea Peoples". All that can be said in the present context is that the Lukka, who participated in the first onslaught on Egypt from the west and allegedly were repulsed by Merenptah, were known as pirates and raiders since the 14th century B.C.E. The Hittite annals place them in southwestern Anatolia, where their name lives on in the classic province of Lycia.[30]

The best-known were the S(h)ardana, mentioned already in some of the Tell-el Amarna letters. They had served as mercenaries for the Egyptian garrison at Byblos around 1375 B.C.E. and later fought with the Egyptian army against the Hittites at Qadesh in Syria. Other groups continued in their traditional role as sea raiders. Thus, S(h)ardana ships fought with the 'Sea Peoples' in the marine attack against Ramses III's navy, while on land, there were other Sardana

---

[26] M.I. Finley, op. cit. p. 58 (1981).

[27] L.W. Taylour, *The Mycenaeans*, Thames and Hudson, London. pp. 15, 161 (1990).

[28] F.M. Combellack, *The War at Troy: What Homer didn't Tell.* Barnes and Noble, New York (1968).

[29] W.W. Hallo and W.K. Simpson, op. cit. pp. 117–120 (1998).
Whether Schliemann's excavations at Hissarlik were, indeed, the mythological ancient Troy has not been proven until now. A Roman period inscription and the recent discovery of a typical Anatolian seal with hieroglyphic inscription could confirm Ilion. The question whether Ilion and Troy are synonymous names for the same or two different sites is equally open.
M.I. Finley, in *Early Greece*, op. cit. surveying the stratigraphy of Hissarlik/Troy, notes that the first destruction, which separates layer VI from VII at around 1300 B.C.E. was due to an earthquake, but not to that related in Homer's *Iliad*. He also doubts that the destruction of VIIa (between 1200 and 1250 B.C.E.) was the Homeric war, because this destruction was by the Achaeians, pre-Dorian Greek people from the mainland. The Achaeians themselves were at that time under attack by a foreign force. Yet, the logical thinking of a contemporary scholar may not be in accord with the folly of history or the fancy of mythology where the bards saw in the destruction of Troy a sin against the gods which brought upon them the later devastation.

[30] N.K. Sandars, "The Sea Peoples" in: *The Penguin Encyclopedia of Ancient Civilization*, A. Cotterel (ed.). Penguin, London. pp. 44–46 (1980).
In ancient Greek, *Lycia* would be pronounced Lukeia.

contingencies that fought with the Egyptian army.[31] They were distinguished by their helmets, which were adorned with two horns. Similar helmets, however, are known from all parts of the Mediterranean Sea, such as some of the warriors depicted on the "Warrior Vase" found by Schlieman in the acropolis at Mycanae and could belong to artistic traditions still found much later in depicting Viking warriors of the early Middle Ages who actually never wore them. This invites a whole string of hypothetical questions, such as whether these warriors were locals or foreigners or even Shardana mercenaries. The vase was found in association with Mycenaean IIIC pottery, mentioned earlier in connection with the Philistines. The Phoenician inscription from Nora on the south coast of Sardinia and dated to the 9th or 8th c. B.C.E. mentiones the letter Sh-R-D-N, obviously meaning 'Sardana' or Sardinia. Whether Sardinia was the original homeland of the Sardana, or a colony founded *en route* like their colonies in Canaan, is still undecided. The uneasy thought about the authenticity of the inscription is itself a problem.

In Sardinia itself, the cultural changes during that time – i.e. the transition from the local Nuraghic II to III – do not indicate important changes. The existing settlements gradually increase in size. The *nuraghes* (central round towers with two or more superimposed corbelled chambers and an uppermost parapet for lookout and defense) became monumental defensive structures serving most likely as sanctuaries with the accompanying rituals as well as the seats of some local warlord. The development of the nuraghes is taken to indicate increased internecine warfare due to population pressure, which may have encouraged emigration but does not exclude invasions from abroad. Except for a few trade items and some pottery shapes found also on the opposing coasts of northern Africa, very little in the archaeological material of the Nuraghic III culture indicates elements introduced from abroad, or showing any signs of immigration nor emigration. We are tempted to assume that the Sardana were actually Sardinians who came from the island of Sardinia itself as well as from the near-by coast of northern Africa with which they shared certain cultural elements.[32] The same applies to the Sekeles or Tjeker who may have been related to the Siculians of antiquity and might, therefore, have come from Sicily. It is, however, equally possible that these islands served only as waystations during a mythic long journey.

About other nations, such as the Teres, Teswes or Tursa, linked perhaps with the Etruscans, or the Denyen, may be the Danuna or Dannoi, we know next

---

[31] T. and M. Dothan, op. cit. pp. 213–214, (1992).

[32] M. Zohar suggests a northern African origin for some of the Sardana and the Sikula, and probably for the Etruscans, too. An influx of northern African cultural elements can be detected on all the islands facing the continent throughout the last centuries of the second millennium B.C.E., but not in Sicily or the Italian mainland where the immigrants probably arrived in small numbers after a sojourn in countries such as Anatolia (mentioned by Herodotus and confirmed by the inscription of Lemnos), Cyprus or the Levant. The Sardana were serving in Byblos as part of the Egyptian army. The term "Phoenician" can only be used in an Iron Age context. The Late Bronze Age population of the Levant was called Canaanite. M. Zohar suggests calling "Phoenicians" only the new people who emerged from the subsequent merger of the immigrant Sea Peoples with the autochthonous Canaanites.

to nothing. Herodotus said that Anatolia was their homeland – and he was probably right, at least for the time being.

The "Sea Peoples" are described invading both by sea and land along the eastern coast of the Mediterranean – seizing control of bridgeheads and attacking cities that were not fortified or could not withstand a siege. There is evidence that the newcomers settled in some places, becoming the rulers. This was generally the case along the southeastern coast after the 'defeat' of the "Sea Peoples" by Ramses III.

An unsolved question concerns the coasts of the Levant in the territory known as Phoenicia. Archaeological information is lacking, largely because the ancient cities of Tyre, Sidon, Beirut and other Phoenician centers are covered by the modern cities. These sites cannot be dug, leaving us in the dark as to whether or not these cities were destroyed during this period. Until now, no destruction horizon of this date has been found at Byblos, which was excavated. Analysis of Ramses III's inscriptions at Medinat Habu led Bikai[33] to conclude that the invaders and the local Canaanites formed a coalition, which had begun as early as the Late Bronze Age, i.e. mid 15th century B.C.E.[34] Indeed, Ramses III's story contains no reference to conquest or destruction of the city-states along the Levantine coast by the 'Sea Peoples'. These cities, especially Byblos, as well as Tyre and Sidon, were for many years, important Egyptian commercial outposts ruled by Egypt either directly or through a local loyal king. Ramses II kept the Lebanese coast within the Egyptian sphere of influence in his treaty with the Hittite king Muwattalis.

Ramses III claimed that he stopped the hordes of the invaders at Djahi (usually meaning the Phoenician coast down to Palestine).[35] This still leaves the city-states on the part of the Lebanese coast that was overtaken by the 'Sea Peoples'. It is difficult to imagine that they sidestepped these strongholds on the way south. This would be possible as they marched along the Syrian coast where the plain may extend to a width of about 20 km. Once at the Lebanese coast, however, they could not have continued, as the coast is not more than a few kilometers wide and, in some places, the mountains reach the sea. This narrow strip is flanked on the east by high mountains (2000 to 3000 meters) dissected by narrow gorges of streams. The streams do not cut through the backbone of the mountain ridges, and thus do not provide passage to the east for circumnavigating the coastal strip. Therefore, the way south could be blocked by city-states situated at natural harbors along the coast, whose navies were probably a match for that of the "Sea Peoples". The attackers would not have been able to proceed unless they conquered the city-states, or circumnavigated them by the sea, or arrived at an agreement with the local people, even having them join their confederation.

---

[33] P.M. Bikai, "The Phoenicians" *The Crisis Years*, pp. 132–141 (1992).

[34] M. Yon, "The End of the Kingdom of Ugarit" *The Crisis Years*, pp. 111–122 (1992). The letters were found during the 18th and 19th season, 1962, by C.F.A. Schaeffer, subsequently published in Ugaritica IV and Ugaritica V.

[35] J.B. Pritchard (ed.) op. cit. p. 186 (1973).

Archaeological excavations all along the coastal plain of Palestine – from Acre in the north to Gaza in the south – seemingly confirm that the "Sea Peoples" had settled there eventually and established city-states. The same may be concluded from the "Onomasticon of Amenope" [36] and from an 11th-century B.C.E. report of the Egyptian Wen-Amon, who was sent to Byblos to buy lumber for the ceremonial barge of Amon. He stopped over at Dor, a port in the middle road between Jaffa and Acre, which was already a land of the Tjeker, one of the tribes that participated in the onslaught and was beaten back.[37] It is thus not logical that Ramses III defeated the 'Sea Peoples' north of Byblos and then allowed them to move southward and settle down. The settling of the 'Sea Peoples' in Canaan was probably a propaganda trick for internal consumption to turn defeat into victory. More likely, the invaders managed to get a few permanent footholds along the coastal plain of the Levant, and then, joining forces with the local Canaanite population, started to "trek" southward towards the fertile delta of the Nile.

Whether or not this scenario is true, it seems that the invaders also secured a bridgehead at Acre. The city has a small natural harbor, very fit for small ships, and was for many years important in securing Egyptian hegemony over the sea routes of the eastern Mediterranean. The Egyptians tried to dominate it, as early as the 19th century B.C.E., and when the local king was reluctant to abide, an "Execration Text" put him under a spell, which did not help the cause. Thutmosis III subdued it and made it an Egyptian post. When Egypt's grip on the Levant weakened in the 14th century B.C.E., Acre was mentioned in the el-Amarna letters as participating in the quarrels between the Canaanite city-states reported to the Egyptian court. The port was re-conquered by Sethi I and Ramses II. Excavations at the tell overlooking the bay unearthed a settlement of 'Sea Peoples'. It contained workshops where Mycenaean III C:1b pottery was found, and nearby was a small portable altar decorated with inceptions of boats similar to what was found in Cyprus. This, together with information from the Onomasticon of Amenope, placed the Sardana in the northern part of the coastal plain of Canaan, leading T. and M. Dothan to conclude that the plain of Acre was the settlement territory of this tribe.[38]

Hallo and Simpson[39] created a scenario, which seems to fit the scanty information we have. The chain of events began with the onslaught of the 'Sea Peoples', which had devastated Cilicia. The inhabitants, mostly Hurrians, fled to the northe across the Taurus, settled in the province of Cappadocia, and thus contributed to the disintegration of the Hittite empire. The Hittite nobility and their retainers living in that region possibly moved southeast into Syria and northern Mesopotamia, where they joined other Luwian and Hittite

---

[36] T. and M. Dothan op. cit. p. 209–219 (1992).

A. Zertal and A. Romano, "El-Ahwat 1993–1996" in *Hadashot Arkheologiyot* 110:32–34 (1999). The authors claim to have found a short-lived settlement that shows relationships in its architecture with ealier Nuraghic villages on Sardinia. Only further excavation can substantiate this claim and the presence of alleged similar sites south of Mount Carmel range.

[37] J.B. Pritchard (ed.) op. cit. pp. 16–24 (1973).

[38] T. and M. Dothan op. cit., pp. 212–214 (1992).

[39] W.W. Hallo and W.K. Simpson, op. cit. p. 120 (1998).

aristocratic families and founded the Neo-Hittite states on the debris of the fallen kingdom of Mitanni and the Hittite empire. The Hurrians, blocked by the Assyrians on the east and invaded by the Aramaeans from the south, appear to have moved northeastward into the mountainous regions around Lake Van. There, they joined other Hurrian-speaking populations and established the mysterious kingdom of Hurri, and perhaps the forerunner of the later kingdoms of Urartu and Armenia.

7.4
A Glance at Ugarit

Trying to understand the chain of events in Eurasia is based on indirect archaeological evidence, except for a few Mycenean Linear B tablets containing economic texts or donation lists, which shed very little light on historical events. The Near East provides a more detailed picture of events due to at least a few written documents as well as archaeological data. The cachets of clay tablets of the city-kingdom of Ugarit, situated on the coast of northern Syria facing the island of Cyprus, has become our prime source of information.

The ancient tell (locally known as *Ras Shamra*, i.e. the Cape of Fennel) is near a natural harbor that has been inhabited since the Neolithic Period.[40] It flourished during the Chalcolithic Period and Early Bronze Age and became an important port connecting Mesopotamia with the eastern Mediterranean islands, especially Cyprus and Crete. At the end of Early Bronze, around 2200 B.C.E., this city was destroyed like all the others. It was soon re-occupied by new settlers, most speaking a Western Semitic language and with a constantly growing sector speaking Hurrian. Soon afterwards, during the Middle and Late Bronze Age, Ugarit became a major commercial port of the eastern Mediterranean. Its locations near the Hittite and Hurrian kingdoms on the one hand, Cyprus and the Levant under the rule of Egypt on the other, guaranteed that each power tried to dominate the city. Ugarit submitted to Egypt's undisputed rule in the 15th and 14th century B.C.E. under the 18th Dynasty, and an Egyptian garrison was placed in the city. During this time, the city reached its economic and cultural acme. When Suppiluliuma incorporated Syria into his rising empire, Ugarit naturally became a vassal state of the Hittites. It tried to free itself from the Hittites, whose rule was stricter than that of the Egyptians as revealed in letters written by the king of Ugarit to the Egyptian pharaoh and found in the Amarna archive. Because the last kings of the 18th Dynasty became increasingly powerless, Ugarit remained under Hittite rule. When Ramses II tried to restore the rule of Egypt and challenged the Hittites, the king of Ugarit was forced to fight on the side of the Hittite king. The ensuing peace treaty between Ramses II and Muwatalis brought a renewed "golden age" to Ugarit.

---

[40] Excavated since 1928 by French archaeologists under C.F.A. Schaeffer; one of the most important sites, from all aspects. During World War II, Schaeffer wrote the first overall documentation of Near Eastern archaeology, *La Stratigraphie Comparée* (Oxford, 1948), which remains unsurpassed to this day in its scope and depth.

During the periods of economic prosperity two large temples were built, one for Ba'al and the other for Dagon. Between them was another building, which served as a library and a school for priests. In its ruins were found clay tablets with writing, most in an alphabetic script using cuneiform signs without vowels. When deciphered, the Ugaritic language was found to be closely related to later Phoenician and Hebrew. Another, yet not deciphered, writing system was found on differently shaped clay tablets, a syllabary known as Cypro-Minoan, and probably used by the resident Cypriot traders. As the marine bridge between the Aegean world and the Near East, Ugarit probably enabled Mycenaean traders and craftsmen to reside beside the Semitic and Hurrian inhabitants, even though no Greek or other Indo-European names were ever found in any of the deciphered documents from the city. The Ugaritic, Akkadian and Hurrian writings included many religious epic tales. In the time of *pax aegyptiaca*, soon after the Egyptian and the Hittite empires entered into a treaty, Ugarit enjoyed a period of prosperity with urban development and a putative migration of villagers to the city. Whether this prosperity was also due to the improved climate, which led to a higher quality of live and a higher birth rate, is a matter of speculation.

By 1200 B.C.E. Ugarit was the center of a commercial empire. Its kings seemed more interested in increasing their wealth than in promoting the prowess of their army. When faced with danger or asked to provide their allies or the Hittite suzerain with soldiers to help fight their wars, they preferred to pay ransom rather than endanger their soldiers. Excavations at Ugarit also reveal that they were not diligent in maintaining the city's walls. Thus, when "the enemy" attacked, Ugarit could not withstand the assault. The urgency and despair of the sudden disaster are apparent in the letters "from the kiln" found in a courtyard of the so-called small palace:[41]

RSL 1 (ex Ugaritica V, 383): Such says the king to Ammurapi, king of Ugarit, as fol bws: Peace be upon you. The gods may guard you in peace! What you have written that enemy ships on the high sea were seen, if this is true that they were really seen, be very strong! Now, you, your soldiers and chariots, where are they? Are they not with you? From the west the enemy will attack you. Surround your cities with walls! Soldiers and chariots bring inside! Wait for the enemy! Be strong, verily!

RS 20.238: To the king of Alasia, my father: Such says the king of Ugarit, your son: I bow down to the feet of my father. Peace be on my father! To your house, your wives, your soldiers, to everything, which is to the king of Alasia, my father, verily,verily, be peace! My father, now - The ships of the enemy arrived! My cities with fire he already burnt down and horrible things he did in the land. Does my father not know that all the soldiers of the lord, my father, are in the land of Hatti, and that all my ships are in the land of Lukka? Until now they did not arrive and the land lies open! My father, this matter you should know! Now seven ships of the enemy have arrived and terrible things they have done to us. If there are more ships of the enemy, please, send me information so I might know.

---

[41] M. Yon, op. cit. in *The Crisis Years*, pp. 111–122.

These letters were probably copies of the correspondence for filing and were never retrieved, fired quite possibly during the very same disaster, which Ugarit befell.

In another letter, unfortunately badly broken, the Hittite emperor wrote to Ammurapi about the food shortage and the disasters that had befallen him, ending with: *"The enemy has risen against me and their numbers cannot be counted whereas our numbers are small (?) Whatever can be found out let me know!"* And in Ras Ibn-Hani, a port across the bay, another letter by the Hittite emperor says: *"I, the Sun, have given the order about Lunadusu whom the Sikalayu had captured, those who live on ships... About the matter of the Sikilayu, I will ask him!"*

The archaeological excavations show that Ugarit did not suffer a general massacre when it was conquered, but was "only" plundered and burned. As good merchants, most people probably bought their way out and escaped to the neighboring hills, never returning to rebuild the city. Was it because "the enemy" remained camped in the area before moving southward along the shore? As the last letter between the Hittite emperor and Ammurapi shows, the danger came from the Sikalayu (Siculians?) *"who live on boats"*. M. Yon suggests Ugarit fell between 1195 and 1185 B.C.E.[42] A letter from an Egyptian vizier to Ammurapi also provides dating evidence. As well as archaeological evidence from neighboring Tell Achana, the ancient Alalah, and Meskene the ancient Emar on the bend of the Euphrates, closest to the Mediterranean. Both cities were destroyed in huge conflagrations dated to the same time.

From Ramses' III inscription and archaeological finds at Cyprus,[43] it appears that Cyprus was also conquered during the same time. Coming from the west, most probably Crete, the raiders first occupied the island, using it as a steppingstone to their bridgeheads on the mainland against the Hittites and Ugarit. From these bridgeheads, they traveled along the coastal plain of the Mediterranean until they reached Egypt. Comparison of the timeframe of the destruction of Ugarit and the attack on Egypt shows that the latter was a consequence of the former and followed within just a short time, or even immediately. The attack, according to depictions and the inscriptions, was by land and sea. The raiders on land came with wagons, women and children. Thus, Ugarit may not have been resettled because the raiders used it, or its vicinity, as their base before proceeding south. Indeed, in Ramses' III inscription he says, *"A camp [was set up] in one place in Amor"*. Amurru was a small kingdom mentioned in documents from Ugarit, and was probably located close to the city.

The lack of Mycenaean IIIC pottery in the ruins of Ugarit,[44] and the failure of the former inhabitants to return to the city may mean that the invaders were staying not too far away. Indeed, at Ras Ibn Hani – a stronghold protecting a harbor about 5 km southwest of Ugarit – Myceenaean IIIC was found in the layers overlying the ruins of the city, which was conquered before the

---

[42] M. Yon, Ibid.

[43] V. Karageorghis, "The Crisis years: Cyprus" in *The Crisis Years*, pp. 79–86.

[44] M. Yon, Ibid.

capital. Summarizing the inventory of pottery in the post-conquest layers at Ras Ibn Hani, which included locally made pottery as well as Mycenaean IIIC, A. Caubet agrees that a faction of 'Sea Peoples' settled it. She raises the question, however, of whether the new settlers were not actually the local population, who had returned after the troubles were over. Yon observes that alterations in the ruins of the capital indicate visits of plunderers who camped there days or months after the destruction, or shepherds squatting in the deserted city.[45] Caubet then asks, "*Where, then, were Ugarit's inhabitants?*" and suggests they probably found refuge in the mountain and plateau villages.

The fact that certain villages preserved their ancient names over the centuries leads one to believe that the local rural population managed to survive due to their inland location. Or, they and others might easily have formed the treks of refugees going south. The present authors suggest that it was the general policy of every conqueror to attack and plunder the urban centers, while leaving the rural population to its own in order to provide food and services for the new masters – the 'Sea Peoples' were no exception. The newcomers, some probably without their women, eventually merged with the local populations, adopted their religion and their language and formed a new people, known by the Greek name of Phoenicians.

7.5
The Aramaeans and the 'Wandering of the Israelites'

The tale of the Exodus of the 'Children of Israel' from Egypt and their alleged wanderings in the Sinai is another piece in the puzzle of the great migrations. At the same time it is an additional opportunity to examine the basic theme of the present book – the interrelation between climate change, and the natural environment compared with the ancient sagas and the Biblical text.

The origin of the Aramaic speaking shepherd tribes in the Syrian desert seem to be less shrouded in mystery than those we have mentioned.

They were with little doubt the direct descendants of the Amorite tribes of the second millennium B.C.E. mentioned above who had maintained their ancestral life-style of herding sheep and goats. By now, they had acquired the newly domesticated camel, which had allowed them, in time, to spread and to dominate the inner flank of the Fertile Crescent. Their first mentioning of the *ahlame aramaya* as enemies threatening the peace came from Tiglath-Pileser I (ca. 1115–1075 B.C.E.). Their emergence and their growing numbers was obviously related with the climate becoming more humid turning the margins of the Syrian desert into arable land and lush pastures supporting a growing rural population. The inhabitants of the neo-Hittite states became increasingly Aramaic-speaking, and so their army. The rulers of these states, stretching over Syria and northern Mesopotamia, have more and more Semitic names. The most famous of these states were Bit Adini, Guzana, Pattina, Zova, and Damascus.

[45] A. Caubet, "Reoccupation of the Syrian Coast After the Destruction of the 'Crisis Years'" in *The Crisis Years*, pp. 123–131.

There is little doubt that some of the earliest Hebrews were a subgroup of one of the many Aramaean tribes. This is reluctantly reflected in the patriarchal narratives by the aversion of the patriarchs to marry their sons to local 'Canaanite' girls and bringing the wives of Isaac and Jacob from Aram-Padan in northern Mesopotamia, the heartland of the Aramaeans. Furthermore, the opening prayer of the Jewish New Year begins with the words: "My Father was a Wandering Aramaean". Considering the later enmity between the two kindred nations this is a remarkable piece of evidence of traditional memories. Remains the question how and when the Hebrew language, a Semitic dialect nearly identical to Phoenician and Moabite and derived from a local Canaanite language, replaced Aramaic as the language of the new Israelite nation.

The exodus story of a people escaping oppression was not unique in the history of the ancient Near East. However, in our case it became the central theme around which other topics were woven. It is an allegory nourished by traditions that may reach back to some of the earliest memories of an "out of Africa" that are lost in prehistory. It may well have conserved memories of the migration into Asia of the first nomadic pastoral tribes speaking proto-Semitic languages, and as such, was shared by other Semitic peoples whose "teachings" were lost (see the recent discovery of the 'torah' of Balaam at Der 'Alla).[46] It certainly retained the memories of "exodi" of other untold families or even tribes throughout the Bronze Ages during appropriate climatic periods. It was transformed during the exilic period into a national epic preparing the remnant of Israel for its 'Second Exodus' from Babylon to the Land of Israel and reminding the Jews of the power of their God who "with an outstretched arm" protected his people then as now – in other words, the Bible is a book which is not history but has made history since then.

As expected, a survey of archaeologists' and historians' writings reveals a tremendous variety of opinions regarding the Biblical stories of the Exodus, the wanderings of the 'Children of Israel', and the conquest of Canaan. At one extreme are the Biblical archaeologists of the classical school, such as Albright, Glueck and Aharoni, and historians such as Malamat, who claim that there is a sound historical basis to that story. At the other are those who deny any historicity and claim that they are pious fabrications and speculations created in the second half of the first millennium B.C.E. when Judaeans returning from exile in Mesopotamia had to re-establish their national identity as Jews and claim the land their own. Most recent scholars, such as Redford's conclusions are between the two extremes. He maintains that only the vague reminiscence of a late tradition of the Hyksos descent into Egypt and its occupation fits the whole fabric of the book of Exodus. All the other narratives, including Jacob's descent into Egypt with his family, his son Joseph's appointment to a high post in the ruling hierarchy, the prosperity and proliferation of their descendants, and their enslavement and exodus, lack any documentary support and were probably adopted as literary topoi from Egyptian novels. Although Redford rejects the extremist socio-economic view that there was no descent of proto-Israelites into Egypt, but only a Canaanite "peasants' revolt", he maintains that

---

[46] J.A. Hacket, "Balaam" *ABD* I: 569–572 (1992).

the historical facts are totally different. According to his model, the Israelites originated from the "Shasu" pastoral nomadic tribes who, according to Egyptian sources, occupied the highlands of Moab and Edom. From the time of the Middle Bronze Age, these people used to move on a seasonal basis into other parts of the Levant, from northern Syria down to Egypt. During the 13th and 12th centuries B.C.E., they crossed the Negev and Sinai, and from there entered and settled in Egypt. The Egyptians reacted aggressively to the Shasu when the hostile nomads endangered their trans-Levant supply routes. Such were Seti I'st and Ramses II's military expeditions against these people in Sinai and as far as northern Palestine and Transjordan. Around 1233 B.C.E., Mernaptah II campaigned against them in the highlands of Ephraim, the 'Land of Israel'.[47]

M. Bietak has a somewhat different approach. He maintains that although the historical truth behind the story of Exodus can never be ascertained, it is possible to explore how the traditions transformed the historical events. Progress in answering this question will, indirectly, help find the historical truth. Bietak also believes that the stories of Exodus were edited and put in final form during the post-exile period. Yet, analyzing the possible routes from Egypt that avoid the *Via Maris* and its armed check-posts (i.e. "the way of the land of the Philistines" Exodus 16:17), he finds few points of conformity between sites mentioned in the Bible and those known from Egyptian sources. One such route is Pithom in Wadi Tumilat, which leads from the Delta to the Sinai Peninsula. He agrees that the Shasu were part of the proto-Israelites, especially so when he states that *"a fragmentary list of toponyms of Shasu Bedouin including a toponym with the T'-s'sw Yhw, which is not unlikely to be the earliest written record about a shrine of Yahweh"*.[48] This would suggest that during the reign of Amenhotep III in the first half of the 14th century B.C.E., a group of nomads worshiping Yahweh are documented by the Egyptians. This would fit the tradition of these proto-Israelites building Pithom and Ramses, as well as of an Exodus that must have occurred before Merneptah documented on his stela his war against Israel at around 1233 B.C.E. The usual fly in the ointment is that Pithom, i.e. Tell Maskhuta, shows no signs of any occupation between the Old Kingdom and the 6th to 5th century B.C.E. but was an important Egyptian border town at the time when the biblical text was put in writing.

The view of these two scholars represents a consensus acceptable to the present authors that the Biblical traditions vaguely and distantly echo some historical events mentioned in Egyptian texts which might have served as building stones of a grand epic blending contemporaneous geographical and ideological element for later scribes.[49]

---

[47] D.B. Redford, op. cit., pp. 408–429 (1993).
  D.B. Redford. "Aegyptological Perspective on the Exodus Narrative" in: *Egypt, Israel Sinai, Archaeological and Historical Relationships in the Biblical Period*, A.F. Rainey (ed.). Tel Aviv University, Tel Aviv, pp. 137–162 (1987).

[48] M. Bietak, "Comments on 'Exodus'" in *Egypt, Israel Sinai*, pp. 163–171. Ibid. 169, quoting R. Giveon, *Les bédouins de Shosu des documents égyptiens*, Leiden, (1971).

[49] See Papyrus Anastasi VI in the British Museum. *Select Papyri in the Hieratic Character from the Collections of the British Museum.*

7.6
The Formative Years of the Israelite Nation

We have seen in the last chapter, that the negative climatic change extending from the 16th to 14th centuries B.C.E. had caused economic distress and foreign rule aggravated by continuous raids of tribes, which might be identified with the Habiru-Shasu, driven by the deteriorating conditions in the desert fringes. They infiltrated the highland region with their herds and, because of their mobility and flexibility, were able to extract their living from abandoned fields and the natural forest by integrating herding and farming. This way of life could be intensified once the climate conditions improved. Their fighting ability, described in the Amarna letters, stood them in good stead when they were involved in the wars between the petty city-states. Thus, some local rulers, for instance the king of Shechem welcomed them, while others, such as Biridaya the king of Megiddo opposed their presence. Both formed coalitions with other city-states which the newcomers probably used to their advantage in subduing and gradually overtaking village by village, and one small town after other. The political struggles seem to have laid the ground for the later kingdoms, Israel in the north and Judaea in the south. Other tribes who had lived for a few generations on the border of the Euphrates or the Nile delta and had acquired a semi-sedentary agricultural lifestyle may have joined the bulk of the settlers. These may have included the lineages of Joseph and the Levites, who may have brought with them a reformed Yahawist version of the former Shadayist monolatric religion. All the tribes seem to have shared a cultural heritage of taboos, such as not eating pig meat. They probably also practiced circumcision, which is of African origin. This distinguished them from the Asiatic peoples from the north, such as the Hittites, Jebusites, and Horrites/Hurrians, and later from the Philistines.

As the right of inheritance to a specific territory or the notion of a 'Promised Land' is a common element in the mythologies of many peoples, it is well conceivable that within some Israelite family traditions memories of a divine promise to this land were preserved. When the climate improved even more, they farmed more and herded less, although they maintained their herds. They had more water sources and could produce food from smaller areas and from reconstructing and tilling the terraces, which earlier sedentary populations had abandoned. The improving climate enabled cultivation of grains and fruits, which could be processed into storable secondary products, such as wine, fig-cakes, and oil. The ratio of sheep and goats to large cattle increased in favor of the latter, as cattle could be used in tilling the terraces. The gradual shift to agro-industrial products seems to indicate the growth of a new tribal elite, which supplied the resources needed for expanding the terraces and converting greater areas of the forested hills into agricultural land. The mention about the Canaanites of Gezer who were allowed to stay once they had consented *"to serve under tribute"* (Joshua, 16:10) might be a hint that the former Canaanite population was drafted to do the heavy fieldwork and terracing. The expansion was peaceful in some places, such as Gezer (which was later destroyed by the Philistines and reoccupied by the Israelites), Gibeon, and Shechem. Others, including Lachish and Hazor, were won by war, as evidenced by the destruction layers of these cities and confirmed by the Biblical narratives.

Herzog attributed the process of settling of related tribes in the semi-arid plain of Beer Sheva to a climate change.[50] This valley, situated along the southern foothills of the Hebron Mountains at the northern margins of the Negev Desert, lies at present near the average annual rain-line of 200 mm, moving south in humid years and north in years of drought. Herzog observed that the pattern of urban settlements went through a series of constant ups and downs. Following the desertion of Early Bronze Age Arad around 2700/2600 B.C.E., there was a break in the pattern of permanent settlements for a period of some eight or nine centuries, coinciding with the warm and dry period that marked the transition from the Early to the Middle Bronze periods. A second gap – of 400 years this time – occurred between the end of the Middle Bronze and the end of the Late Bronze and again this was during a warm and dry period. From the start of the Iron Age around 1200 B.C.E., the archaeological survey shows a clear pattern of gradual establishment of permanent communities. Observations at the archaeological sites in the valley, along with the time series of proxy data discussed above, indicate that improved climatic conditions, rather than political reasons, were responsible for the early stages of settlement. Dendrochronological findings, as well as pollen found in the layers of ancient Beer-Sheva, do indicate that the climatic conditions at the end of the Late Bronze period and early Iron Age were more favorable than today.[51] The inhabitants of this valley re-established an agro-pastoral society the economy of which was based on field crops and herds. The ratio of cattle to sheep and goat bones found in the layers of the early Iron Age, also attest to an improved climate in this region. The introduction of cattle clearly indicates wetter climate conditions, not only because cattle need more water and pasture than sheep and goats do, but also because it indicates the possible use of draft animals to plough. Thus at Tell Masos, the percentage of cattle bones is 22% in a layer dated to the 12th century B.C.E.; rising to 31% in an 11th century B.C.E. layer; then declining to 20% in the 10th century and dwindling to 14.6% in the 6th to 7th centuries B.C.E.[52] There is an extraordinary conformity of these percentages with the paleo-climatic curve (Fig. 8).

The passage from a pastoral to an agricultural society can be seen in the nature and pattern of buildings at Tell Masos. From the end of the Late Bronze to the beginning the early Iron Age, i.e. from around 1250 to 1200 B.C.E., the

---

[50] Z. Herzog, "The Valley of Beer-Sheva, From Nomadism to Monarchy" in: *From Nomadism to Monarchy, Archaeological Aspects of Early Israel*, N. Na'Aman and I. Finkelstein (eds.), Yad Ben Zevi and The Israel Exploration Society, Jerusalem. pp. 215–242 (1990).

[51] N. Liphschitz, "Dendrochronological Studies in the Negev" in: *The Negev and its Past Climate, Settlement Patterns and Nomadism (Abstracts)* Ben Gurion University, Beer Sheva, p. 2 (1987).
A. Horowitz, "Pollen Analysis of an Early Iron Age Sample from Beer-Sheba" in *Beer-Sheba II*, Z. Herzog (ed.), Tel-Aviv. pp. 118–119 (1984).

[52] E. Tchernov and I. Drori, "Economic Patterns and Environmental Conditions at Hirbet-el-Mesas during the Early Iron Age" in: *Ergebnisse der Ausgrabungen der Hirbet Mesas (Tel Masos) 1972–1973*, V. Fritz and A. Kempinski (eds.), pp. 213–222 (1983).

Fig. 8. Percentage of cattle bones at Tell Masos (Herzog 1990) correlated to precipitation rates based on data from the Soreq Cave (Bar-Matthews et al., 1998)

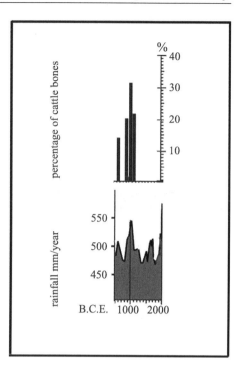

people lived in tents or sheds, as only floors and hearths were found, along with store pits pointing to the harvesting of field crops. The high water table in the gravel beds probably allowed these ancient dwellers to excavate shallow wells for watering their livestock and may even have been used to cultivate small vegetable gardens. Buildings characteristic of a more permanent settlement were built later in the early part of the 12th century, and included structures whose architectural features were typical of that used by Egyptian governors. This led Herzog to deduce that once the region enjoyed a more favorable climate and economic sustainability, the Egyptians became interested in controlling it.

Tell Beer-Sheva was settled a few decades later (Plate 10). The first evidence of occupation consisted of tent floors and sheds, as well as storage pits. In the next stage, dated to the 11th century B.C.E., remains of simple stone buildings were found and a deep well was dug. During the same time, the upper section of Arad was resettled in a similar pattern.

The trend changed in the 10th century. Some strategically favorable sites that had undergone the transformation from temporary hamlets to permanent buildings were now turned into administrative centers and military strongholds of the Judaean kingdom – especially Beer-Sheva and Arad. Water supply projects were constructed within the walls to ensure supply in times of siege. At Beer-Sheva the large shaft resembles the water projects of Megiddo and Hazor, which will be discussed later. At Arad the water system is based on the ancient well in the valley with water reservoirs built within the walled

Plate 10. Air Photo of Tell Beer Sheva, Negev (Courtesy I. Zaharoni R/Col/IDF)

citadel on top of the hill. The reservoirs were probably filled with water brought by donkeys and carriers from the well in the shallow valley below.

Herzog found that the favorable climate conditions of the 11th century B.C.E. had promoted the expansion of farming communities into the Negev's more arid highlands. He also attributes the desertion of these posts in the 10th century to the worsening of the climate, rather than to the anthropogenic causes suggested by Meshel and Cohen. Indeed, the Soreq Cave curve shows a steep decline in humidity at the end of the 10th century, but its impact in the Negev may have started earlier.

In the Beer-Sheva valley proper, the deteriorating climate caused many farmers to desert the permanent settlements and return to a more pastoral way of life. Part of the population was able to move to the administrative-military centers, engage in other types of livelihood and gain a measure of security, which had become a key concern. The worsening climatic conditions probably led to severe pressure from the desert tribes, such as the Amalekites (I Samuel 27:8), against whom defensive and offensive measures had to be applied. The Negev sites experienced rejuvenation in the 7th century when the climate became humid again, again seen on the Soreq Cave curve. There is little doubt that the success of the royal initiative of king Uzziah (Azaria) who *"built towers in the desert and digged many wells: for he had much cattle ..."* (II Chronicles 26:10) was due to the improved climate.

Herzog further suggests that the ethnic identity of the people inhabiting the Beer Sheva Valley during the early Iron Age can be regarded as Israelite – their ethnicity the outcome of an integration of various ethnic groups who settled there because of a better climate and, therefore, better living conditions. The traditions of each group influenced those of its neighbors, and *vice versa*, until they were all drawn together by the *maelstrom* of the new Israelites cultural dominance.

In this respect one can understand the inscriptions found at Kuntillet Ajrud in the Sinai bordering the Negev,[53] which express adoration of a variety of deities, from Yahweh and his consort Asherah to El and Ba'al of the Canaanites. This should not surprise anybody as the Bible itself contains many echoes of the El, Ba'al and Asherah being worshiped side by side with Yahweh. Moreover, "non-conventional religion" is typical of peripheral zones. The present authors, who both lived for a time with the bedouin "Moslems" of the Sinai, noted the shallowness of belief in the official religion of the establishment – the spirits rule their lives.

The weak cultural ties of those living in the border zone of the desert may have been a reason for maintaining local temples at Arad – (Plate 11) a small version of the temple in Jerusalem – and at Beer Sheva to represent the official cult of the royal house of David. These temples seem to have functioned from the 10th or 9th century until the end of the 7th century B.C.E. when king Josiah enforced his radical Deuteronomistic reform.

Plate 11. The Holy of the Holies of the Israelite temple at the citadel of Arad (Photo A.S. Issar)

---

[53] R. Cohen, *The Settlements in Har'HaNegev*, Ph.D. thesis. Hebrew University, Jerusalem (1986, Hebrew with English abstract).

Z. Meshel, "Did Yahweh Have a Consort? The New Inscriptions from Sinai" *BAR* 5/1:24–34 (1979).

W.G. Dever, *Recent Archaeological Discoveries and Biblical Research*, University of Washington Press, Seattle. pp. 134–155 (1990).

Changes in the distribution of settlements, due to climate change, are illustrated also in the highlands of Hebron stretching north of the Beer-Sheva valley. Even though, because of their altitude, these highlands receive much more precipitation than the southerly and easterly foothills of Judaea, their rough topography discouraged widespread settlement. The coastal plain further to the west, on the other hand, was heavily populated and well protected by a dense chain of fortified urban centers ruled by Egyptian overlords. Thus it was into the central and western highlands where the new people penetrated when the semi-arid zone to the east and south became too arid.

A. Ofer's survey of the highland region from Jerusalem in the north to the foothills descending to Beer-Sheva in the south[54] shows that the region began to be settled in the Early Bronze humid period. Evidence of permanent settlements is lacking during the very dry phases of the Intermediate Bronze period, except for seven cemeteries. We agree with Ofer that these cemeteries are evidence that more mobile pastoral nomads replaced the permanent population. Permanent settlements reappear during the Middle Bronze period, but the nomads remained in the region as well, as indicated by cemeteries located outside but in the vicinity of the settlements. In the warm and dry Late Bronze period, permanent settlements again disappeared, except for one called Devir, which was built as a fortified city with its wall encircling an area of 6 hectares. At that time, four big cemeteries continued to serve as burying places for the nomadic population.

With the beginning of the Iron Age, about ten new settlements, including Devir, appeared, representing a continuation of the Late Bronze settlement. Two of them were in places where only cemeteries had existed before. From the end of the 11th century to the end of Iron Age IIA, the number of settlements in the region increased significantly to about 17, extending into the semi-arid regions in the south and east. This area thrived and became densely populated during the next five centuries until the destruction of Jerusalem by Nebuchadnezzar and the end of the Judaean kingdom in 586 B.C.E. Unlike those of previous periods these settlements were not restricted to the close vicinity of springs and were able to expand because of the development of impervious coating for cisterns which could be filled by the sufficient quantity of rainfall during this time.

The same general pattern of increase and decrease in the number of permanent settlements was seen also in the highland region extending north from Jerusalem to Shechem known as "the land of Ephraim".[55] The number of settlements here is even more significant as the variety of geographical characteristics within the region enabled a complementary type of agriculture with either grains, vegetables, vineyards, olive plantations or livestock grown in each sub-region. About 87 permanent settlements were mapped during the

---

[54] A. Ofer, "The Judaean Hill Country – from Nomadism to a National Monarchy" in N. Na'Aman and I. Finkelstein (eds.) op. cit. pp. 155–215 (1990).

[55] I. Finkelstein, *The Archaeology of the Period of Settlement and Judges*, pp. 307–314 (1990).
I. Finkelstein "The Iron I in the Land of Ephraim – A Second Thought" in N. Na'Aman and I. Finkelstein (eds.) op. cit. pp. 101–131 (1990).

Middle Bronze II, i.e. 1800–1550 B.C.E. This number declined towards the end of the period, and the few that remained were heavily fortified. The Late Bronze saw an almost total desertion with only five settlements remaining in greatly reduced areas. Then, in the early Iron Age the number of smaller settlements increased abruptly reaching 115. During the centuries of the Israelite kingdom, they grew in number to 190 – a peak not matched again until the Roman period, which also was mostly a cold and humid period (Fig. 9).

Similar waves of settlement are also apparent in the hill country stretching from Shechem to the Valley of Jezreel in an even greater intensity than observed further south. Archaeological surveys show that during the Middle Bronze the center of the region was densely populated and dominated by the city-state of Shechem. This capital of the local kingdom was heavily fortified and played an important role in the politics of Canaan. There were about 135 settlements in the region, mainly situated near springs. At the beginning of the Late Bronze period, Shechem was destroyed twice, probably by the Egyptians. Later in the period the number of settlements declined to 33. At that time Shechem was rebuilt and was ruled by the Labayu dynasty, frequently mentioned in the Amarna letters. These letters also tell of the involvement of the Habiru tribes in the fights among the local city-states, which heralded the beginning of the Israelite take-over of Canaan. A new wave of settlements marked the

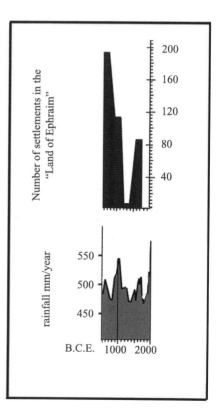

Fig. 9. Number of settlements in the "Land of Ephraim" (Finkelstein 1990) correlated with historical precipitation rates based on data from the Soreq Cave (Bar-Matthews et al., 1998)

next stage in the Iron Age, increasing the number from the 111 reached in the first part of the period. A. Zertal maintains that the new settlers were Israelites who settled peacefully beside as well as apart from the Canaanite cities, which had survived the Late Bronze crisis because of their location near the main springs. His conclusions about the ethnicity of the new settlers are based primarily on the new and characteristic type of pottery in the new settlements, and also on the discovery of a new cult place on Mount Ebal overlooking Shechem.[56] He holds that the archaeological findings generally conform to the Biblical story of the settlement of the Israelites in this region described in the book of Joshua. Further support comes from the lack of pig bones in the new settlements in the hill region. These bones are regularly found in excavations of contemporaneous settlements in the coastal plain, and they are also abundant in the earlier Bronze Age settlements in the hill region.[57] This seems to be a striking evidence of the occupation of the region by a new population whose taboo code was different from that of the aboriginal one. Most likely, the new population had been present in the region, moving along the margins of the agricultural area and sustaining itself mainly on herds and occasional on agriculture typically practiced by semi-nomads. At some point they began to settle, mainly in the hill country, and became increasingly sedentary, practicing more agriculture until it became predominantly an agricultural society with a strong urban commercial economic basis. Could the new population simply have been descendants of the previous inhabitants, who had been forced to abandon agriculture because of the aridity of the Late Bronze Age and then returned when the climate improved? This reasoning does not explain the new types of pottery nor the "pig bones anomaly" and does not agree with the tremendous amount of "proxy-proxy data" in the Biblical stories about this period, which does conform to the archaeological data.

All this does contradict the general conclusion that differing and often conflicting stories, including the traditional "one-drive-conquest," represent a synthesis compiled by scribes some centuries later and based on oral traditions. However, the Habiru-Shasu settlement hypothesis presents a problem in that the early Iron Age was probably a humid period. One would expect that pastoral people who had penetrated into sown land during a dry period would return to the green open lands of the desert once the climate improved. Instead, these people settled in a region that demanded a high investment of effort to turn it into an agricultural land, as will be discussed later. In examining these considerations, one cannot refrain from entertaining the notion that the social heritage of the new settlers included, along with the taboo of eating pig and other traditions, a legacy that this land was promised to their ancestors by their tribal god. Whether the drive to realize this promise come true may have been spurred by a fresh wave of kin families or clans that had lived for a while in

---

[56] A. Zertal, "In the Land of the Perizzites and of the Giants – The Israelite Settlement in the Hill Country of Manasseh" in N. Na'Aman and I. Finkelstein (eds.) op. cit. pp. 53–101 (1990).
   A. Zertal. "Has Joshua's Altar Been Found on Mt. Ebal?" *BAR* 11:42 (1985).

[57] I. Finkelstein, "The Great Transformation: The Conquest of the Highland Frontiers and the Rise of the Territorial States" in N. Na'Aman and I. Finkelstein (eds.) op. cit. pp. 365 (1990).

Egypt and acquired the habits of a sedentary agricultural population is another question.

The archaeological evidence and the pollen evidence from the Sea of Galilee suggest that clearing the natural forest enabled this settling initiative. Again, this process is echoed in the words of Joshua to the house of Joseph: *"But the mountain shall be thine; for it is wood, and thou shall cut it down"* (Joshua 17:18). A. Zertal emphasizes the significance of the water supply problem in settling the region of the Hills of Manasseh whose easterly part is semi-arid (less than 400 mm of rain per year).[58] He disagrees with W.F. Albright's suggestion that the Israelites were able to settle the region because of the invention of the impermeable plastered water cistern[59] as the archaeological survey found no such cisterns in most Israelite sites in this region of permeable limestone. Zertal concludes that the Israelites had to obtain their water by peaceful means from springs dominated by the Canaanites, and that they occupied the region in a process of quiet infiltration. He further suggests that the large earthenware pithoi, characteristic of the Israelite settlements and found in abundance there, was used to bring water from the springs by

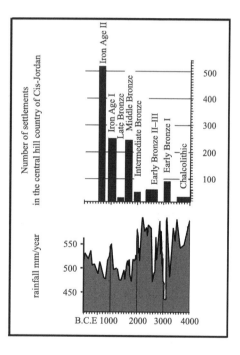

Fig. 10. Number of settlements in the central hill country of Cisjordan (Finkelstein 1998) correlated with historical precipitation rates based on data from the Soreq Cave (Bar-Matthews et al., 1998)

---

[58] A. Zertal, "The Factor of Water and its Influence on the Israelite Settling of the Hill Country of Manasseh" in: *Settlements, Population and Economy in Eretz Israel in Ancient Times*, S. Bunimovitz, M. Kochavi and A. Casher (eds.). Tel Aviv University, pp. 126–145 (1988).

[59] W.F. Albright, *From Stone Age to Christanity*. Doubleday, pp. 212, (1957).
    See also Y. Aharoni, "The Settlement of Canaan" in: *The World History of the Jewish People*, Vol. III: Judges, ed. B. Mazar, Massada Publishing Co., Tel Aviv. p. 98 (1971).

donkey. He thus emphasizes how the importance of ensuring their water supply influenced the manner and pace of the Israelite occupation of the mountainous regions of Canaan. Whereas in the more southern parts of Canaan the relative scarcity and weakness of the cities enabled the Israelites to gain control of the water resources, the Hills of Manasseh required a more gradual takeover. Here, the occupation progressed stage-by-stage, first with the peaceful acquisition of water rights from the Canaanites and later through military and political domination. This model does not consider the possibility that, because of the more humid climate from the 12th to 10th centuries B.C.E., there were many more small springs in this region of more and less permeable rocks. Such perched springs also appear in the present time during a more humid spell. This climate may explain Zertal's observation that the settling process started in the belt between the semi-arid eastern zone and the more humid central region, where the Canaanite towns survived near larger springs. For the new settlers, who adopted a rural dispersed pattern of communities, the quantities of water from the small springs were sufficient.[60] They may well have needed the pithoi to fetch the water from the springs, most often located at the bottom of a deep creek. Unlike Herzog, who attributed settlement of the semi-arid zone of southern Canaan to climate change, Zertal does not consider climate as a factor in the Israelites' expansion to the semi-arid zone of central Canaan.

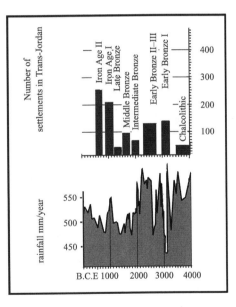

Fig. 11. Number of settlements in Trans-jordan (Finkelstein 1998) correlated with historical precipitation rates based on data from the Soreq Cave (Bar-Matthews et al., 1998)

---

[60] L. Marfoe concluded that a farmer in southern Lebanon just before the 1975 civil war needed a minimum of one cubic meter per year to survive! L. Marfoe, *Between Qadesh and Kumidi: a History of Frontier Settlments and Land Use in the Biqa', Lebanon*, unpublished dissertation, University of Chicago, (1978). ibid "The Integrative Transformation: Patterns of Sociopolitical Organization in Southern Syria" *BASOR* 234:1–42 (1980).

The role of climate in settling the Lower Galilee[61] is also ignored, even though it experienced the same settlement pattern. This omission is especially conspicuous in Finkelstein's[62] impressive, but anthropo-genically biased, analysis of the process of settlement of the entire highland region. This is best illustrated by overlaying Finkelstein's diagrams of settlement oscillations in the central hill country on the climate change diagram. The correlation between the precipitation levels and the high and low periods of settlement is notable (Figs. 10 and 11).

---

[61] Z. Gal, "The Iron I in the Lower Galilee and the Margins of the Jezreel Valley" in: N. Na'Aman and I. Finkelstein (eds.) op. cit. pp. 34–53 (1990).

[62] I. Finkelstein, "The Iron I in the Land of Ephraim- a second thought" in N. Na'Aman and I. Finkelstein (eds.) op. cit., pp. 101–130 (1990).
I. Finkelstein, "The Great Transformation: The Conquest of the Highland Frontiers and the Rise of the Territorial States" in T. Levy op. cit. pp. 349–368 (1995).

CHAPTER 8

The Age of Iron and Empires

From ca. 900 B.C.E. to ca. 800 C.E.

"To open the canal I sent an Ashipu priest and a Kalu priest ... and I presented gifts to Enbilulu Lord of Rivers and to Enemibal (Lord who digs canals) ,The sluice gate like a flail was forced inward and let in the waters in abundance. By the work of the engineer its gate had not been opened when the gods caused the water to dig a hole therein. At the mouth of the canal, which I had dug through the midst of Mouth Tas I fashioned six great stele with the images of the great gods and my royal image in a gesture of obedience. Every deed of my hands I wrought for the good of Nineveh I had engraved thereon. To the kings my sons I left it for the future."

[Sennacherib's inscription on the rock near the inlet of his water project to supply water to Nineveh]

"This is the story of the tunneling through: Whilst (the workers lifted) the pick each towards his fellow and whilst three cubits to (be) tunneled (there was heard) the voice of a man calling his fellow, for there was a way in the rock on the right hand and on (the left) :And on the day of the tunneling through, the workers struck each in the direction of his fellows, pick against pick, And the water started to flow from the source to the pool twelve hundred cubits. A hundred cubits was the height of the rock above the head of the workers."

[Rock inscription near the outlet of the Shiloah tunnel in Jerusalem].

The paleo-climatic proxy-data (Fig. 3a) show that after the spell of cold and humid climate around 1000 B.C.E. came a warm and dry period with a low peak of humidity around 850 B.C.E. Then it steadily improved reaching favorable conditions from ca. 300 B.C.E. until the 3rd century C.E. After a short interval of worsening from ca. 350 to 450, the climate improved again reaching a positive peak around 550 C.E. It was followed by a time of crisis reaching a climax around 900 C.E. The humidity curve from lake Van shows a generally drier period in this phase, but still the differences between the peaks of humidity and aridity are less extreme than those characterizing the previous millennia. The evolution of the technologies of water extraction, transport and utilization helped in mitigating the impact of average negative climate changes. However once the change was extreme as in the case of the last centuries of the first millennium C.E. a major socio-economic crisis followed and in its wake a major change in the political and religious pattern of the Near East.

In the following chapter we will survey the general pattern of the history of the region, highlighting from time to time the role of climate in deciding the scenarios, which developed.

Similar to the arsenic and antimony copper ores of the Chalcolithic and Early Bronze Age, high quality iron ores are found in abundance in the northern and northwestern regions of the Near East and beyond. The archaeological evidence indicates that iron production started in central and eastern Anatolia, where it was well known by the second half of the second millennium B.C.E., and from there it spread to the other parts of the ancient Near East. The oldest existing Near Eastern iron artifact is a dagger presented by a Hittite ruler, found in the tomb of Tutankhamon (1336–1326 B.C.E.) and now in the Cairo Museum. In addition, a letter from the Hittite emperor Hattusilis III (1275–1250 B.C.E.) to a neighbor, probably the king of Assyria, suggests that the Hittite empire had a monopoly on the manufacture of iron. In the letter, the emperor says he is postponing the transport of good iron requested by his neighbor, but compensates him with a gift of an iron dagger blade.[1] The collapse of the Hittite kingdom following the invasion of the 'Sea Peoples', and probably others from the north, led to the disintegration of this monopoly. With this, the skilled iron manufacturers most likely migrated to other parts of the Near East, carrying the Hittite and Luwian language, as well as Anatolian technologies, southwards to Syria and even Canaan.

We have seen already, that this migration may have been aided by the colder and more humid climate conditions in the rain-fed belt forming the inner flanks of the Fertile Crescent skirting the Syrian Desert. The prosperity had a positive impact on the growth of the sociopolitical entities of Assyria and the various neo-Hittite and Aramaean states. The city state of Assur was already a consolidated state with a long history. Equipped with new and mass produced iron weapons, Assyria began to gobble up the other kingdoms one by one, forming in due time the Assyrian Empire, which was to dominate the entire ancient Near East in the centuries to come.

Egypt began its final decline. The kings of the 20th, and last, dynasty of the New Kingdom all bore the name of Ramses, from III to XI, recalling the great king of old, but to no avail. Around 1080 B.C.E. the Third Intermediate Period began. Some of the kings of the 21st to the 25th Dynasties were of foreign origin – Libyans and Nubians – and others were local priest-kings. Later historiography streamlined various groups competing among themselves and ruling at the same time. The arts still flourished, and some of the large tombs on the West Bank of Thebes/Luxor were built during this period. But the economy declined due to a series of low Nile floods and political instability. Once again, a cold-humid climate spelled prosperity for the Near East, except for Egypt, which, fed by the Nile, depended on tropical and subtropical climate systems.

---

[1] O.R. Gurney, *The Hitties*, Penguin Books, London, p. 83 (1952).

The Assyrians ousted the Nubian kings of the 25th Dynasty. The revival of the 26th Dynasty, with its capital at Sais (thus, the Saitic Period) followed a short Assyrian occupation. Some of the most accomplished and sophisticated pieces of late Egyptian art were produced during this dynasty, under the kings Necho and Psammetich, from 670 B.C.E. until the Persian conquest in 525 B.C.E. Egypt remained under foreign rule, except for the short local rule of the 28th, 29th and 30th Dynasties from 400 to 350 B.C.E. The Persians then returned, followed by the Greeks, and thereafter, Egypt was to remain under foreign domination until the 20th century C.E. changing its language and twice its religion.

Anatolia, like Greece, entered a "Dark Age" lacking historical documentation; it was probably darker for us than for the local populace, who might have enjoyed a few centuries without kings and priests. With the establishment of the Greek city-states on the coast, the Phrygian and Lydian kingdoms in the center and Urartu in the east from the 9th or 8th century B.C.E. on, new political entities and civilizations were born that owed remarkably little to those of the Bronze Age.

8.1
The Aramaeans Occupy Center Stage

It should be underscored that so far there is no external historical evidence for events taking place during the late 11th, 10th, and early 9th c. B.C.E. in the southern Levant. We, therefore, have to rely only on the biblical traditions written down half a millennium later with all what this implies. The archaeological sequence seems to have been established based on the development of local pottery types common already in the preceding Late Bronze Age and continuing into the Iron Age without a major break. During the brief period of the United Monarchy, the Israelite kingdom became a regional power underscored by the legendary king David's campaigns against his neighbors including the Aramaeans in Syria. The rise and fall of the Aramaeans paralleled and was interwoven with that of the Israelite kingdoms. We have seen already that the benign climate favored the formation of these independent kingdoms, especially among populations concentrated around the springs emerging from the Syrian-Taurus-Zagros Mountains' arch. Thus, at the end of the second millennium B.C.E., this string of Aramaean states was stretching from the borders of Assyria and Babylon in the East to Phoenicia and the Israelite kingdoms in the west and south. Although these states shared the same language and religion they never united to form a single political entity to rival the other states of the Near East. At certain periods, however, they did form a kind of confederation headed by one of their states. At the beginning of the 10th century B.C.E., Hadad-Ezer, the king of Aram-Zova, located in the Beqa between the Lebanon and Anti-Lebanon, headed one such confederation, which fought and defeated the Assyrians. The Aramaeans could have obliterated Assyria as an independent state, but Hadad-Ezer retreated, probably under pressure from David's armies on the southwestern borders of Aramaean kingdom. Thus, the first successes of the Israelites sowed the seeds of their destruction generations later.

During the 9th and 8th centuries B.C.E., Aram-Damascus was the leading Aramaean kingdom and was, for most of that time, the chief rival of the northern Israelite kingdom. Aram-Damascus gained independence from Israeli sovereignty during King Solomon's reign (Samuel II, 8:5) and later expanded in all directions. Its capital, Damascus, rivaled the Israelite capitals of Jerusalem and Samaria. It is situated in an oasis created mainly by the two rivers of Barada and Awaj (Biblical Amana and Parpar), which emerge from the highly permeable limestone rocks of the Anti-Lebanon mountain range. The high altitude of these mountains guarantees the perennially flow of the springs, and thus the water supply of the oasis. During more humid periods, the rivers maintain a high flow for a longer period of the year, due to storage of water in the snows on the mountains and in the aquifers. The abundance of water in such times enabled the area of irrigation to extend into the desert.

Yet Damascus owed its prosperity not only to extended agricultural areas, but also to its location on the desert border in the belt of semi-arid areas with an annual precipitation of 200 to 400 mm. During the more humid periods the city was the market for the rain-fed agricultural region, whose economy was based on grain and livestock. In addition, because it was situated on the main caravan road connecting Arabia with the other parts of the Near East, its markets, and later bazaars, became a pivotal point of transit. Thus, it is not surprising that at the beginning of the 1st millennium B.C.E., when, according to the biblical tradition, Jerusalem became a flourishing capital, once the Aramaeans gained their independence, Damascus followed suit. It became the capital of a strong kingdom that in a rather short time became the major rival of the Israelites kingdoms. The wars that followed weakened all parties involved. In the meantime Assyria gathered force and organized a highly efficient army, with which it put the entire Near East under its dominion.

8.2
The Empires of Assyria and Babylon – A Brief and Brutal Performance

After the death of Tiglath-Pileser I, the Aramaeans began to strike back. During the reign of Tiglath-Pileser I's grandson (966–935 B.C.E.), who bore his grandfather's name, the Aramaeans, headed by Hadad-Ezar, the king of Aram-Zova, crossed the Euphrates and reached the Tigris. They might have subdued Assyria completely, but, for some unknown reason, they did not complete their conquest, thus enabling Assyria to survive and later turn the tide. One possible reason for the Aramaean halt may have been King David's attack on them from the south and west after he had consolidated his kingship over the tribes of Israel.

Assyria's renascent began towards the end of the 10th century with the persistent campaigns of the Assyrian king Ashur-Dan II (934–912 B.C.E.) against the Aramaeans, whose easternmost kingdoms encroached on the Assyrian border. His heirs continued these campaigns and subdued not only most of the Aramaean kingdoms but also large tracts of land in the northern highlands.

Shalmaneser III (858–824 B.C.E.) expanded the range of conquest southward to the Israelite kingdoms but was halted in 853 B.C.E. at Qarqar by a coalition led by Ahab, king of Israel. He had more than 2000 chariots and 10,000 warriors at his disposal and represented the largest military force in the region. The defeat of the Aramaean kingdoms – as well as their failure to unite – may have been prompted by economic destabilization of their kingdoms in the semi-arid regions bordering Assyria, resulting from a negative climate change beginning in the 9th century B.C.E. Assyria was not seriously hurt by the change, because by this time the Assyrian kings had developed a highly effective war machine, which they proudly portrayed on the walls of their palaces, and were less reliant on agriculture than on plunder. Their army included cavalry fighting beside war chariots and infantry composed of slingers, archers and storm troops. They used armored machines to besiege a city, built huge siege ramparts and destroyed the city's limestone wall by setting fire to trees piled against it, thereby melting the limestone into lime. The army became an efficient tool to exact from the conquered nations tribute in cash and kind. Those who did not surrender were cruelly punished. Many were deported to provinces of the Assyrian empire and other people were settled in their place. Thus, Sargon II (721–705 B.C.E.) vanquished the northern Israelite monarchy in 720 B.C.E. and deported the local population to northern Mesopotamia, replacing them with people from Kuta in Babylonia. Ration lists from Niniveh document the fact that Israelite warriors were integrated into the Assyrian army giving rise to the myth of the 'Lost Ten Tribes' scattered all over the then known world.

His son Sennacherib (704–681 B.C.E.) tried to do the same to the kingdom of Judaea after its king, Hezekiah, had rebelled. Sennacherib conquered and destroyed many cities in Judaea, among them Lachish, the conquest of which is illustrated in a relief in the king's palace at Nineveh. He laid siege to Jerusalem, but the capital did not surrender. The city was protected in the first place by the immense walls, built by Hezekia and which were excavated recently in the Jewish Quarter in the Old City of Jerusalem. Secondly the city's water supply was assured by Hezekiah's tunnel, in Hebrew *niqbah*. This was part of the siege water supply system of Jerusalem, which King Hezekiah perfected. It included the ancient spring of Gihon's emergency siege system, which was instigated already during the Middle Bronze Age around 1800 B.C.E. (see Appendix I).

Hezekiah diverted it through a tunnel to the Pool of Siloam, further down the valley but still inside the city walls (II Kings 20:20, II Chronicles 32:30). The story of the tunneling was found inscribed on the wall of the tunnel. The tunnel was cut from both ends simultaneously and the diggers met with a difference of not more than 30 cm, even though the tunnel meanders greatly, which may be due to its following a karstic solution channel. It is about 533 m long, and the difference in height of the floors between the inlet at the spring and the outlet at the pool is about 35 cm and carries water to this day (see Appendix I). The Siloan inscription is the longest Hebrew text known from the pre-Exilic period presently kept in the Archaeological Museum of Istanbul.[2]

---

[2] A. Issar, "The Evolution of the Ancient Water Supply System in the Region of Jerusalem" *IEJ* 26:130–136 (1976).

After entering a truce with the king of Judaea, who paid a heavy tribute, Sennacherib's son, Esarhaddon (680–669), invaded Egypt in 671 B.C.E. His son Ashurbanipal (668–627) continued the success of his ancestors, defeating and subjugating his opponents – including his own brother, whom his father made king of Babylon. Two years after Ashurbanipal's death, the Chaldeans, aided by the Medes, overtook Babylon.

Sennacherib himself had constructed one of the most impressive water projects of the ancient world, probably inspired by the example of the Urartians. He built a new capital, Nineveh, where he planted many gardens inside and outside, and to irrigate them he constructed a system of canals about 220 km long for carrying water from the foothills of the Anti-Taurus Mountains. For the farmers and gardeners he imported from Egypt the well-sweep device (*shadoof*) needed for pumping water from canals.[3]

The Assyrian aggression had a rather surprising effect on the countries in the northern highlands of eastern Anatolia. Here, a conglomerate of peoples speaking languages related to Hurrian (and probably ancestral to some of the modern Caucasian languages) had formed several principalities – the lands of Biaini (Van). In order to survive the constant threat of Assyrian predations, they rallied behind King Aramu and his successor Sarduri I, who established his capital at Van on the lake with the same name.

The Assyrians called this kingdom Urartu (the original pronunciation of Ararat).[4] Around 800 B.C.E., the kings Ispuini and Menua began building one of the most elaborate irrigation systems of the ancient Near East. Artificial irrigation and water control systems had a long history in the south Caucasian countries. For instance, at the third millennium B.C.E. site of Moghrablur in Armenia, stepped dams and sluices of mud-bricks faced by reeds and wooden panels reduced the velocity of rushing water from the mountain which was then led through simple channels onto the fields. The southern slopes of Mount Aragatz are dotted with small and large artificial ponds, remnants of stone dams dated to the 15th to 14th c. B.C.E. These barrages were linked to each other and to a complex network of canals controlling and directing the run-off water from gradually melting snow on top of the mountain to the bottom of the Ararat

---

Y. Shiloh, Underground Water Systems in Eretz Israel in the Iron Age, in: *Archaeology and Biblical Interpretation, Glen Rose Memorial Volume*, L.G. Perdue, L.E. Toombs and G.L. Johnson (eds.). Knox, Atlanta (1987).

D. Gill, "Subterranean Water Supply Systems of Biblical Jerusalem: Adaptation of a Karst System" *Science* 254:1467–1471 (1991).

R. Reich and E. Shukron, "Light at the End of the Tunnel" *BAR* 25/1:22–33 (1999).

R. Reich and E. Shukron "Jerusalem, the Gihon Spring" *Hadashot Arkheologiyot- Excavations and Surveys in Israel (ESI)* 109:116–118, Hebrew, English version 77–78 (1999) Ibid. "Jerusalem, City of David", ditto, 112:102–103, Hebrew, English version 82–83 (2000).

A. Frumkin, A. Shimron, J. Rosenbaum, "Radiometric dating of the Siloam Tunnel, Jerusalem" *Nature* 425 (2003).

[3] G.E. Sandstrom, *Man the Builder*. McGraw-Hill Books, New York (1970).

[4] The consonantal sequence '-R-R-T of the Hebrew text can be pronounced in various ways. In one of the Dead Sea Scrolls, a fuller (plene) writing is spelt '-W-R-R-T and thus confirms the pronounciation *Urartu*. In the 4th century C.E. the original pronunciation was forgotten, and a simple /a/ was inserted into the consonantal sequence, resulting in its mispronunciation as *Ararat* in common use today.

Valley. Similarly, the Urartians dammed major rivers into large reservoirs, directing the overflow through rock-cut channels and built aqueducts around the flanks of mountains, in some cases crossing rivers by bridges. In the valleys, particularly the plains around the Van, Urmia and Sevan lakes and the Ararat Valley, plantations were irrigated with these waters, often transported by interlinked aqueducts over large distances. Inscriptions in Urartian cuneiform found scattered in the landscape describe the grandeur of the gardens, giving rise to the later myths of the 'hanging gardens of Semiramis'.[5] Some of the old dams and portions of the aqueducts are preserved to this day, including the famous Menua Canal south of the town of Van which, nearly three millennia later, continues to irrigate the fertile plain to the east of the lake.[6] For a century the wealth and political power of Urartu influenced the cultural development of the eastern Mediterranean countries. Urartian bronze artifacts are found all over the ancient Near East,[7] including hundreds of ceremonial cauldrons in Delphi and Etruria. In addition, the rock-cut chamber tombs in the St. Etienne compound of the *Ecole Biblique* and the so-called Garden Tomb of Jerusalem, dated to the Iron Age, show architectural elements in common with the royal tombs in the Rock of Van.

After the defeat suffered by the Assyrinans, Nineveh was sacked and burned, its magnificent water supply system fell slowly into ruin, as the city was not reconstructed and the country quickly became rural and impoverished. The local rural society could not maintain the huge, sophisticated system without subsidies from the royal treasury, enriched by tribute and booty exacted from the subdued nations. Possibly contributing to the project's decline was the dependence of the water supply on the diversion of the flow of springs in the foothills of the mountains of eastern Anatolia instead of on a major river. The project was built in the time of Senacherib around the beginning of the 7th century B.C.E. – a humid period when the springs were at a high level. At the end of the century the climate became drier, probably causing the springs to fall below the level of the intake sluices. This would have required re-planning and changing the diversion system, a task that could not be executed in the absence of a central Assyrian government.

The climatic deterioration, which affected northern Assyria, may have pushed the Chaldeans (Kaldu or Kasdu in Akkadian, Kasdim in Hebrew) from the desert fringe in the south of Mesopotamia or the 'Sea Lands' in contemporaneous documents. They had become the masters of Babylonia and under King Naboplassar (625–605 B.C.E.) and joined forces with the Medes, from the semi-arid mountains and plains of northern Iran, with the purpose of overthrowing the Assyrian empire. Nabopolassar's son, Nebuchadnezzar II (604–562 B.C.E.) delivered the final blow to the Assyrian empire, annexing it in 612 B.C.E. to his empire, which we call neo-Babylonian. When the kings of Judah revolted,

---

[5] O. Belli, Urartian Irrigation Canals in Eastern Anatolia, Arkeoloji ve Sanat Yayinlari, Istanbul (1997).

[6] G. Garbrecht, "The Water Supply System at Tuspa (Urartu)" *World Archaeology* 2/3:306–312 (1980). M. Salvini, "Il Canale di Semiramide" *Geographia Antiqua* 1: 67–80 (1992).

[7] R. Merhav, ed., *Urartu - a Metalworking Center in the First Millennium B.C.E.* The Israel Museum, Jerusalem (1991).

relying on Egyptian promises of help, Nebuchadnezzar besieged Jerusalem. In this first rebellion the king of Judah surrendered and was deported to Babylon, while Jerusalem was spared. In a second rebellion, however, Jerusalem was besieged for two and a half years before it was finally conquered, burned to the ground and its nobility deported in 586 B.C.E. The book of lamentations on the destruction of the city describes the terrible famine in the besieged city, but is silent on the water situation. It seems that the *niqba* project functioned well, guaranteeing the city water during the siege, but nevertheless, did not save it from destruction. The difference in exile practices between the Assyrians, who scattered their captives among their own populace, mostly deportees themselves, and the neo-Babylonians, who settled them in compact and socially homogeneous communities on sites from "Before the Flood", *till abubi* (Tel Aviv) in Akkadian, allowed the conservation of their traditions and enabled the formation of Judaism.

During his reign, Nebuchadnezzar II had transformed Babylon with construction of elaborate temples, palaces and public buildings, such as the Gates of Ishtar. The city also became famous for its 'hanging gardens', irrigated by water lifted to the roofs of the palaces. Herodotus, who visited Babylon about a century later, was impressed by its beauty and included the gardens in the "Seven Wonders of the World." He also described the city's elaborate canal system, designed to make the city invulnerable in case of siege. The canals were of no help, however, when Cyrus the Great besieged Babylon.

According to Herodotus, he diverted the canals using their beds to enter the city.[8] All the tremendous effort invested in creating this wonderful city crumbled after a few centuries, but the intellectual edifice of systematic knowledge cultivated by Babylonian priests-savants in the fields of astronomy, mathematics, anatomy and medicine, survived. These fields of knowledge were initially developed in connection with the tasks of the priesthood to inquire from the gods their plans with regard to the future of the realm. It involved astrology, based on astronomical observations that were recorded and used to predict astronomical events. In the later Hellenistic period this knowledge was misapplied to a futile modern astrology. Information about anatomy was acquired while divining according to the state of the innards of sacrificed animals, especially the liver. Much of the knowledge amassed by the Babylonians was transmitted through the ages by savants of Persian, Greek, Roman, Jewish and Arab origin and became cornerstones in the scholarship of the Western World.

The fall of Babylon soon after the death of Nebuchadnezzar II is one of the curiosities of Near Eastern history. For some unexplained reason, his descendants' reigns were short, with Nabonaid being the last king of Babylon. He was a native of Harran in northern Mesopotamia, from where he introduced the cult of the Moon-God Sin into Babylon, declaring this god the sole god of Babylon instead of Marduk, and not only of Babylon but of the universe. The priesthood of Marduk accused him of the heresy of monotheism, which looked to them outrageously ridiculous. Nabonaid invaded Arabia and established a centre for

---

[8] Herodotus, *The Histories* (Translated by Aubrey de Sélincourt). Penguin Books, Harmondsworth, p. 90 (1954).

the worship of Sin at the oasis of Teima about 800 km southwest of Babylon, where he stayed for over a decade with his son Belshazaar acting as viceroy. The downfall of Babylon was accelerated by these religious disturbances causing the worshippers of the traditional Babylonian god Bel/Marduk to rebel. Thus when Cyrus II of Persia (559–529 B.C.E.) invaded Babylon in 536 B.C.E. he hardly needed to divert the city's canals in order to conquer the empire by 539 B.C.E.

8.3
The Persian Empire and the First Unification of the Ancient Near East

With the conquest of Cyrus the Great, the Near East experienced a new era of a more tolerant and, in many ways, enlightened system of government. Cyrus himself adopted the reformed quasi-monotheistic Persian religion initiated by Zoroaster, a Medean who probably lived and preached during the early part of the 1st millennium B.C.E. The tenets of this religion were the belief in Ahuramazda as the supreme entity representing light, truth and moral behavior, while on the opposite side stood the antagonist Ahriman (the "godfather" of the Judaeo-Christian Satan or devil), representing the essence of evil. The story of mankind was understood as the constant struggle of light symbolizing good against darkness symbolizing evil. The end of this struggle was believed to end with the triumph of light. The adoration of Ahuramazda as a supreme entity symbolized by light constantly struggling against the forces of evil, which were symbolized by darkness, had strongly influenced the similarly inclined Judaeans. This struggle became a basic theme in the Judaeo-Christian mystical thought and was dealt with at length in Jewish kabalistic and Christian theological writings.

Whether predisposed by the example of Zoroaster himself or by the spirit of his teaching, or maybe by sheer political wisdom, Cyrus II – most justly called the Great – adopted a policy of tolerance for the numerous religions and cults of the people of his empire. The return of the Judeans, or now Jews, to their ancestral homeland is the best-known example of Cyrus's tolerance. It was encouraged by the fact that he was already planning to conquer Egypt and in case of hostilities he clearly preferred a population loyal to Persia occupying the bottleneck on the road to Egypt. Both Ezra and Nehemiah, the leaders of the community of returning Jews, were Persian officials, and Jews later served in the Persian army post of Elephantine/Aswan.[9]

The important role of Persia, located in the Iranian plateau, in shaping the history and culture of the Near East and beyond, requires a special study. This plateau, south of the Alborz mountain range is, in general, a vast, harsh desert. Except for its most southeastern part, it belongs to the Mediterranean climate system and receives its rains in winter. Its aridity is due partly to the distance over the continent that the storms bringing humidity from the Atlantic and

---

[9] The Elphantine papyri written in Aramaic and dated to the early 5th century B.C.E. is one of the most important historical sources of Jewish life in this period. See Bezalel Porten's 1992 survey in *The Anchor Bible Dictionary*, vol. 2, pp. 445–455, which includes an exhaustive bibliography.

the Mediterranean must travel. In addition, the Zagros mountain chain along the western rim of the Iranian plateau is a barrier to the rainstorms from the west. Fortunately, the interior mountain ranges traversing the Iranian plateau capture the humid air masses that do succeed in crossing the border mountain barrier, thus trapping the humidity brought by the westerly cyclonic system. Were it not for these interior high ranges, the storms crossing the barrier would lose their humidity when forced to descend into the low plateau.

Many springs emerge from the limestone aquifers of the Zagros mountain ranges. The spring water is used to irrigate the fields and orchards in the valleys and the intermountain basins and to supply potable water for the urban centers. They were the source of water for Pasargade, the ancient royal capital of Persia, and Persepolis. Until today, these springs supply the water for cities in the Fars province, such as Kashan, Shiraz and Isfahan. In contrast, the mountain ranges of central and northern Iran are mostly composed of impermeable metamorphic and igneous rocks. The springs in such rocks are usually relatively small and their flow does not reach the wide intermountain valleys, but percolates into the gravely riverbed. The water from the rains and melted snow pours down the mountain slopes in torrential floods. Some of it infiltrates the vast alluvial fans built of gravel and sands. These fans form *bajadas* (apron-like covers) that stretch all along the foothills of the mountains and form regional sponge-like aquifers in which the water flows slowly towards the center of the intermountain basins.[10] This water, along with some other floodwater, flows into the inland basins and evaporates to form huge saline lakes and marshes. The Iranian farmer, now and most likely also in the past, used the water from the higher altitudes in many ways. The floods, as well as the water from the numerous springs in the mountain valleys, were channeled into terraced fields along the banks of the mountain's rivers. Yet the period of flow of the floods is rather short, seldom extending into the long dry and hot summer (April to September). To the purpose of ensure water for drinking and irrigation during this dry season, the Iranians tapped the groundwater reservoirs of the alluvial fans. To avoid pumping, they made use of the force of gravity by tunneling into the alluvial. These tunnels, or subsurface galleries, are called *qanat* in the Persian-speaking provinces of Iran, *kharez* in the Turkish speaking parts, and *foqqara* in the Arabic-speaking ones. Issar believes that this ingenious method was invented in the mountain valleys of Iran. It may have started with building galleries into consolidated rocks to ensure the flow of mountain springs. Later people moved downstream to the wide alluvial fans emerging from the riverbeds coming from the mountain ranges. Here, galleries were dug in the gravel beds of the rivers to catch their subsurface flow after the surface flow disappeared. Such galleries still function in many

---

[10] National Iranian Oil Company, Geological Map of Iran, 1:250,000 with explanatory notes, 1959.
A. Issar, "The Groundwater Provinces of Iran" *Bulletin of the International Association of Scientific Hydrology* 14/1:87–99 (1969).
D.N. Lerner, A.S. Issar and I. Simmers, *Groundwater Recharge: a Guide to Understanding and Estimating Natural Recharge*, Heinz Heise Verlag, pp. 33–36 (1990).
L. Lockhart, "Persia as Seen by the West" in: *The Legacy of Persia*, A.J. Arberry (ed.), Clarendon Press, Oxford, p. 325 (1953).

of the mountain riverbeds. Later, when agriculture expanded into the wide plains of the plateau, the experience gained in the narrow valleys was applied to the alluvial fans. This system was probably invented in the pre-Achaemanid period, since during the time of Cyrus II it was already the basic system for supplying water to the towns and villages in the central and eastern part of Iran. Polybius, the Greek-Roman historian (204–122 B.C.E.) who visited Ecbatana (Hamadan), the ancient capital of the Medes, provided the first written report of such a water system.

From a neo-deterministic standpoint, the invention of the *qanat* system can be seen as adding extra resiliency to the water supply system, and consequently to the survivability of human society in the arid Iranian plateau. This is vividly demonstrated to the modern traveler who drives great distances in the arid plains of Iran, especially during the summer, when temperatures exceed 45 °C. The traveler will arrive at a small town or village surrounded by orchards and will be invited to sit under the shade of a tree beside a pool supplied from the canal, *jube,* fed by a *qanat.* In Old Persian an orchard was called *pairidaeza* "formed around the pool", which became "paradise" in Western languages.[11]

The greater resiliency the *qanat* system enabled to change Iran from a desert with a few scattered oases and scattered fertile valleys supporting mainly a population of pastoral tribes to an agricultural society. The ruling upper classes were the former tribal chiefs, who had become landlords, and the warriors maintaining the values and sports of a tribal warrior stock with archery, riding, hunting, and excelling in war, which continued to play an important role in their way of life. These traditions, which continued to characterize Persia's ruling class until recently, are described in the great national epic, the *Shahnameh* of the distinguished Persian poet Firdousi (935–1020 C.E.).

Thus when Cyrus the Great started his campaigns after forcing unity between the Medes and Persians, he had at his disposal an elite army of horsemen who fought according to the traditions of the Indo-Aryan warrior tribes, excelling in riding and mounted archery. He could also raise an army of infantry from the peasant society. With the development of irrigation methods based on diverting dams from the rivers, as well as the *qanat* system, the army was backed by a country turned from a desert dotted by oases to a rich agricultural economy supplied by water from natural and artificial rivers. In thirty years, Cyrus the Great formed an empire stretching from the Hindukush Mountains to the Mediterranean Sea and including the Anatolian plateau previously ruled by Croesus, the king of Lydia. With the victory over Croesus, the Persian Empire reached the Aegean Sea and conquered the Ionian cities in western Anatolia. Control over this vast empire was achieved through local governors who maintained close connections with the central administration at the royal court through informers (the 'king's ears') and other means of communication. Paved roads had relay stations where horses could be changed, and messages by fire signals could be transmitted from one end of the empire to the other usually by the help of artificial hills constructed along the roads.

---

[11] J.H. Liffe, "Persia and the Ancient World" in A.J. Arberry (ed.), op. cit., pp. 1–39 (1953).

The heirs of Cyrus the Great extended the empire, which at its zenith reached from the borders of Libya in the west to India in the east. With the expanding Persian administration also spread the system of *qanats*. When the son of Cyrus II, Cambyses II (529–522 B.C.E.), conquered Egypt in 525 B.C.E., his people dug a *qanat* in the Western desert depressions, which form an outlet for the fossil water aquifers underlying the Western desert. The conquest of Egypt required crossing the Sinai desert and securing a supply of water for the invading army. During years of average precipitation, adequate water was made available from shallow wells in the El-Arish oasis and some wells dug in the sand dunes along the Sinai coast. These, however, did not meet the needs of the army, either because of its size, or perhaps also because this was a period of low humidity, which can be seen from the paleo-precipitation curve of the Soreq Cave. This water shortage called for non-conventional engineering methods, which Herodotus describes most vividly: Cambyses made a pact with the Arabian king by which the king would supply the Persian army with water carried in camel-skins that were loaded on all the king's camels. Furthermore, the Arabian king constructed a pipeline of cowhides and other animal skins stitched together to carry water from a river in Arabia that runs into the Red Sea – a distance of twelve days' travel. Regardless of whether the pipe was fact, fiction or exaggeration, one can glean from this story that solving water supply problems was integral to the planning, operation and logistics of the Persian army. After conquering Egypt the Persians, according to Herodotus devised another method to secure water supply for their posts along the road crossing the Sinai desert. Taking advantage of the fact that Egypt had to import all its wine from Phoenicia and Greece, the Persians ordered all the wine jars, once emptied, to be filled with water and returned. The water was emptied into containers at the desert stations.

A lack of appropriate planning and logistics played havoc in another campaign, however, this one against the king of Ethiopia, Nubia of today. According to Herodotus, Cambyses sent a group of spies pretending to be good faith missionaries to the king of Ethiopia. The Ethiopian exposed the trick and told the spies his opinion of the Persian king. Cambyses, furious of having been told the truth to his face, a practice to which, absolute sovereigns were not used to, tried to invade Ethiopia. This time however, maybe because of his anger, Cambyses did not plan ahead, and his invasion southward ended in great failure, without seeing any combat. According to Herodotus, the army had no food and was reduced to cannibalism. Interestingly there is no mention of thirst, which may tell us that the Persian army marched along the Nile. Another part of the army was sent westward to conquer Libya. On the way to the oasis of Siwa (Ammonium) it disappeared in the sands of the Sahara. The Libyans' story was about the Persian army, covered by sands driven by a severe sandstorm while having their lunch. Cambyses did not live long after the Ethiopian campaign. He had to rush back to Persia to quash a rebellion by a *magos* pretending to be the king's brother and died.[12]

---

[12] Herodotus, *The Histories*. pp. 176, 177, 185.

Darius I (521–486 B.C.E.), an army officer who was crowned after putting down the rebellion, succeeded Cambyses. He married Cambyses' daughter, thus becoming the dead king's son-in-law. In 499 B.C.E., the Ionian cities in western Anatolia, aided by Greeks ships from the Greek mainland, rebelled against Darius I. Darius subdued the rebels and later invaded the Greek mainland. He began his campaign in Thracia and moved northward into the vast plains of southern Russia, hoping to inflict a decisive blow on the Scythians, either to secure his northern borders or to control the gold trade from the Ural.[13] The Scythians avoided a full-scale war, engaging instead in the first guerilla warfare reported; that is, sudden attacks by their horsemen, while their main army and their folk retreated. At the same time they adopted a scorched land policy, *"blocking all the wells and springs, which they passed on the march and stripping all the country of all green stuff, which might serve as forage"*.[14] At last Darius was forced to retreat, leaving all the wounded and sick behind, even though his army had seen no full-scale combat.

8.4
Hellenism Dominates the Near East and Unites East and West

Colder and more humid conditions started around 300 B.C.E. It benefited the Levant causing the desert margins to become more hospitable. On the other hand, the climate change being rather moderate, it did not drastically influence the tropical and subtropical system, and therefore, the Nile flow, though somewhat reduced, was not strongly affected.

A long series of wars between the Persians and the Greeks, among the Greeks themselves, and within the members of the Persian royal family set the stage for Philip II, the king of Macedonia, to gain control of the other Greek cities and for his son to challenge the Persians. It was Alexander III, so-called The Great (336–323 B.C.E.) who crossed the straits into Asia, defeated the King of Persia, and became ruler of the Persian Empire and Greece. After his death at age thirty-three, Alexander's vast empire, stretching from India to the Danube and to the oasis of Siwa, fell apart. His generals who inherited his empire did not pursue his vision and began a series of wars, which ended with a subdivided Hellenistic-oriented Near East. The wars between Alexander's successors, known as the wars of the Diadochi, continued until 200 B.C.E. When they ended at last the eastern and northern parts of the Near East, including Mesopotamia, Syria, part of Iran and part of Anatolia emerged as the Seleucid Empire, while Egypt became the kingdom of the descendants of Ptolemy. The general Lysimachus controlled Anatolia and Thrace and the descendants of Antigonus ruled parts of the Greek homeland.

Once the political situation stabilized and the economy improved, both the Seleucids and the Ptolemies could undertake the major cultural transformation of the Near East to a new cultural structure, Hellenism, which embraced the

---

[13] R. Girshman, *Iran*. Penguin Books, Harmondsworth, pp. 146–152 (1954).
[14] Herodotus, *The Histories*, p. 281.

upper levels of society. They adopted Greek as their language and began imitating the literature, the arts, and the architecture of classical Greece. Initially, Macedonians and Greeks were encouraged to immigrate and settle in existing cities, intermarry with the local people and participate in public life. New cities were built and populated by a mixture of new immigrants and natives. These cities were awarded the status of a *polis*, which, in many cases, involved changing of the name of the city; thus Acre became Ptolomeis, Amman was renamed Philadelphia and Gaza became Seleucia. This status conferred upon the inhabitants various privileges, including the right to mint their coins and to exercise some degree of autonomy in the local government affairs. These cities became centers of a new heterogeneous Hellenistic culture and religion, a profound cultural transformation that was to retain its influence until the Arab conquest of the Near East in the 7th century C.E. The economy of most of the cities was unchanged and based on agriculture and commerce. The difference is found less in quality than in quantity: The favorable climate conditions fostered the spread of agriculture over extensive regions, including the desert fringes, and the establishment of trade routes to Arabia and China through Persia and central Asia. Thus, in spite of the outbreak of intermittent warfare between the Seleucids and Ptolemies – ending only with annexation of the Near East by the Roman Empire – the Hellenistic period was one of economic prosperity and its cultural influence lasts into our days.

Some time around 250 B.C.E., tribes related to the Scythians from the plains east of the Caspian, locally known as the Parni, invaded Iran. They established a kingdom that would become known as Parthia and that was destined to become the main opponent of the Seleucid, and later the Roman, empires. The invasions coincided with a swell of colder climate that perhaps made the grasslands of central Asia less hospitable and the plains of Iran and the Syrian deserts more desirable.

Around the same relatively humid period, the Nabataeans,[15] semi-nomadic pastorals, who originally arrived from eastern Arabia, had settled in the formerly Edomite territories of Transjordan, the Negev and northern Sinai deserts. From the beginning, these people were not just nomadic shepherds; they were expert camel and horse breeders, practiced agriculture as their principal or secondary occupation and, for many generations, controlled the trade routes to southern Arabia and its ports with goods from Africa, southern Asia and even the Far East. They spoke an Arabic dialect and adopted Aramaic as their written language. The Roman scholar Strabo describes them as being ruled by a royal family, yet maintaining a principle of personal liberty that forbade enslavement of others. They worshiped a pantheon of deities with Dhu-shara, a mountain god, as the principal deity, the sun and moon and other astral bodies adopted from Mesopotamia. As they expanded northward they made the city of Petra their capital. One of their first recorded wars was with Antigonus, one of the diadochs, who attacked Petra in 312 B.C.E. Antigonus was victorious at first, sacking the city, but his army was destroyed when it was stranded in

---

[15] N. Glueck, *Rivers in the Desert*, pp. 191–242. ibid. *The Other Side of the Jordan*. The American School of Oriental Research, Cambridge, MA (1971).

the desert, burdened by the enormous weight of their booty and unable to find water. After the Seleucids emerged as the rulers of the Levant around 200 B.C.E. the Nabataeans were left to themselves to develop the commercial routes that would profit the Seleucid government as well. This peaceful cooperation led to a partial Hellenization of the Nabataeans. When the Seleucid government weakened, the Nabataeans expanded northward into Syria.

After the middle of the 2nd century B.C.E. the Jews attained their independence after a series of wars against Antiochus IV Epiphanes (175–164 B.C.E.). They were headed by the priestly family of the Hasmoneans, who had risen to leadership in a rebellion against the Seleucids following a period of religious oppression and an attempt of forced Hellenization. The Jewish state expanded to its greatest limits during the reign of its king Alexander Janneus (103–76 B.C.E.), extending to the Galilee and Golan in the north and the Negev in the south, where it confronted and expelled the Nabataeans.[16] Janneus'/Yannai's conquest of Gaza blocked the Nabataeans' access to the Mediterranean, forcing them to move towards Moab, Syria, and Damascus in the wake of weakness of the Seleucid kingdom. Their trade routes were now through Syria to the coast of Phoenicia. After occupying Damascus, the Nabataeans attacked the Jewish kingdom again (85 B.C.E.) with the war ending in a truce between Janneus and the Nabataean king Aretas/ Hartaat III. In 76 B.C.E. Janneus died and was succeeded by his wife Salome Alexandra/ Shlomzion. The time of her reign is considered to be the "Golden Age" of the Hasmonean kingdom. After her death in 67 B.C.E., the Nabataeans became involved in a conflict between her sons Hyrkanus, who was the great priest and king of the Jews, and his brother Aristobulus who was supported by the army. To gain Aretas III's support, Hyrkanus promised to return to the Nabataeans twelve cities that Janneus had taken from them (Antiquities 14, 29–33). The combined forces of Hyrkanus and Aretas defeated the army of Aristoblus and put Jerusalem under siege.

During the reign of Aretas IV (9 B.C.E.–40 C.E.) the Nabataean kingdom reached its zenith. In addition to Petra, their capital in Transjordan, other cities of their realm in the Negev and the Hejaz flourished and prospered.[17] A network of fortresses connected Petra via these cities to Gaza, which served once more as their main port on the Mediterranean. The increasing demand by the Roman market for frankincense, spices, and perfumes from Arabia and India strengthened the Nabataean economy, and the kingdom continued to thrive until the middle of the 1st century C.E.[18] The majority of archaeological

---

[16] Josephus Flavius, *Antiquities of the Jews* 13, The Complete Works of Josephus, translated by W. Whitson. Kregel, Grand Rapids, Mich. p. 374 (1981).

[17] The cities in the Negev were: Obda (Hebrew Avdat, Arabic Abdeh); Mampsis (Hebrew Mamshit, Arabic Kurnub); Sobata (Hebrew Shivta, Arabic Subaytah); Ruhaiba (probably) (Hebrew Rehovot; Arabic Ruheibe) Elusa (Hebrew Halutsa, Arabic Khalsa); Nessana (Hebrew Nizanna, Arabic Auja el Hafir).

[18] A. Negev, *The Nabataeans and the Provincia Arabia*. Walter de Gruyter, Berlin (1977).
D.E. Graf, "Romans and the Saracenes: Reassessing the Nomadic Menace" in *L'Arabie Pré-islamique et son Environnement Historique et Culturel. Travaux du Centre de Recherche sur le Proche-Orient et la Grèce Antiques* 10: 341–400 (1989).
Concise summary is D.E. Graf, "Nabataeans" in *ABD* vol. 4, pp. 970–973.

remains of Nabataean buildings, arts and inscriptions are from the period of Aretas IV. It was a relatively peaceful period between the Nabataean and the Judaean kingdoms, as both prospered from the trans-Arabian trade. The Romans soon recognized the potential profits of this trade, and tried to seize the caravan routes. With Aretas' IV death, the Roman pressure increased. During the reign of Malichus/Malho II (40–70 C.E.) the Nabataean kingdom became a vassal kingdom of Rome, and after the death of Rabbel/Ravaal II (70–106 C.E.) the last Nabataean king, it finally became a Roman province.

The source of the Nabataeans' prosperity began and continued to be trade overland. When maritime trade in Greek hands through the Red Sea encroached upon a substantial portion of the caravans' trade, agriculture became increasingly important in their economy, especially after the first part of the 1st century C.E.. The Nabataean agriculture was a remarkable achievement, as they cultivated the belt between the desert and the sown land. Against all odds, they managed to develop a flourishing agriculture based on a unique, sophisticated system of water harvesting and irrigation, which enabled them to excel in the production of olive oil and wine. The latter was of special quality and was exported to many Mediterranean countries. Still, ingenuity and the subsidy of rich merchants, and even government, would not have been sufficient to develop such an intensive agricultural economy if the climate had not become colder and more humid. This benign weather regime characterized most of the period during which the Nabataeans, under their own kings or as Roman citizens, participated in the history of the Levant from the 3rd century B.C.E. to the 7th century C.E. The peaks of humid climate during the 1st century B.C.E. and again in the 6th century C.E. coincided with the zenith years of Nabataean prosperity. Thus the combination of optimal climatic conditions and human ingenuity makes the Nabataean culture a classic model for the study of neo-determinism.

8.5
Under the Boot of Rome and the Beginning of Christianity

In 63 B.C.E. the Roman general Pompey began wiping out the "pirates" who dominated the sea-lanes of the eastern Mediterranean. Taking advantage of the war between the Hasmonean brothers, he brought Judaea under the control of Rome. Towards the end of the century and during the civil wars after the murder of Caesar, the Nabataean kingdom was drawn into the war between Marc Antony and Octavian/Augustus on the loosing side. On the other hand, Herod, the son of Antipatris, an Edomite officer in Janneus' Hashmonean army, played the Roman card cunningly and was awarded the crown as king of Judaea in 37 B.C.E.

Like all the other contemporaneous rulers, Herod's reign (37–4 B.C.E.) was marked by economic prosperity combined with tyranny, of continuous plotting and assassination of anyone who appeared a threat to his rule. Among those murdered were all the descendants of the Hasmonean dynasty, including his own wife Mariam and their two sons. Herod indulged in a stupendous activity,

Plate 12. The street pavement of Herod's port at Caesaraea under water (Photo A.S. Issar)

bringing a booming economy to his country. As Josephus Flavius tells us he was personally involved in the planning and control of his construction projects.[19] In his kingdom he rebuilt Sebasteia and founded Caesarea and its artificial port and aqueduct. Part of the pavement of the port is now under water (Plate 12) therefore, one can conclude that during Herod's time the sea level was lower, due to a global colder climate regime and glaciation. The uninterrupted slopes of the Roman aqueducts (Plate 13) show that there were no tectonic displacements since then along the shoreline thus the change in sea level is because of climate change and not tectonics.

But even in the best of years there were reversals. In a story reported by Josephus Flavius, three successive years of drought and starvation occurred some time before the reign of Herod (67–62 B.C.E.). They ended when a righteous man by the name of Onias (in Hebrew called Honi the Circler) drew a circle around himself and told God that he will not leave it if rain does not come. The rain came with such intensity that it filled all the cisterns of Jerusalem and caused floods all over, forcing the people of Jerusalem to escape to safety in high places – the lesson to be learned is that one should be more precise when asking God for help. The story continues when Onias was killed in 65 B.C.E. during the fight between the two Jewish factions, the Sadducees supporting Aristobulus, and the Pharisees on the side of Hyrkanus. Onias refused to pray against Aristobulus, and his party and was stoned by the hostile party. This brutality caused a strong wind, which destroyed the grain crops all over the country and was followed by an awful drought. This time there was nobody to ask for rain, which teaches us how dangerous it is when political and religious fanatics take the law in their hands.[20] Flavius mentions another drought in the

Plate 13. The northern Roman aqueduct of Caesaraea (Photo A.S. Issar)

---

[19] Josephus Flavius, *Antiquities* p. 15.
[20] Babylonian Talmud, Taanit, 19:71; 23:71.
    Josephus Flavius, *Antiquities* 14:22–24.

thirteenth year of Herod's reign (24–25 B.C.E.).[21] The famine forced Herod to import grain from Egypt, which made him popular with the Jewish population to whom he distributed the food for free. At about 46–47 C.E. another famine occurred during the reign of the Roman emperor Claudius.[22]

The Judaean desert, although situated in the shadow of the rain brought by the westerlies cyclonic storms, nevertheless sustained human life during periods of optimal climate conditions, especially when precautions were taken to catch the winter floods and divert them into cisterns. With such cisterns, enough water was collected at Herod's desert fortresses, including Masada, to enable the Jews to endure many months of siege.

The members of a sect known from the famous Dead Sea Scrolls who considered themselves "The People of Light" used cisterns filled with diverted floodwaters to guarantee the supply for their ritual baths, an important part of their purification rites. These people chose for their shelter a desolate hilltop, today called Khirbet Qumran, on a bluff over the Dead Sea shore. It was believed that here, from around 150 B.C.E. to 70 C.E., they wrote their sectarian scrolls. When they anticipated disaster with the approach of the avenging Roman legions, they sealed the scrolls in jars and hid them in nearby caves.[23] Lately, this traditional interpretation of Qumran is losing credibility among archaeologists and historians alike who claim that there was no connection between the scrolls and the inhabitants of the site. Most scrolls were actually carbon-dated to the 3rd and 2nd century B.C.E. It has been suggested that the place may have been the manor of a wealthy nobleman, a factory, or a trading station for the production of a priceless cosmetic ingredient made of "balm" or "balsam" – an as yet unidentified plant allegedly growing only in the vicinity of the Dead Sea and Jericho. Other sites similar to Qumran are coming to light now, including cells of pre-Christian hermits at Nahal Qidron South, which open entirely new vistas on life in the Judaean desert before the revolt of the Jews against the Romans in 66 C.E. These differences of opinion, however, do not diminish the importance of the water supply and intricate water storage at these sites.

Herod has also built a series of fortresses, such as Masada, as places of refuge for him and his family. In Jerusalem he rebuilt the temple in monumental scale, which can be deduced from the remnant of the walls supporting the podium on which the temple stood, known today as the Temple Mount. In addition he built numerous other temples, pagan and Jewish, public urban features such as columnaded streets, baths, markets, roads, water supply projects, not only at home but also abroad. All these ambitious projects could not have been financed without an agricultural infrastructure that could prosper despite

---

[21] Ibid. 15:299–314.

[22] Ibid. 20:51–53.

[23] R. de Vaux, *Archaeology and the Dead Sea Scrolls*, Schweich Lectures, British Academy, Oxford University Press, London (1959). The main stumbling block is preconceived ideas: The monks of the French Ecole Biblique who excavated Qumran immediately recognized the site as innately related to their own monastic ideals rooted in the Middle Ages. They recognized every detail, from the *scriptorium* to the *refectory*, from their own viewpoint. Lately, a lively debate has reopened with more secular explanations gaining authority.

excessive taxation, and the resources of an international trade network which could be tapped.

Herod's death, followed by direct Roman administration and the ensuing steep and long decline of the economy of the country was, beside religious frenzy the major reason for the revolts of the Jews in 63–70 C.E. This first Jewish war ended with the destruction of Jerusalem and devastated the country around it. More oppression led to rebellions of Jews in Cyprus and Egypt and to the second Jewish war 133–136 C.E. and, finally, the expulsion of the Jews from their country and the beginning the Jewish diaspora. Other parts of the Near East continued to prosper and there can be little doubt that the favorite climate as well as the *Pax Romana* during most of the period, were two complimentary positive factors.[24]

As in former cold periods, the colder and wetter climate that now benefited the Near East was not favorable for the people in the Eurasian-steppes, resulting in a series of invasions from these regions beginning in the first and second centuries C.E. Roman sources speak of Germanic people constantly pushing south and west in an effort to enter the Roman Empire across the Rhine and Danube borders. By 200 C.E. these incursions seriously threatened the Roman provinces along these rivers. It clearly was not simple *Wanderlust* or the temptations of civilization and wealth that pressed them. Rather, they faced a more serious peril from the east - the nomadic pastoralists of the Eurasian steppe, who always have been and would become again an important political factor for the next one and a half millennia.

During the first millennium B.C.E. many former-farming peoples in central and eastern Asia increasingly relied on their herds for sustenance in favor of agriculture. Traditional pastoral nomadism, however, is an economic dead end. Unlike farmers and horticulturists, who could intensify or extend their economic base with terracing, water procurement or alternative crops, these nomads' subsistence was restricted to their sheep, goats, cattle, and horses. Once a given area reached its carrying capacity, which occurred in a relatively short time, the ever growing population had no choice but to try to occupy their neighbors' lands. If the invaders succeeded, the conquered people then had to try to invade their neighbors in turn, with the weaker neighbors falling in the same domino pattern seen a thousand years earlier. Even the smallest climatic and, therefore, economic or political crisis in the heartland of the nomads multiplied this phenomenon.

The nearest outlet was the rich lands of China, which were protected by the well armed and trained armies of the Han dynasty who withstood the onslaught of the invading Yue-chi and Hsing-Nu nomads, successfully deflecting them to the west. The Yue-chi were most likely some Indo-European people, such as the Sarmatians, Cumani, or Alani, and the Hsing-Nu probably Turkic Huns.

It is the first time that we have historical sources about the pattern of ethnogenesis of these nomadic peoples. In this context is of some importance to clarify this phenomenon, which is not unlike the formation of other tribes and

---

[24] D.D. Anderson, "The Impact of Rome on the Periphery: The Case of Palestina – Roman Period (63 B.C.E.–324 C.E.)" in: *The Archaeology of Society in the Holy Land*, pp. 446–468.

tribal leagues based on fictitious lineages or descent. By promising booty and good prospects to "deliver the goods", any charismatic and opportunistic war lord would be able to gather around him followers from the most heterogeneous groups: Siberians and Tungids from the farthest east, Mongols, Turks, Chinese, Tibetans, Indians, and a plethora of peoples of Indo-Iranian origin, such as Soghdians, Scythians, or Sarmatians. Depending on the situation and some initial success, other peoples were attracted and would also have been thrown into the melting pot of this warrior comradeship. By sharing a basic common language, necessary for internal communication and passing of commands, by installing occasional commemorative or festive days and rituals, and assimilating the traditions of one or the other group by the entire community, a new people was born. If successful, the charismatic leaders of old were soon glorified in various forms, were endowed with special powers or intimate relationships with their gods, and often became the eponymous ancestors of that tribe or nation. From the names of their leaders, recorded by their literate enemies, we learn that Iranian dialects were the primary languages in this process during the early part of the first millennium C.E., whereas Turkic languages prevailed in the second half.

In most cases, these tribal commonwealths would be quite short-lived, their members reshuffled into other configurations, only their memories would often live on in some oral traditions and the epics of their bards. All the available evidence suggests that countless similar but only imperfectly recorded 'fission and fusion' processes have taken place among the semi-nomadic or nomadic tribes of the ancient Near East, whom we so confidently give generic names such as Amorites, Aramaeans, Shasu, Habiru/Hebrews, Israelites, etc., without knowing anything concrete about the stages the process of their agglomeration and what these names might have meant at a particular point in time.

According to the paleo-climate proxy-data time series, the climate started to change around 50 C.E., although it remained above the average of the precipitation line until around 250 C.E. Thus, despite the many wars on the eastern borders against the Parthians and the two destructive and bloody rebellions of the Jews against the Romans, the first and second centuries C.E., in general, were periods of relative economical affluence.

Among the prosperous Syrian Desert cities was Palmyra, situated about 250 km east of Damascus and now in the middle of a parched desert. This city is dependent on groundwater that is recharged from a shallow marshy lake called Sabkhat el-Mouh. At present, approximately every ten years or so, when the rain is above average, the lake fills and causes the shallow groundwater table in the wells of Palmyra to rise, perhaps for a decade. When the rainfall stops, the groundwater becomes brackish and then increasingly saline. During most of the Roman period amounts of rainfall greater than today's average supported Palmyra's economy. Surviving custom-laws show that the territory surrounding the city was settled with a large number of small villages and extensive cattle pastures.[25] In 258 C.E., when the Roman emperor Valerianus was defeated and taken prisoner by the Persian king, Shapour I, Palmyra's governor Septimus

---

[25] Personal communication to Issar by Prof. Em. M. Geyh, Hannover, March, 2001.

Odeinat defeated the Persians and drove them back. He declared himself king and, because of the anarchy in Rome, became ruler of the Near East. After he was killed, his wife Zenobia and his son declared themselves emperors of Rome, but the rival general Orleanus defeated them and destroyed Palmyra. It is quite probable that Palmyra was not restored due to the steadily deteriorating climate in the Near East, which had reached a low point at end of the third century and discouraged any initiative of reconstruction.

The poor climate magnified the impact of the anarchy in the empire's central government, the heavy taxation and, of course, the devastations brought by endless internal and external wars. In 220 C.E., a new dynasty, the Sassanids, had gained rule of Persia and waged war on Rome with renewed vigor. All these together profoundly influenced the economy of the Near East and the Mediterranean countries, which fell to one of its lowest ebbs. The third century C.E. was a terrible time for the Roman empire: 22 emperors came and went, the economy was grossly mismanaged, taxation was crushing, and inflation and famines wiped out what was left of the empire's wealth. In 251 the Goths crossed the Danube, and a few years later the Franks poured into Gaul.

The disastrous situation did not change much even after Diocletian – an able army officer and excellent administrator – became emperor in 285 C.E. He immediately made peace with the Persians and divided the responsibilities of governing the empire with another outstanding officer, Maximinianus. Both adopted two other officers, Galerius and Constantine, to share in governing different parts of the empire. The military and civil reforms saved the empire from collapse and improved the security of the empire, yet the economy continued to deteriorate. In 301, because of the rising inflation, Diocletian issued an edict fixing prices of goods, payments and salaries and prohibiting changing trades. Despite all these measures, the situation continued to deteriorate, famine had become commonplace, and, according to Gibbon, a third of the empire's farmers perished, while the number of administration officials and army personnel had grown to unprecedented size, devouring what commodities were left.[26] No wonder that religions from the east, such as the Egyptian mystery cults and Christianity, which promised redemption and a better life, at least after death, were enthusiastically accepted by those who suffered most, and often cynically exploited by those in power, above all Constantine, to further their own interests.[27]

The basic problem in the region, the climate's low ebb during the early 3rd century C.E. could not be surmounted under any circumstances. The problematic location of a Roman villa at 'Ein Fashkha near Qumran, on the north eastern shore of the lake, dated to 2nd to 3rd century C.E can now

---

[26] M.D. Herr, ed., "The Roman Byzantine Period" in: *The History of Eretz Israel*, vol. 5, 48 (1985).
  F. Millar, *The Roman Near East 31 B.C.E. – AD 337*. Harvard University Press, Cambridge, Ma. (1993). A classic survey is N.H. Haynes N.H. and H.St.L.B. Moss (eds.), *Byzantium: An Introduction to East Roman Civilization*, Oxford University Press, London (1969).

[27] Despite it currently somewhat being out of fashion and subject to controversies, Edward Gibbon's classic *The Decline and Fall of the Roman Empire*, first published in 1776, remains one of the greatest works in the world's historical literature and is the most essential and awarding reading for any discussion of this period.

be explained not only by the data already discussed but by new geological observations which show a sharp drop in the level of the Dead Sea.[28] As usual, during warm periods, the level of the Dead Sea dropped while the level of the oceans came up, and indeed, at ca. 300 C.E. the rising sea covered Roman villas built in southern England on the seashore.[29]

Information about a series of further drought years from around 400 to 550 C.E. was obtained from a tree-ring study of trunks of cedar trees (presumably from Lebanon) that were re-used for the ceiling of El Aksa mosque in Jerusalem.[30]

Aggravating factors were created by the voracious consumption of wood for the purpose of cement making, fuelling the gigantic bricks kilns and heating the public baths. It brought about the destruction of the natural forests, which caused higher rates of surface water flow and thus soil erosion and as a result high rates of infiltration. Infiltration of rainwater into the subsurface, aided by the decrease of transpiration of the trees (up to ten times more than in forested area) produced high levels of ground water creating swamps, ideal breeding grounds for malaria-carrying mosquitoes. The history of malaria in the Mediterranean countries has still large lacunae. Evidence of malaria is found in some prehistoric skeletons but appears massively in its present shape around the 5th century B.C.E. on the island of Sardinia, brought there most probably by the Carthagenians colonists and remained endemic on that island until the 3rd century C.E.[31] The proliferation of the swamps was instrumental in the diffusion of this disease throughout the Mediterranean countries.

In the meantime more German tribes penetrated the west – Goths, Vandals, Suevi, Alemani and Langobardi – with Attila's Huns in pursuit. Rome was sacked in 410 and 455 C.E. The Huns were defeated in 451 C.E. at Chalons, but a host of other mostly Turkic speaking tribes kept invading from the east, Avars, Petchenegs and Bulgars, as well as the Indo-Iranian Croats and Serbs, conquered and joined forces with the Slavs who had penetrated the Balkans in the early 5th century and, within a few centuries, 'Slavicized' their former masters. These new intruders were halted by Byzantium, however, which formed a barrier protecting the eastern Mediterranean and the Near East.[32]

Here the economic situation began to improve during the 5th century, perhaps in part because of the administrative and religious changes taking place in the eastern provinces of the empire, now better known as the Byzantine

[28] Y. Enzel, D. Sharon, R. Ken-Tor et al., "Rainfall variations affecting recent changes in the Dead Sea level etc." *Proceedings "Science For Peace Conference"* Hebrew University, Jerusalem (2002).

[29] F.H. Thompson (ed.), *Archaeology and Coastal Change.* The Society of Antiquitaries of London, p. 70 (1980).

[30] S. Lev-Yadun, N. Liphschitz, and Y. Waisel, Chronology of Rings of Cedar of Lebanon Trees from the Roof of El Aksa Mosque. *Eretz Israel* 17:92–96 (1984, Hebrew).

[31] P.J. Brown, "Malaria in Nuraghic, Punic and Roman Sardinia – Some Hypothesis" in: *Studies in Sardinian Archaeology*, M. Balmuth (ed.), The University of Michigan Press, Ann Arbor. pp. 209–236 (1984).

[32] V.A. Fine, *The Early Medieval Balkans – A Critical Survey from the Sixth to the Late Twelfth Century,* The University of Michigan Press, Ann Arbor (1991).

empire.[33] Interestingly, this economic improvement parallels an improvement in the climate – a fact hardly ever mentioned by modern historians. With Christianity now the official and dominant religion of the empire, Palestine in general, and Jerusalem in particular, became the focus of pilgrimages, and the construction of churches and monasteries created work and provided income. Many of the pilgrims settled down not only as hermits in the desert regions but also as farmers in the hill country. They repaired the old Israelite terraces and planted the very same crops – olive and fig trees and wine grapes. This new prosperity was facilitated by the more humid climate spurring immigration of entire communities from Cappadocia in central Anatolia, for instance, where conditions were less than ideal due to the cold. The settlement pattern in Palestine during the Byzantine period was one of the densest the country has seen until the modern period. The agricultural infrastructure flourished and, as before, brought about an equal flowering of the urban centers. The prosperity continued during most of the 5th and 6th centuries despite many internal religious tensions among the various Christian sects, such as the Monophysist heresy, or the rebellion of the Samaritans almost causing their annihilation by the Byzantines as a people and as a political entity. The so-called Roman Law formulated by Justinian (518–565 C.E.) which served as the basis of all the later European laws, deprived the remaining Jewish inhabitants of their lands, legalized their persecution, and planted the seed of traditional Western anti-Semitism.

The investigations carried out by Izhar Hirschfeld of the Hebrew Universty, Jerusalem, also pointed to a period of about 300 years of substantial increase in precipitation, which started at ca. 300 C.E., ie. The early Byzantine period, causing the expansion of the settled area in the Middle East.[34]

After the death of Justinian, the empire experienced a few more strong convulsions and loss of territories. In the east, the Persians managed to conquer a large part of the Levant only to lose it again to the Byzantine emperor Heraclius (610–641) who even tried to invade Iran. This war ended with a peace which returned all the Levant and large parts of the Fertile Crescent to Byzantine rule.

8.6
The Arabs and Islam Emerge from the Desert

While the Christian Byzantine and Zoroastrian Persian emperors were fighting each other, the heretics within their own fold, and all their neighbors, a new faith evolved in the oases of Arabia where Mohammed preached yet another

---

[33] 'Byzantium' (and 'Byzantine') is a late medieval term with some derogatory connotations that still survive in English. The Greeks of Constantinople call themselves *Rhomaioi*, i.e. Romans, to this day. The invading Turkish Seljuks called Anatolia '*Rum*' i.e. Rome, and themselves as '*Rum Selchuk*'. M.D. Herr, ed., *The Roman Byzantine Period*, p. 40.

[34] Y. Hirschfeld, "A Climatic Change in the Early Byzantine Period? Some Archaeological Evidence". *PEQ* 136/2: 133–149 (2004).
Y. Hirschfeld, "The Expansion of Rural Setllememt During the Fourth-Fifth Centuries in Palestine" in: *Les villages dans l'Empire byzantin (IV-XV siècle)*. J. Lefort, C. Morrison et J.P. Sodini (eds.), Lethiielleux, Paris, pp. 523–537 (2005).

version of monotheistic religion, Islam. It demanded submission to principles allegedly written by *Allah* (i.e. God' himself, in the Holy Book of the *Qur'an*. Mohammed's tirades against the old gods infuriated his own people and he was forced to escape from his home in Mekka to the Jewish town of Yathrib, or Medina, in 622 C.E. in a move called *hijrah*, which is considered the beginning of the Islamic calendar. Many of Mohammed's earliest admonitions were taken from Judaism, such as one fast day per year modeled on *Yom Kippur* and the direction of prayer towards Jerusalem, the first *qiblah*. Eventually, Mohammed managed to turn the tide, returned to Mekka and made it the center of his new religion, which he purged of any Jewish aspects in order to appease his people. One of its main commandments is *jihad* – holy war against infidels who do not accept Islam as the only true religion. In the first skirmishes against the Persians and Byzantines in 629, 630 and 632 C.E., the Arabs were defeated. After Mohammed's death in 632 C.E., his successors, the caliphs, took it upon themselves to fulfill the *jihad*. The third invasion began in 634. In a crucial battle in 636 C.E. on the Yarmuk River, a confluent of the Jordan, the Byzantine army was defeated and all the Levant fell under the rule of the new Moslem empire.

The rapid growth of the Islamic empire and collapse of the Byzantine rule in the Fertile Crescent and of the Persian Empire had many roots. For one, the extended bleeding of both the Christian kingdom and its Zoroastrian opponents as well as the internal wars for the throne helped by the Armenians and other non-Greek minorities, worked to the benefit of the Moslems. The religious zeal of the Moslems and their belief in the expansion of the realm of Islam as a divine commandment stood them in good stead as did their well-known tolerance for hardship. In addition, new war tactics based on light cavalry and camel riders, enabled them to travel long distances and attack their less mobile opponents from every direction. Against them stood armies that practiced the old and conservative warfare with mercenaries commanded by corrupt local leaders. Moreover, the garrisons stationed at the *limes* along the border of the Arabian Desert were often recruited from local Arab tribes who saw no reason to defend the frontier against kindred tribes from Arabia. It is interesting, however, that the invasion from the deserts of Arabia in the 7th century C.E. coincided with the beginning of a phase of severe desiccation which reached an extreme peak towards the end of the 1st millennium C.E. One might question whether this change, first felt in Arabia, may have heightened the fervor of the religious zeal fuelling it with despair.

The 8th to the 10th centuries C.E., following the Muslim conquest, saw a general movement of Arab tribes and their settlement into the Fertile Crescent. The local farming population was obliged to pay a certain toll to the new invaders, who for the most part, did not practice agriculture, but relied for most of their sustenance on their herds. Papyri found at Nezzana in the Negev reveal that about forty years after the Arab conquest, the local populace, which still included a majority of Christians, had to supply wheat and oil to the *Banu-Saad* bedouin. The caliphs of the Umayyad dynasty (636 to the mid-8th century) had encouraged Arab immigration, but maintained the Byzantine settlements and preserved its Christian Greek-speaking administration. Thus, the Arab conquest did not cause the disappearance of settled life in the Negev. According to the papyri at Nezzana, this town remained unchanged at least to the end of the 7th Century and it was not until the first half of the 8th century when the cities of the Negev were abandoned and remained as ghost towns ever since. While some pottery and Umayyad coins at Nezzana are dated as late as 750, none of the Greek papyri is dated later than ca. 700 C.E.[35]

The Ummayads choose Damascus as their capital and constructed the earliest Islamic buildings on the Temple Mount of Jerusalem, a site the Byzantines considered unclean. At its foot, sumptuous palaces merged Byzantine and Persian elements into a new art form, initiating later Muslim art and architecture. Many members of the new ruling class often came from the sedentary horticulturists living in the oases of Arabia who had long recognized the potential of agriculture and the value of a well kept irrigation system: They introduced the Iranian qanat system into the Arava Valley and, last but not least, it was they who brought Near Eastern irrigation methods and a wide variety of plants and trees to northern Africa, Sicily, and the Iberian Peninsula, thus laying the foundation of the "Golden Age" of Spain. By and large, despite a steadily deteriorating climate, the Ummayad rule benefited the people not only of Syria-Palestine, but also of their entire realm.

There is no question that the immigration of a pastoral non-agricultural population contributed significantly to desertification of the Levant. Yet, information from the time series of proxy-data shows that a severe climate change undoubtedly forced the pastoralists in the desert margins to enter and occupy the sown land. The burden of the autochthonous farmers was further exacerbated by taxes the Moslem governors imposed on the non-Moslems, from which the believers were exempted. Thus desertion of the villages and towns in the semi-arid belt of the Levant accelerated during the 8th and 9th centuries, simultaneous with a rapid process of aridization, clearly evidenced by the isotopes and pollen curves as well as the levels of the Dead Sea.[36]

The desertification of the coastal plain was amplified as a result of the invasion of the sand dunes caused by an increased supply of sands to the Mediterranean Sea. The recent investigation of the Hellenistic-Byzantine cities of Herakleon and Canopus on the canopic branch of the Nile delta have shown that they have been deeply covered by sands brought by the Nile starting from the 8th century C.E.[37]

During this period, most of the Byzantine cities of Transjordan, including Heshbon, Medeba, and Garussa (Jeresh), were abandoned, often with a *coup de grace* of several severe earthquakes along the Rift Valley, and reduced to impoverished villages below the cultural level of the early Neolithic period. Recent investigations based mainly on soil profiles analyses brought to the concluion that the desertion of Abila and the other Decapolis cities was due to

---

[35] N. Lewis, "New Light on the Negev in Ancient Times" *Palestine Exploration Quaternary* 80:102–117 (1948).

[36] Y. Enzel, et al. "Science For Peace Conference".

[37] Stanley et al. *Nature* (2001).

a series of droughts.[38] The same was true for towns and villages in the Golan Heights, some Jewish, others Gentile. In Cisjordan most settlements on the fringe of the Judaean desert or in the rain-shadow of the mountains, such as the Jewish towns of Sussia and Maon south of Hebron or Korazim in the eastern Galilee, were deserted.[39]

Moreover, the center of power, which was reliant on provisions and was, therefore, interested in maintaining the economy, agriculture, and commerce, had moved from Damascus to Baghdad in 750 C.E. with the take-over of the political power by the new Abbasid dynasty. Situated now far away from the capital and becoming an outlying provinces, Syria-Palestine began its decline whereas Mesopotamia saw a revival. The later history of the Near East is a classic example of this process. The combination of a climatic crisis with the ascendancy of the Abbasids in the east, who could rely on river-fed irrigation, and the rising power of the Fatimids in northern Africa and Egypt, the "gift of the Nile", turned the Levant into a neglected no-man's land serving as a battlefield between these two Muslim empires. They would soon be joined by a flood of Turkish tribes and, worst of all, semi-barbarian invaders from Europe known as Crusaders from the west, topped by the Mongols from the east.

Thus ended the prosperity that characterized the ancient and the Hellenistic and Roman-Byzantine periods in the Near East. Numerous ghost towns covered by desert silts and sand, elaborate palaces, shrines and theaters – many built of marble imported from Greece and Italy – attest to the descent from the heights of prosperity. To blame this desertion solely on human misbehavior is to disregard the extraordinary creative genius of people whose achievements form the basis of modern civilization. The paleo-climate proxy-data time series stands in defense of *Homo sapiens* of the Near East who withstood the hostile forces of nature and of men, to succumb only when the combined forces of these adversaries reached beyond the limit of resiliency of their socio-economic system.

---

[38] B. Lucke, *Abila's Abandonment*, M.S, Thesis BTU, Cottbus, Germany, Yarmuk University, Irbid (2002).

[39] *NEAEHL* 290–299, 367–377, 552–556, 774–779, 929–934, 969–971, 1096–1101.

CHAPTER 9

Crusaders, Mamluks, and Ottomans on the Eve of the Era of Industry (ca. 800 C.E. to the Present)

"This blessed basin of water was built by the order of our master the greatest sultan, king and very honorable Khan, who holds the nations by their neck, the sultan of the Turks, Arabs and Persians. The Sultan Suleiman, son of Sultan Selim Khan. May Allah make his kingdom and rule eternal.
On the date of the tenth of the month of Mukharemin the year nine hundred forty three."

(Inscription on drinking basin (sybil) on the Pool of the Sultan (Birket el Sultan), near the Jaffa Gate of the Old City of Jerusalem, built in the time of sultan Suleiman the Magnificent – 1520–1566 C.E.).

During the last quarter of the first millennium C.E. tensions between the various political, ethnic and sectarian factions within the Islamic world led to constant warfare. The campaigns swept all over the Near and Middle East. Raiding Arab bedouin tribes and Turkish nomads intensified the deleterious conditions of the warm and dry climate. Fortunately, after the beginning of the second millennium C.E. the global climate started to change, becoming cooler and more humid, as evidenced by the various paleo-climatic data already discussed (Figs. 3 & 3a). This general positive trend was interrupted from time to time by dry periods.

An archaeological underwater survey showed a drop in the level of the Mediterranean Sea on the order of magnitude of 1.5 meters during a cold spell, a fact which can be deducted from the depth of floors of homes of the Crusaders in Acre (Akko, Ptolomeis) during this period. Also, a the lowest part of a tunnel built by the Knights of Saint John to enable safe passage from their stronghold to the port is now flooded by seawater.[1] Further evidence for a humid climate during the period is found in the rings of an old pine tree *Pinus nigra*, recently discovered in the Troodos Mountains of Cyprus (Fig. 12). A correlation was found between the average width of rings of other trees in Cyprus and the average of rainfall in Nicosia during the last 76 years. As

---

[1] Z. Goldmann, "Acco" in *NEAEHL* vol. 1, 24–27 (1993).
This tunnel dates from the Roman Hellenistic period, (a cold period) and was improved by the crusaders. When Issar visited this site in May 2000, he learned that the archaeologists who carried out the excavation were not aware of the paleo-climatic significance of the flooding of the tunnel. The uninterrupted slope of the tunnel shows that it is not faulting which may have caused the flooding.

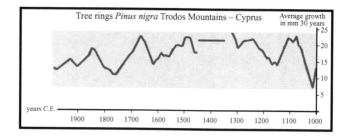

Fig. 12. A dendro-chronological diagram of *Pinus nigra* from Cyprus (Kypris 1996)

Cyprus is situated in the trajectory of most of the cyclonic lows that bring rain to the Near East, it can be assumed that this dendrochronological information can serve as a general indicator of climate for the Near East on the whole. Unfortunately, although more than 900 rings could be counted in the tree trunk, the central part has decayed, leaving a gap of unknown time between the first 380 and the last 570 years.[2] Correlation of these data from Cyprus with other dendrochronological time series throughout the Near East[3], as well as the information from the Soreq Cave, enables to reconstruct the climate changes over the time span of the last 1000 years. This correlation shows that after the hot and dry phase of the last part of the 1st millennium C.E. the climate improved reaching a cold and wet peak around 1500 C.E. known as the Little Ice Age in Europe, while between 1700 and 1800 C.E. it became warm and dry again.

9.1
The Crusaders' Interlude

The cooling at the dawn of the 2nd millennium C.E. coincided with an invasion of the Near East by people from the northern and temperate zones of Eurasia: Mongols and Turk in the Far East, Turks spearheaded by the Seljuks in Central Asia and the Near East, and the Normans in Western and Mediterranean Europe. We are by now in the full light of historiography which fuels the debate whether the changing climate has been the primary cause of the invasions or not. As always, it depends on the personal viewpoint of the historian and his or her interpretation of the sources. Nevertheless, in our opinion it may well have influenced the course of history in some way. The direct motivation for the Crusades, appear to have been religious zeal and the call of Pope Urban II for Christian warriors to leave their homes and join the crusade against

---

[2] D.C. Kypris, "Cyclic Climatic Changes in Cyprus as Evidenced from Historic Documents and One Century's Rainfall Data" in: Angelakis and Issar op. cit pp 11–128 (1996). The exact date of the fall of the tree is unknown and is assumed to be around 1990.

[3] N. Liphschitz, Y. Waisel and S. Lev-Yadun, "Dendrochronological Investigations in the East Mediterranean Basin" *Proceedings of the Pacific Regional Wood Anatomy Conference*, University of Tsukuba, Japan, (1984).

Table 5. A historical timetable of the Near East, from 0 to 1500 C.E.

years C.E.	Egypt	Syria-Palestine	Mesopotamia	Anatolia
		The Near East becomes part of the Ottoman Empire, end of Middle Ages		
1500				1453 fall of Constantinople
1400		Mamluk Period	Il-Khanate	Ottoman
1300		Mongol Invasions		Empire
1200	Ayyubids	Zengids		Seljuks of Rum and Armenian Kingdoms
1100		Crusader's conquests	Great Seljuks	
1000		Fatimid Period		
900				Middle Byzantine Period
800		Abbasid and Ummayyad Dynastic Caliphtes		
700				
600		——— Spread of Islam ———		Early Byzantine Period
500	Byzantine Empire		Sassanids	
400				
300		——— Spread of Christianity in Near East ———		
200	Roman Empire		Sassanids	Roman Empire
100			Parthians	
0		70 Jerusalem destroyed		

the "infidels". This summon did not fall on deaf ears, but instead, led the leaders and masses of Europe to adorn themselves with the sign of the cross and embark on one of the most romanticized bloodsheds in history. It started with massacres and plunder of their own Jewish communities, sacrificing on the way thousands of Christian children and peasants who were enthralled by their fervent spiritual leaders and was followed by hordes of warriors, peasants, knights, and noblemen. Their goal was to liberate the tomb of a Jew from Nazareth who had preached peace and love from the hands of the infidel Moslems.

The religious fervor may have been bolstered by socioeconomic conditions that resulted from the sequence of warm and cold climate alterations, beginning with the warm spell at the end of the first millennium C.E. that swept Europe, enabling the Norse seafarers to reach and settle Iceland and Greenland.[4] In the more temperate areas of Europe this warm period brought prosperity to agricultural communities and a dramatic rise of population. The increasing number of noble families who did not inherit land reached a peak in the 11th century. Thus, the Crusades served also as a channel to absorb the surplus population, in addition to the drive to clear the forests and to drain the marshes of Europe to create new lands.[5] Some historians estimate that about a tenth of Europe's population, which at that time had reached about 50 million people,

---

[4] H.H. Lamb, *Climate, History and the Modern World*, Methuen & Co. London, pp. 162–173 (1982).

[5] Prawer, *The World of The Crusaders*, Quadrangle Books, New York (1972).

was sacrificed between 1095 and 1291 in the expeditions to win the Holy Land. Although this estimate may be exaggerated, the Crusades were a major factor in defining Europe's future, as they reestablished the borders on the shores of the Mediterranean.[6]

In 1097 the Christian armies began the conquest of the Levant through Anatolia from the north and marching south and east. With the conquest of Jerusalem on July 15, 1099, they realized their goal. Christian spirit of charity was sorely lacking when the triumphant bearers of the cross, massacred not only all the Moslem and Jewish inhabitants of the city, but Christians of other persuasions as well in a killing orgy of unprecedented gruesomeness even for their time. The Frankish feudal Kingdom of Jerusalem and other principalities and earldoms similar to those of Europe were established in the Near East from Edessa/Urfa to Antioch, Tripoli and Cyprus. The first Kingdom of Jerusalem survived for about a century until Salah ed-Din, a Kurdish warrior and former governor of Egypt, vanquished it after defeating the Frankish army in the battle of the Horns of Hittin in the Galilee.

This battle, which represents a major turning point in the Frankish venture in the Near East, clearly demonstrated the military superiority of Salah ed-Din over his Christian opponents. It illustrates equally well the critical role water has played in shaping Near East history. The battle of the Horns of Hittin was fought on a hot summer day – July 4, 1187. The fate of the Frankish army, numbering about 18,000 infantry soldiers and about 1200 mounted knights had already been sealed July 2, by the king of Jerusalem, Guy de Lousignan (1186–1192). De Lousignan decided to ignore the advice of the ruler of the Galilee, Reymond of Tripoli, not to counter-attack Salah ed-Din's army, which had already conquered the major Crusaders' forts in Transjordan and was besieging Tiberias. Reymond believed that the Frankish army would not be able to sustain itself and would have to disperse. He warned against the danger of the long march to Tiberias without an adequate source of water for the army. Unfortunately for himself and his kingdom, the king of Jerusalem instead heeded the advice of the head of the Order of the Templar Knights to attack Salah ed-Din's army in order to lift the siege of Tiberias. Thus, on that very hot day of the third of July, the Frankish army began its march from the springs of Zippory/Sepphoris in the Galilee towards Tiberias. As Reymond had predicted, the army and its animals suffered from thirst during the long hot day. Neither did the knights' heavy armor lighten their burden or that of their mounts. As a result, the commanders decided to change direction and head towards the springs of Hittin. Night fell before they reached the springs, and they made camp, thirsty and tired. During the night the Moslem army positioned itself between the Frankish camp and the springs. The next day, the thirsty Franks were surrounded and had to fight without refreshing themselves or their horses. It is no wonder, therefore, that they were defeated, with all their nobles and knights either being killed or taken prisoner, and no reserve army to continue the fight. On October 9, 1187 Jerusalem surrendered to the Moslem leader. This

---

[6] F. Braudel, *A History of Civilizations.* Penguin Books, Harmondsworth. pp. 309–310 (1995).
    A. Maalouf, *The Crusades through Arab Eyes*, Schocken Books, New York (1984).

time there was no massacre; most of the Christians who could afford to pay a small ransom were allowed to leave the city, while the poor were sold as slaves.[7]

The Christian world could not accept the loss of the Holy City and new crusades followed. As by now the Seljuks blocked their way over land, they had to hire ships of the Italian city-states of Venice, Pisa and Genoa for a handsome sum. Most of the campaigns ended in defeat, while some regained territories of the Levant, only to be terminated by the Mamluks, a new phenomenon in the Near East. Mamluks were originally "slaves"[8] of mostly Turkish and Circassian origin. They were purchased as young boys, adopted by their masters, educated and trained as warriors or administrators. Anyone who showed ability could rise in the ranks. In a short time they became the backbone of the armies of the descendants of Salah ed-Din, the Ayyubide dynasty, and in 1250, they themselves took over the government of Egypt. After their sultan, Kutuz, defeated the invading Mongol army at 'Ein Jalud (Biblical 'Ein Harod) in the Jezreel Valley in 1260, his inheritor Baibars[9] (1260–1277) started an offensive against the last Crusaders strongholds in the Near East. His followers carried on to the last stage when the Sultan Al-Malik el-Ashraf Halil (1290–1293) conquered Acre/Akko, the crusaders' last fortress, and thus ended the Christian endeavor to gain control over the "Holy Land".

In spite of incessant wars, the economy of the Levant flourished during the time of the Crusaders' rule. Pilgrims, soldiers, colonists and merchants from all the European countries found their way to its ports. At the same time, capital was invested in building castles, hostels and rural and urban strongholds. Recent archaeological investigations show that the rural Frankish colonization thrived. They learned irrigation methods not known in Europe, renovated aqueducts and built watermills and windmills.[10] These findings are contrary to the previous notion that the crusaders were predominantly an urban society.[11] The flourishing agricultural economy was most certainly made possible by the increasingly humid climate, abundant water for irrigation and only rare droughts.

---

[7] A. Jones, T he Art of War in the Western World. Oxford University Press, Oxford, pp. 138–139 (1987).

[8] An important distinction must be made between the Western concept of slavery, based on Roman law, and slavery in the ancient, as well as the more modern, Near East. In the West, the slave was a dehumanized chattel without any rights and could be worked to death or simply killed at will. In the East, according to earlier laws, including the Islamic shari'ah, killing a slave is regarded as murder, as a slave never lost his or her humanity. The Arab word 'abd, translated as "slave" in English, actually means "servant". It wasn't until the 18th century that slavery in the West and East became comparable. Mamluk means "belonging" to another person who expects absolute loyalty. From the 8th century we hear of desperate families in central Asia offering their children for "sale" or better "adoption" into the Muslim society in Iran and Afghanistan. It was expected that the children, boys and girls alike, would be treated decently, fed, clothed and provided an education. A mamluk had certain rights of inheritance from his master in return for loyalty and obedience.

[9] Succession in the Mamluk's regime was not from father to son but according to the merits of the man as a warrior among his mamluks.

[10] R. Ellenblum, "Settlement and Society Formation in Crusader Palestine" in: T. Levy. op. cit., pp. 502–511, (1995).

[11] J. Prawer, "Colonization Activities in the Latin Kingdom" in: Crusader Institutions, Oxford University Press, Oxford, pp. 102–142 (1980).

9.2
The Slaves who Became Sultans – Mamluks, Mongols and Turks

The Ayyubides and the Mamluks who took their place adopted a policy of destroying the cities all along the coastal plain of the Levant, from Tyre in the north to Gaza in the south, to keep the Crusaders from landing and re-conquering Palestine. As a result, the coastal area became barren and sparsely populated for centuries.[12] Sand dunes began to bar the outlet of rivers from the mountains, creating marshes in which malaria-transmitting mosquitoes bred. The letters and annals of pilgrims to the Holy Land during the Mamluk rule describe an impoverished country. Cities bordering of the desert, such as Gaza were subject to raids by the bedouin tribes. The governor of Gaza had to organize punitive sorties in order to relieve the pressure they exerted on his city.

The period of the Crusades also saw attempts by the Mongols to conquer the Near East. In 1256 they invaded Iran and destroyed Alamut, the stronghold of the Shi'ite-Ismaelite assassins. Two years later they conquered and destroyed Baghdad, which was during the times of the caliphs Harun ar-Rashid (786–809 C.E.) and Al-Mamoun (813–833 C.E.) a center of Islam culture. The Abassid caliphate spiraled into a steep decline. The Mongols pushed on and tried to conquer the Levant but were, at first, repulsed by Baibars, the sultan of the Mamluks. In another attempt, during the reign of Sultan Kala'un, they managed to conquer Aleppo in Syria and defeat the sultan in the battle of Homs, but were forced to retreat because of internal problems. Their following attempt, near Damascus in 1303, ended in defeat.

Besides devastating entire communities, the Mongols destroyed many water projects the inhabitants had built and maintained for millennia. The effects of these Mongol invasions on the Near East are felt until the present. In Iran, they destroyed the qanats, on which the besieged towns depended for their water supply.[13] In Mesopotamia, now 'Iraq, they destroyed the irrigation canals, mainly in the southern part of the country. The horrendous massacres and destructions eliminated the cultural superiority of Mesopotamia in Near Eastern history once and for all.

The arrival of the Turks in the Near East had already signaled economic distress due to a function of demographic pressure accentuated by climate fluctuations in the steppes of northeastern Asia. As in Europe, the warm climate that came at the beginning of the second millennium C.E. had caused a surplus of herds, and therefore, increase of population. As already mentioned in the last chapter, pastoralism is an economic dead end, a fragile system allowing no intensification, only expansion into new territories. Chinese records from the 6th century B.C.E. on show the similar numbers of

---

[12] M. Rosen-Ayalon, "Between Cairo and Damascus: Rural Life and Urban Economics in the Holy Land during the Aybbid, Mamluk and Ottoman Periods" in: *The Archaeology of Society*. pp. 512–523.

[13] While surveying groundwater resources in the Khorassan province of eastern Iran in the early 1960s, Issar was shown quite a number of tracks of ancient qanats leading to ancient ruined towns. All were, without exception, destroyed by the Mongols and have since remained as ruins.

cattle, and therefore people, as the Soviet census of 1928.[14] Other data from China suggest that the Mongolian expansion may have been the outcome of a combination of socioeconomic impacts. A warm and rainy climate in the last centuries of the first millennium C.E. brought prosperity and overpopulation in the Mongolian steppes. When the climate started to change early in the 2nd millennium C.E. a series of cold and dry years may have threatened starvation. This sequence of climate changes we can infer from the records of the Chinese Imperial Court, which enumerated the endowments and reports from the various provinces, including the dates of the blossoming of the plum tree, which was celebrated by the Chinese. This tree, normally widespread in the Yellow River valley, disappeared from this region after the Song Dynasty (960–1279), presumably indicating a colder climate. The cold climate during the 12th century also affected the cultivation of *litchi*, a Chinese tropical fruit widely grown in southeast China at least since the Tang Dynasty (618–907). In this region *litchi* were eradicated twice by cold weather, in 1110 and 1178. Other Chinese historical documents describe a warm period from 650 to 1000 and then a cold period from 1050–1350. These changes can be inferred from analysis of the flood records of the Yellow river, as high floods coincided with warm climates and *vice versa*. The longest wet period was from 810 to 1050, whereas the time from 1050 to 1270 was of low frequency of floods. These dry periods in northern China, especially in the latitudes 30 – 45 °N coincided with a much colder climate, with strengthening of the Siberian anticyclone and a southward shift of the polar front.[15]

By the middle of the 14th century, the Black Death arrived, decimating the Mamluk empire, although to a lesser degree than Europe. The drier climate may have helped to limit the spread of disease in the Near East, but the relative sparing was more likely due to habits of personal hygiene, as Moslems wash their hands and feet at least five times a day before praying. Unlike Europe, the classical tradition of bathing was never given up in the Islamic world. More stringent hygienic practices probably accounted for the fewer deaths from plague among the Jews in Europe as well, but this led to accusations of poisoning the wells and consequent persecution by their Christian neighbors. In the Near East there is no report of blaming the infidels, even though the people were aware of the connection between the death and the water sources. Many people fled from the densely populated cities where drinking water came from pools or cisterns and camped near springs. The plague undoubtedly contributed to a religious fervor and much land and buildings were donated to the *wakf*, an authority in charge of consecrated property for the common good.

Needless to say, the plague damaged the economy of the region. The population of the urban settlements declined while many rural settlements and

---

[14] A.M. Khazanov, *Nomads and the Outside World*, Cambridge University Press, Cambridge (1984).

[15] Zhang Peiyuan and Wu Xiangding, "Regional Response to Global Warming: A Case Study in China" *World Laboratory and CCAST Workshop Series,* vol. 5, ed. YinoYuan Li, London, Gordon & Breach Scientific Publishing 26–41 (1990).
Fang Jin-qi, "The Impact of Climatic Change on the Chinese Migrations in Historical Times" in: *Global Change and Environmental Evolution in China,* ed. Liu Chuang, Zhao Songqiao, Zhang Peiyuan, and Shi Peijun, Hohot, China, Editorial Board of Arid Land Resources, pp. 96–103 (1990).

their fields were abandoned. Invading bedouin tribes settled various parts of the Levant, filling the vacuum left by the thinning population. Their sheikhs began to dominate the trade routes, exacting tribute from all who traversed their territories.

Timur Lang was the last Mongol khan to attempt to invade the Near East. In 1400 he invaded Syria but had to retreat because of a war he was waging on another front against the Ottoman Turks in Anatolia. Although he defeated the Turks, Timur did not pursue another Near East invasion. He died in 1405 and his empire disintegrated due to conflicts between his heirs.

9.3
The Ottoman Centuries: Peace and Stagnation

Ural-Altaic and Turkic speaking tribes had begun to encroach into Europe and the Near East as early as the first half of the first millennium C.E. New Turkish tribes came from the steppes east of the Caspian and settled along the eastern borders of the Byzantine Empire. Whether their southwesterly migration was related to the colder climate that persisted in the region during this period remains to be investigated. Some of the tribes (e.g. the Khazars and probably the first Seljuks) adopted Judaism. The Seljuks soon adopted Islam and due to their military prowess became the elite regiments and palace guards of the various Moslem potentates, gaining power and influence throughout the eastern Moslem world. Their *khan* Tughrul conquered Baghdad in 1058 and became the ruler, or *sultan*, of the Realm of Islam,[16] opening the gates of the Near East to ever-increasing numbers of Turkish warriors and their families. From the 10th century the Seljuks ruled the eastern Moslem states from Afghanistan and Persia to Anatolia, with its center at Konya. The ruthless Mongol assaults during the 13th century and their conquest of the eastern Turkish states caused many inhabitants of these regions to flee to the Near East. The Seljuks had enlisted other Turkish tribes to wage constant war against their Christian neighbors to the west. After the decline of the Seljuks in the wake of the Mongol advances, a Turkish *emir* (leader) near Bursa in northwestern Anatolia by the name of Osman (1281–1326) proclaimed himself sultan. He and his son Orhan competed with other emirs but were able to consolidate their power. The victorious Battle of Kosovo in 1389 led to the occupation of the Balkan Peninsula and remains etched in the memory of the people of the region to this day. The Ottomans[17] strengthened their military might by developing the *devshirmeh* – a method of recruitment similar to that of the Mamluks, in

---

[16] Until this time, the head of the Abbasid dynasty was both the spiritual (*khalifa* i.e. caliph) and secular ruler, the *sultan*. After 1058, the Abbasids who escaped to Egypt after the Mongols conquered Baghdad retained a virtual caliphate until the Ottoman conquest of Egypt in 1517. The various states of the medieval and more recent periods were created mostly by Turks and ruled by dynasts of Turkish sultans. In modern Arab historiography, this was the moment *"when history went wrong"*.

[17] The name "Ottoman" is derived from an 18th century "overcorrect" mispronunciation of the name of the founder, *Osman*, based on the Arab pronunciation *'Othman* (with a sound like the English "th") but difficult for Persian and Turkish speakers who prefer the /s/ sound.

which a professional army is built from converts, mainly of Christian children, who are converted to Islam and educated to be outstanding fighters as well as accomplished administrators. In due time these Sipahis and Janissary (> *yeni chari* = new army) soldiers learned how to use black powder for siege and personal guns. In 1453 Constantinople fell to Mohammed II (or Mehmet), the Conqueror (1451–1481)[18] who renamed the city Istanbul and made it his new capital. After the expulsion of the Jews from Spain and other Christian countries in 1492, they were welcomed to settle in the Ottoman Empire.

The Ottomans continued their attacks first to the west, and by 1528 they were outside the walls of Vienna. The year 1517 has seen the demise of the Egyptian Mamluk Empire, which had deteriorated from government corruption and internal conflicts. The local rulers in the Levant, many of them bedouin sheikhs, contributed to this disintegration with their continuous internecine warfare. The Mamluks were conservative fighters who viewed the use of firearms with contempt, regarding guns as the weapon of cowards afraid to face their enemy in short-range combat. Their arrogant refusal to adopt military reforms was a boon to the Ottoman sultan Selim I, who, after defeating the Safavid Shah of Persia in 1514 headed south to fight the Shah's allies, the Mamluks. Their fate was decided in three devastating battles, the last at the gates of their capital, Cairo, where the Mamluk horsemen never managed to engage in face-to-face combat, as the Ottoman guns destroyed them at long range.

The defeat of the Persians and the Mamluks brought the Near East, including Arabia, under the rule of the Ottoman Turks, whose empire reached its zenith under Suleiman the Magnificent (1520–1566), son of sultan Selim I. Suleiman penetrated deep into the Balkans by conquering Belgrade and Hungary. He also besieged Vienna a second time, but again failed to take the city. He forced the Knights of St. John – the Hospitallers, to evacuate the island of Rhodes, to flee to Malta and to become pirates. He further extended his empire to include Libya, thus making the eastern Mediterranean a Turkish sea. In the east he conquered the western part of Persia, including 'Iraq, and reached into the Caucasus and the plains of southern Russia. The empire stretched south into Sudan, gaining footholds on the coast of eastern Africa.

In addition to his military prowess, Suleiman was an effective legislator. He published the *Kanoon Nameh* (hence his Turkish title *Kanooni*, i.e. the law giver), a codex of Islamic law and administration for the Ottoman Empire. His court was also a center for cultural activity where jurists, architects and poets were welcomed. "Ottoman" has become a cultural concept irrespective of individual ethnic or religious background. When Selim conquered Cairo, he got possession of Islam's most sacred relics – the sword, the mantle and other personal items belonging to the Prophet Mohammed – and, to the deep resentment of the Arab world, declared himself caliph. Thus, Istanbul replaced Cairo and Baghdad as the political and cultural center of the Islamic world. The Sultan favored Jerusalem, ordering reconstruction of the walls and gates

---

[18] To guarantee against palace wars of inheritance, Mohammed II enacted a law to kill all the sultans' brothers borne to the many women of the sultan's harem. This did not prevent many intrigues designed for the survival of these women and their servant eunuchs.

through which the contemporary visitor enters the Old City. He also reconstructed the water supply system from the springs south of the city and the drinking basins in the city.[19]

The economy of the Near East in this period initially benefited from a favorable climate. Tree ring data show that the second half of the 16th century was relatively humid. It brought a short period of economic revival and prosperity that was manifested in a growing number of villages and expanding area of farming in the Fertile Crescent.[20] However, it did not last for long. The proxy-data time series and tree ring data available from various parts of the Near East[21] show a phase of warming, and thus drying, beginning in the first half of the 17th century. This turn of fate led to a general trend of desertion of the villages and decline of agriculture not only in the central and southern parts of the coastal plain of Syria-Palestine but also in once prosperous Mesopotamia. This may have been also the period when the sand dunes completed their conquest of Palestine's coastal plain.

At this same time, the deterioration of the Ottoman Empire accelerated. The first seeds of degeneration had been sown in the higher levels of Ottoman society during the reign of Suleiman. It became customary for the sultan not to have one legal wife but to sire offspring from many concubines given by their families to the royal harem. The resulting intrigues of the harem led, in Suleiman's lifetime, to the rebellion and execution of two of his sons. Later, the death of a sultan resulted in the massacre of all his sons except one who was chosen as his successor and one 'reserve'. In 1570 the Turkish navy suffered its first major defeat at the battle of Lepanto, demonstrating to the European nations that the 'invincible' Turkish army could, in fact, be beaten. The main cause of economic decline, however, was the system of tax farming, i.e. the leasing of the rights of taxation of farmers by some influential magnate. It fostered corruption, oppression of the farmers and injustices very similar to the tax farming system of the Roman period that contributed to the disasters of the 3rd century C.E. Thus, the seeds of deterioration of the empire had begun to germinate in the time of Suleiman's successors, just after their conquest of the Near East and northern Africa and the establishment of administrative regions. In each region they installed a local governor who was the official chief leaser and responsible for collection of the local taxes. Each region was subdivided into smaller units governed by a sub-tax leaser and thus a local officer for the collection of taxes (*Pasha, Wali, Bey*). The governors' performance was defined by his success in collecting the amount of taxes expected to reach the treasury in Istanbul. Because these tax amounts were fixed, the leaser exacted his personal profit margin by demanding a sum beyond the official levy, a practice that was enormously destructive to the local economy. Furthermore, the governors were commissioned for one year only, not knowing if they would maintain their post the next year, and the land, being government property, was leased for one

---

[19] See introductory paragraph to this chapter.

[20] A. Cohen and B. Lewis, *Population and Revenue in the Towns of Palestine in the Sixteenth Century*, Princeton University Press, Princeton (1978).

[21] Liphschitz, et al., op. cit. (1984).

year only as well. There was, therefore, no long-term incentive for either the local governors or the farmers to develop the land. Instead, the system became more and more corrupt, easily accounting for the gradual degradation of the economy of the Near East during the Ottoman rule. As such, the worsening climatic conditions were not the principal cause of the empire's decline, but helped to expedite its deterioration.

The discovery of the sea routes to the Far East around Africa by the Portuguese Dias in 1488 and the discovery of the Americas in 1492 by Columbus were the death knell of the Ottoman empire and disintegration of all the other political and economic systems of the Near East and beyond. The Portuguese and Dutch, followed by other Europeans, usurped the commerce by colonization and unprecedented exploitation. Trade along the "Silk Road" or "Incense Road", which had for millennia linked China, India and the coasts of the Indian Ocean with the Mediterranean world virtually dried up, ruin the livelihood of merchants, local seafarers and caravan-farers and contribute to the growing isolation and neglect of the Near and Middle East.

The combination of these developments hastened the impoverishment of south-western Asia, especially after the first half of the 17th century during which the climate became very dry. This may have been the main factor that led the bedouin tribes from the plains of the Arabian and Syrian deserts to invade the sown lands once more.[22] The Ottoman government tried to control the incursions of the nomads by building forts and settling army cavalry and infantry at these sites. These units were also expected to escort the pilgrims on their way to Mecca. However, as time progressed, the central government in Istanbul could not afford to maintain the fortresses and pay the soldiers, and the government of many parts of the Levant fell into the hands of the local sheikhs and petty warlords.

Istanbul's power over the vast Ottoman Empire reached a nadir during the 18th and 19th century. In the Near East, local *beys*, *sheikhs*, and *pashas*, such as Ali Bey el-Kebir in Egypt and, Sheikh Dhaher al-Omar and then Ahmed Jezzar (the Butcher) Pasha in Palestine, became *de facto* independent rulers. They maintained their own armies and even conducted foreign relations quite independently. Their one concession to Istanbul was to deliver a portion of the taxes they squeezed from the local inhabitants. Jezzar Pasha, originally a Bosnian, had an army strong enough to stop Napoleon Bonaparte at the walls of Acre/Akko, his capital and fortress, helping to thwart Napoleon's plan to include the Near East within the French circle of influence.

In the tumult of the period, Egypt became independent in a similar fashion when an Ottoman officer of Albanian origin, Mohammed Ali (1769–1849), first fought the French and then had declared himself governor of Egypt. At first he collaborated with the central government in Istanbul by suppressing the revolt of the Moslem fundamentalist Wahhabites in the Arabian Peninsula and by fighting against the Greek rebellion in the Balkans. However, when the Sultan rejected his demand to become also governor of Syria and Palestine, he sent his stepson Ibrahim Pasha to conquer these lands and then advance northward

---

[22] M. Maoz (ed.). *Studies on Palestine during the Ottoman Period.* Magnes Press, Jerusalem (1975).

to defeat the main Ottoman army at Konya. This argument convinced Sultan Mahmud II (1803–1839) to agree to Mohammed Ali's terms. Mohammed Ali then demanded that his post be made hereditary, and once again, his demand was rejected. The Ottoman army was sent to battle the Egyptian army supported by French arms. The Egyptians prevailed, with the surrender of both the Ottoman army and navy. Now, Mohammed Ali could have seized the entire empire were it not for the intervention of Britain, Russia, and Prussia, who opposed a greater influence of the French in the Levant through their Egyptian ally. Mohammed Ali was forced to withdraw but was granted hereditary rights.

Even after the local governors of the Levant achieved *de facto* independence from the "Sublime Porte"[23] of Istanbul in all military and diplomatic activities, the Near East nevertheless remained part of the Ottoman Empire. It retained, in particular, the conservatism and corruption of the administrative cadre while ignoring advances in science and technology. Moreover, the most important social ideology introduced by the American and later the French Revolution – that government exists for the benefit of the people and not the reverse – did not penetrate the consciousness of the rulers nor most of the subjects of the Ottoman Empire. Thus, this huge empire did not participate in the social, scientific and technological achievements of the industrial revolution in the Western world from the 17th to the 19th centuries. During this period the Ottoman Empire became 'the Sick Man on the Bosphorus' owing its survival solely to the conflicts of interests among the European nations. Yet, with the colonial expansion of Europe on one side and the evolution of nationalism and democracy in those nations on the other, winds of change stirred the Ottoman Empire. The defeat of its army and navy led to the independence of the Balkan nations, the Slavs and the Greeks, teaching the Turks a painful lesson. In 1839, bending to pressures from the European governments, Sultan Abdul-Hamid I (1839–1861) instituted a series of administrative and socioeconomic reforms. Along with other concessions granted to European companies, he introduced technological innovations, including telegraphs and railways, and admitted colonies of Europeans, especially to the 'Holy Land', Palestine.

Towards the end of the 19th century, the socioeconomic conditions of the empire had improved somewhat as a result of administrative and judicial reforms that were enforced after the Egyptian authority withdrew. Yet these were short-lived, as Abdul-Hamid II (1876–1909) reverted to autocratic rule.

Still, because of concessions[24] the Turkish government had to make under the terms of various treaties, a growing number of European nationals were allowed to settle in the Levant. In Palestine, many members of the clergy of various Christian religious institutions settled, as did religious layman who wanted to restore the settlements abandoned after annihilation of the Crusaders' states. One such group was the German Templars, who established

---

[23] The 'Sublime Porte' was the entrance to the Grand Wezir's palace and seat of government, similar to the 'White House' in Washington or '10 Downing Street' in London.

[24] Since the beginning of the Ottoman rule of the Levant in 1535, the consulates of France, and later those of other European countries, attained special rights which were known as 'capitulations'.

a few agricultural colonies all over Palestine and settled in the suburbs of the main cities where they set up mechanical workshops.

At the same time, Jewish immigrants from Europe began to arrive and strengthened the native Jewish population, with the ultimate goal of reviving Jewish independence in their 'Promised Land'. This immigration, which later became organized as the Zionist movement, aimed also to change the main occupation of the Jewish people from merchants to farmers. Using modern methods, the new settlers completely transformed the local agriculture. Exceptional progress was achieved in orange orchards irrigated by pumped groundwater. Arab landowners adopted the new agriculture which then spread over vast areas of the coastal plain, the soil and climate of which were well suited for this fruit tree. A chance mutation of the orange fruit (locally known as *Shamooti* and abroad as *Jaffa*) produced an excellent taste and quality for export. The development of this intensive type of agriculture was central to the agriculture of Palestine until late in the 20th century.

Thus despite the rather warm and dry climate of the first part of the 19th century, orchard agriculture thrived because of groundwater-pumped irrigation. Originally, the water was raised from hand-dug shafts by traditional animal-driven wooden pumps. This method could not satisfy the water requirements of the growing number of new orchards in the coastal plain. Thus, in the second half of the 19th century, the shafts were drilled deeper by machines, and the pumps were powered first by steam and later by kerosene engines. The more efficient centrifugal pumps replaced the piston pumps enabling to draw more water from increasing depths.

The reforms introduced by Abdul-Hamid I were nullified by Abdul-Hamid II (1876–1909), an act leading to the inevitable end of the autocratic rule of the sultans and the darkest period in the history of the empire. He was the first to use religious fanaticism to further his ends – and with disastrous results in the first wave of the atrocities committed on the Armenians by Muslim fanatics. A popular movement known as 'Young Turks' rebelled in 1908 and a year later was able to force Abdul Hamid II to abdicate. He was followed by Mehmed V (1909–1918), the last Ottoman emperor and puppet in the hand of the nationalist 'Union and Progress Party' led by Enver Pasha. The new regime insisted on keeping the empire intact and even enforced measures in order to "Ottomanize" its numerous nationalities. On one hand, this ended discrimination between Muslims and people of other religions. At the same time, however, the attempt to integrate the empire's various ethnic groups into a single nation led individuals to be deprived of their rights, creating an environment ripe for rebellion and countermeasures, including the indiscriminate massacres and deportation of the Armenians during the ensuing wars. Other minorities suffered, too, but to a much lesser extent. In trying to maintain their empire, the Ottomans joined the German-Austro-Hungarian axis during World War I. It was defeated, largely by the British based in Egypt who were securing their hold on the route to India through the Suez Canal. With this defeat, the Ottoman Empire collapsed in 1918.

CHAPTER 10

An Epilogue

The Near East Enters the Era of Technology and Global Change

After World War I, the Near East was divided into zones of influence between Britain and France. Egypt remained under British control. Britain obtained the mandate over Iraq and Palestine on both banks of the Jordan, France over Syria and Lebanon. In Turkey, Greeks, Italians, Armenians and others tried to occupy large parts of Anatolia, until an army officer named Mustafa Kemal Pasha stopped them in battle. In 1923, he established a modern democratic independent republic and was awarded the title Ataturk (Father of the Turks). None of the other Muslim nations of the region followed his example, which are still ruled by conservative autocratic and often oppressive regimes.

The region enjoyed rapid technological and economic progress in the period between the wars. This was especially true for Palestine, the destination of a major immigration of European Jews who had embraced the Zionism, a movement to establish a "Jewish National Home" in Palestine. The movement was encouraged by the promise of the British government during the First World War to help in fulfilling this goal. In addition, anti-Jewish riots in Russia, Poland, and later, Germany persecutions brought many Jews to seek refuge outside Europe. Urbanization and agricultural growth was aided by modernization of the administration and public services by the mandatory governments and the new immigrants. Thus from the end of World War I, technological innovations played a prominent role in Palestine's socioeconomic development. In agriculture, for example, the dependency on rain-fed crops waned, supplanted by irrigated agriculture based on pumped groundwater. The new technologies considerably reduced the potential of climate fluctuations to determine the country's fate.

Another important factor to emerge in Near East economy and politics was the discovery of oil fields in its eastern part, bringing the region into international focus. This was especially true for the United States and its oil companies, whose interest was to push the British out of the region. Thus in the decades following World War II, the countries of the Near East became independent, among them Israel. This caused a series of political conflicts that still remain to be solved. Among the reasons (or the pretense for the reasons) for these conflicts are issues of water rights. These issues were central in the tensions over the last few years between Turkey, Iraq and Syria regarding the water of the Euphrates and Tigris, part of which Turkey plans to use for hydroelectric power and irrigation.[1] The water of the Jordan River is the subject of debate among Syria, Israel and Jordan, and historians undoubtedly will argue about whether this issue led to the outbreak of the 1967 war. Another, as yet unresolved issue between the Israelis and Palestinians, concerns the sharing of the joint groundwater resources.[2]

The debate over these conflicts and their possible solutions do not take into consideration the Global Change monitored by atmosphere, ocean and earth scientists all over the world. In general, it can be said that industrial and technological advances in the Near East have weakened the link between climate and the welfare of the people. Yet, oddly enough, in the wake of the industrial development of the world, helped in no small way by the oil flowing from the Near East, this link may be strengthened again. A new crisis of global warming, this time wrought by human deeds responsible for increasing the quantity of greenhouse gases – especially carbon dioxide – in the atmosphere and the rise of temperature, may have severe consequences for the region. That it may endanger the economy of many countries in the region is a foregone conclusion that can be drawn from the present book, namely, that a warm period is equated with a reduction in precipitation. This conclusion supplies information wanting in that provided by the global climate computerized models. Because of the small regional scale of the Near East, these models cannot predict the exact impact of the global climate change on this region.[3] Taking into account that, with the exception of Israel, agriculture still plays a significant role in the economy of the Near East, a severe reduction in precipitation will most certainly, damage the economy of most countries in this region.[4]

A conclusion that can be drawn from the history of this region is that such a crisis resulting from a shortage of water resources has been averted or mitigated in the past by human resourcefulness and innovation. This is especially true in modern times when these efforts are based on scientific research, engineering, and agronomic innovations as well as the education of farmers to

---

[1] M.Tomanbay, "Turkey's Water Potential and the Southeast Anatolia" in *Water Balances in the Eastern Mediterranean*, D.B. Brooks and O. Mehmet (eds.), International Development Research Center, Ottawa pp. 95–112 (2000).

[2] D. Hillel, *Rivers of Eden: The Struggle for Water and the Quest for Peace in the Middle East.* Oxford University Press, Oxford (1994).
N. Kliot, *Water Resources and Conflicts in the Middle East.* Routledge, London, (1994).
S. Altout, "Water Balances in Palestine, Numbers and Political Cultures in the Middle East" in: *Water Balances in the Eastern Mediterranean*, Brooks and Mehmet (eds.), pp. 59–84 (2000).

[3] Intergovernmental Panel on Climate Change (IPCC), Special Report on The Regional Impacts of Climate Change, An Assessment of Vulnerability, Chapter 7 – Middle East and Arid Asia (WMO – UNEP (2001).

[4] An exception is Egypt, which may enjoy higher Nile flow due to a stronger easterlies regime. Hopefully, the volume will not exceed the capacity of the Aswan Dam.

adopt new technologies. Needless to say, scientific and technological advances, together with education, are interdependent processes that can succeed only when the vicious cycle of poverty and ignorance, religious fanaticism, and aversion to innovation, is broken.

The authors hope that this book will contribute to an understanding of the problems and processes this region had experienced in the past, and be of some help to the reader to draw her or his historical as well as practical conclusions about the future.

Appendix I

Groundwater Use and the Evolution of Groundwater Exploitation Methods as a Function of Climate Changes

In semi arid regions like the Middle East groundwater is of crucial importance for the survival of human societies. Some of the most outstanding monuments of antiquity were emergency water supply systems built with an amount of labor second only to the foritifications of a town or city. In order to understand the reason for this fact we present here a short survey of the principles of groundwater flow and storage, especially in such regions.

The water infiltrating the subsurface immediately after the rains, or from the water of the floods running over permeable gravel layers, starts its voyage in the subsurface in a vertical direction through the pores, fractures, and channels in the soil or rock, until it reaches an impervious layer at a shallow or deep level. In the upper layers, it may be drawn back to the surface by upward capillary pull and evaporate, leaving the salts in the soil, the thickness of this zone depends on the grain size of the soil. Below the capillary zone there is the root zone, in which the water may be drawn back upwards by the roots, and be transpired. Some trees like the acacia may have roots a few meters deep. The water, which does not transpire accumulates in the subsurface on the impervious layers. The pores and fractures above these layers become saturated. The water in the saturated rock is termed groundwater. The direction of movement in is towards the outlets, either springs or the sea. The plain, which separates the saturated part of the rock from the overlying non-saturated rock, is called the water table.

The velocity of the flow of water in the saturated part of the subsurface is directly proportional to the transmisivity[1] of the rock and gradient of the water table. Order of magnitudes may range from a few hundreds to a few meters per year. Thus there exists a retardation factor between the time during which the rainwater starts its underground flow to its emergence in springs or dripping into caves. For example, the retardation factor of springs flowing out from fossil aquifers may amount to a few thousand to tens of thousands of years. In annual recharged regional aquifers, where the outlet is far from the recharge zone, it may amount to hundreds of years. For local aquifers, retardation may range

---

[1] For more information with regard to the laws deciding the flow of groundwater see D.K. Todd *Groundwater Hydrology*. John Wiley & Sons, Chichester (1980).

from a few years to a few decades, depending on the hydraulic conductivity, storage and geological structure.

The water table slope or gradient indicates the direction in which the ground-water will flow. This gradient is very low, a few meters per kilometer and the flow is very slow, from a few to a few hundred meters a year. The reason for this can be grasped if one can imagine the subsurface as composed by billions of minute retarding dams, which cause the water to slow down. Only under special circumstances, mostly in limestone rocks, where big solution channels exist the velocity of the flow is high and in some cases, where large caves exist the groundwater flow will be by subsurface rivers.

In places where the water table meet the surface, either along a channel of a river or along a cliff formed by fault, a spring emerges. The quantity and the duration of flow is a function of the permeability of the rock and quantity of the water stored in the saturated layers.

Perennial springs in semi-arid and arid zones, however, are rather scarce, and their water may not suffice to supply the demands of an expanding population. Thus, when the people had to leave the overcrowded spring and river valleys, they had to locate another source of water for themselves and their livestock. The needs of men are rather small and often the water accumulated in a pothole in a riverbed or in a hole excavated into the previous ground sufficed for a hunting party or a traveler crossing the desert from one watered valley to the other. But after goats, sheep and cows, and the beasts of burden and transportation like donkey, horse, and camel were domesticated and the forage in the vicinity of the spring or river was not sufficient, men had to look for a dependable and adequate supply of water away from the spring. It is natural that he went looking for the water, which was hidden underground.

It is quite obvious that the first wells were dug in riverbeds in which the flow had dried up. It is quite natural even for a primitive man, seeing the water disappear, and seeing the moist soil, to excavate a hole to find the disappearing resource. The excavation of a shaft to a small depth is not a difficult task and even primitive societies can maintain themselves along river-beds which continue to carry water in the subsurface during the dry seaedouinsday this technique is used by bedouins in the deserts of Sinai, and this type of shallow well is called a Tamila.

When the water table below the riverbeds drops, reaching the supply of water becomes more difficult, as the excavation has to go deeper. The walls have to be protected and the material, which is excavated has to be carried up from deep below the surface. In the deserts of the Levant one can find all stages of the evolution of the techniques of excavation. The nomads satisfy themselves by a Tamila or a deep hole, which they excavated, hauling out the material by a rope and a basket. The walls of the hole are seldom protected. The digger goes down into and out of the hole by digging small holes in the walls of the shaft. As communities became settled, first near springs, and their population extended beyond the supply capacity of the springs, especially during dry seasons or years, the excavation of wells became a prominent feature characterizing the urbanization phase in the Middle East. At a certain stage these urban center had to defend themselves by walls. The fortified stronghold obviously preferred the high ground. For ensuring the water supply during times of siege and even peace, tunnels to spring outlets, or deep wells were built as will be later described.

The type of tunnels and wells depended on the hydrogeology of each region, which is decided by the geological, hydrological and climatic factors. The frame and backbone of the Fertile Crescent is built of folded chains of mountains composed mainly of limestone and dolomite of Mesozoic age (Jurassic and Cretaceous). The limestone and dolomite rocks are very permeable due to extensive dissolution of the rocks by water. The high altitude of the mountains of Taurus, Zagros and Lebanon and the fact that these mountains rise in the path of the rainstorms coming from the sea brings to the interception of these storms. The mountainous regions receive much precipitation, part of which is stored as groundwater. This water give rise to the very large springs feeding the large rivers, which cross the dry plain of Mesopotamia. The flow from these springs forms the seasonal as well as multi-annual base flow of these rivers. The control of the short term flow of flood water, but mainly the diversion of the water of the long term base flow, demanded the organization of the infra structure, physical as well as social, on a regional scale. This brought about the establishment of kingdoms on a regional scale. (The Nile basin has its regulating storage in the swamps of the Sudd).

In the western part of the Fertile Crescent, which is mountainous, the population inhabited the valleys, each of which was supplied by local springs. In these valleys independent city-states were established and the local inhabitants developed various methods for the most efficient exploitation of these springs. Because of the high permeability of the limestone rocks, the seasonal as well as multi-annual storage capacity is depended on the thickness of the strata. Thus in case of a series of drought years the regional aquifers will continue to flow and thus mitigate the impact of the climate change, while small springs will dry up, causing severe socio-economic crises.

While the excavation of wells in a riverbed or above and behind a spring is obvious and needs no sophisticate hydrological understanding, wells excavated in the Coastal Plain to reach the groundwater table evidence an advanced stage of intelligence. Such a dug well of 1.5 meters in diameter and the depth of about 6 meters[2] (Nir and Eldar 1987) was dug during the Pre-Ceramic Neolithic period at Atlit-Yam, south to Haifa, along the shoreline of the Mediterranean Sea. Today the edge or the well is at the depth of 10 meters under the sea level, however when it was dug by the ancient inhabitants the height of its edge was about 6 meters above sea level, which was then 16 meters below the present level. The well was deserted around 5500 B.C. when the warming up that occurred between the pre ceramic to the ceramic periods reached to its pick and brought a rise of the sea level. From the structure and the depth of the well can be learned that the inhabitants had an idea about the existence of a groundwater table and that the diggers worked in the water and were probably been helped pumping the water with buckets.

---

[2] Y. Nir and I. Eldar, "Ancient wells and their geoarchaeological significance in detecting tectonics of the Israel Mediterranean coast line region" *Geology* 15:3–6 (1979).

During the Chalcolithic Period, (4500 to 3300 B.C.E.), which was mostly a cold and humid period the inhabitants of the deserts of the Negev and Sinai exploited the floodwaters that flooded the river sh(eds.) They developed irrigation methods that are known to day as water harvesting, which will be discussed in more detail later on. As the area was arid in the summer time there was a need to secure for the inhabitants and their livestock an ordinary supply of drinking water. For this purpose the ancient inhabitants used to dig shallow wells in a shape like a cone to ground water level that existed in limestone a layer of clay[3]

During the Early Bronze Age (3300–2200 B.C.E.), which was cold and humid, the rivers' diversion projects in Mesopotamia spread into the desert. In the same time there exists interesting evidence from which can learnt about the roll of the wells in Sumer, although the water supply to the cities was based mainly on water canals. The evidence comes from the war tale between the two cities Kish and Uruk/Erech that was translated by the Sumerolog Samuel Noah Kramer (Kramer 1963) written on a clay tablet. It tells the story of the messengers of Aga king of Kish who arrived at Erech and demanded its surrender to their king. Gilgamash the king of Erech urges the people of his city to be prepared for siege, once the enemy diverts the water canals.

As already mentioned, the cold and humid conditions explain that there is not much evidence of water engineering during the Early Bronze Age in most other parts of the Middle East, such as Syria-Palestine or Anatolia. A water supply system that was used as a well and also as a system for enrichment of ground water was found in the city of Arad in the more arid southern part of Israel. The city had been laid out in such a manner that the built up area created a collecting area for rain water and as all it's streets were leading towards the lowest point of the city where, in times of rainstorms, a pool was formed. The water enriched the shallow perched groundwater table and supplied the city with water during the dry season. Due to the limited recharge area the reserve of this perched groundwater table was quite limited, and during the warm dry periods, which followed during the second half of the third millennium B.C.E. it dried up and the city was deserted.

The studies of Avner (op. cit.) showed that like in the Chalcolithic Period so also during the Early Bronze Age settlements in the central part of the southern Negev Desert such as the valley of Uvda flourished. These agricultural settlements used methods of damming and water harvesting for irrigation and excavated wells into the perched groundwater table.

During the transition period from the Early to the Middle Bronze, which is termed the Intermediate Bronze Period, in which the climate was very dry, Arad and its water system were deserted, not to be re-settled until thousand years later in the Iron Age when another humid period occurred.

After the extreme aridity that occurred during the Intermediate Bronze Age, the Middle Bronze Age (2200–1200 B.C.E.) has been characterized with a more temperate climate and the need for more water and more secure water reserves

---

[3] U. Avner. "Settlement, Agriculture and Paleoclimate in the Uvda Valley, Southern Negev Desert, 6th–3rd Millennia B.C.", in: Issar and Brown, op. cit. pp. 147–202. (1998)

for the fortified towns and cities arose. One of the first written evidence for a project of canals conveying water from springs is that of Yahdun-Lim, the king of Mari on the Euphrates (1800 B.C.E.) He boasts that he brought prosperity to his people by digging canals, which eliminated the use of buckets in his country.

Although the climate during the Middle Bronze Age improved and the gradual revival of the urban centers, still it did not returned to the humid conditions of the Early Bronze. Thus in order to ensure the supply from the springs during the dry seasons and especially during spells of droughts the inhabitants developed the spring system by galleries dug into the rocks (Fig. 13).

As the outlets of these springs were at the bottom of the hills on top of which the towns were built, and in order to guarantee the supply also during time of siege the inhabitants dug shafts inside the wall to reach groundwater and in those cases that this target was not reached, meaning that they did not succeed reaching the ground water, a tunnel was lengthened from beneath the shaft towards the nearest well in order to secure the water supply in time of siege. The system of shafts and tunnels is found at a number of urban centers from the Middle Bronze Age, like Jerusalem, Hatzor, Gezer, Megido, Taanach, and Gibeon. In many cases, the Iron Age Israelites adapted and even developed this system. Never the less, it is reasonable to assume that the development of the system has to be ascribed to the Middle Bronze Age.

The Late Bronze Age was characterized by a warm and dry climate, which caused the desertion of many settlements along the border between the sown land and the desert. This enabled the nomad tribes and their herds to penetrate

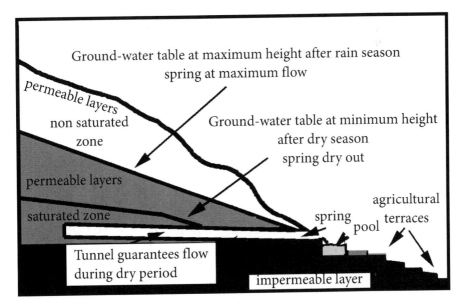

Fig. 13. Section of a spring tunnel

into the once fertile area. Once the climate improved they settled down and starte – farming the land.

The Iron Age 1200–580 B.C.E. was again a cold and humid period. The Assyrian kings Sargon and especially his son Sancherib brought the technology of water supply to their'cities to very high level. Sancheribs's projects compromised of tunnels, channels, and bridges.

During this period the systems of siege water supply to the fortified cities became very sophisticated. Shafts and tunnels, inside the walls, were dug to reach the groundwater table, thus avoiding the need to reach the outlet of the springs. In places, tunnels diverting springs into a walled section of the city, like the Niqba (Tunnel of Siloam) of Jerusalem, were excavated.

In the water system of Gibeon north of Jerusalem (modern El-Jib), one can see the evidence for the change of the climate that caused the drying of the upper system during the Late Bronze. This town is built above a spring that is emerging from the limestone rocks. In the inside of the wall, dated by the excavator J.B. Prichard to the Iron Era[4], a graded tunnel of two levels was dug with two slopes: the upper one is more moderate and the lower, diverted from the upper one by 90 degrees, is steeper and reaches to the fountain's pool room from where the water tunnel emerges. Issar[5] suggested that the upper part was excavated during the Middle Bronze period as a tunnel from inside the wall of the town and reached an high groundwater feeding a small spring emerging outside the town. The outlet of the fountain was eventually hidden and the inhabitants had an access to the fountain through the tunnel. When the upper spring has been dried out at a certain stage, most probably due to the climate change of the Late Bronze Age, arose the need of deepening the tunnel in order to reach the water level of the lower spring.

Inside the town Prichard (op. cit.) exposed a large shaft with a system of spiraled stairs quarried in the limestone rocks which enabled the water carriers to go up and down. Again one can see two stages: the upper shaft with the spiraled stairs with a depth of 10.7 meters, and the lower graded tunnelgoing down to the depth of 13.7 meters from the bottom of the shaft towards a cave-room which is at the level of the lower spring. Also in Prichard's opinion the deepening beneath the bottom floor of the pool with the spiraled stairs was necessary because of the drop of the ground water level. There can be little doubt that the primary function of the installation was an emergency water supply, but like in many other contemporary civilizations, in particular the Hittites, these water sources served a religious purpose, as for instance the libation of water as part of a ritual in honor of Baal or Adad to ensure the safe supply of water. This tradition lives on in the Jewish festival of Succoth (Simhath Beth haShoevah) when water from the Gihon Spring is carried in a procession onto the Temple Mount. The very name of Gibeon is derived from "Gev" i.e. a water hole.

---

[4] J.B. Pritchard. *The Water System of Gibeon*, University of Pennsylvenia Press (1961).

ibid. *Gibeon, Where the Sun Stood Still.*, Princeton University Press, Princeton (1962).

[5] A. Issar. "The Evolution of the ancient water supply system in the region of Jerusalem" *IEJ* 26:130–136 (1976).

Our knowledge of past climatic changes can be of help in dating the ancient water supply project at Gezer. All archaeologists agree that Maccalister [6], who excavated this site at the beginning of the century, was deeply influenced by his faith and endeavor to prove the Biblical stories. Examining his findings shows that the engineering principles of the digging of a tunnel from inside the wall in order to gain access to a spring outside the walls is identical to the upper part of the water project at Gibeon. The slope of the tunnel was calculated according to the difference of the height between the land surface inside the walls to the height of the spring on the outside close to the "Southern Gate". The reason that tunnel is not perpendicularly to the wall probably conforms with the will to secure a suitable slope as well as to preserve some existing structures. It is assumed that an access entrance to the spring should be found by continuing the horizontal tunnel cleaned by Maccalister. If this assumption will prove to be correct, the tunnel have had to be quarried during the Middle Bronze Age when the groundwater level was relatively high and a spring would have emerged at the foot of the hill on top of which the town of Gezer was built. Macalister found a pothole dug at the beginning of the horizontal part of the tunnel confirming that the ground water level fell during the beginning of the climatic crises of the Late Bronze Age. With the worsening of the climate and the continuing fall of the groundwater table which probably became brackish, the tunnel was abandoned. The salinity of the water, meaning the bad taste of the water in the pothole, as reported by Maccalister is due to the presence of salts in the layers of chalk of Eocene age building this region. The many pits all over the mound, were most probably for collecting rainwater to secure domestic water supply because of the salinity of the spring water which may have been used mainly during periods of siege. Maccalister describes a shaft with a spiral stairs system similar to that of Gibeon.

Another engineering solution was by a vertical shaft quarried in the rock from where the water was lifted with the help of leather bags or buckets and ropes. Such a well was found at Tell Lachish, located within the area of the city fortification. Like the others, it was built in the Middle Bronze Age and remained in use till the end of the Iron Age. The depth of the well is about 44 meters and its diameter about 2.4 meters quarried into the chalks of Eocene Age. The groundwater table outcrops in the valley at the foot of the mound and the inhabitants understood the hydrological meaning of these seepages.

The development of the agricultural terraces on the slopes enabled also the exploitation of the small-perched springs flowing mainly during the spring. The research carried out by Zvi Ron[7] showed that only a small percentage (about 0.5%) of the whole agricultural terraces area was irrigated by the springs. Nevertheless, a huge effort was invested in the development of these springs by digging tunnels into the limestone and dolomite rocks. In many cases, if not in most of them, the tunnels were following natural solutions channels or large caves, and the economical investment to improve the system show

[6] R.A.S. Macalister. *The Excavation of Gezer.* v. 1 and 2 (1902–1912).

[7] Z. Ron. "Development and Management of Irrigation Systems in Mountain Regions of the Holy Land" *Transactions Inst. British Geographers* 10:149–169 (1985).

that it was very worthwhile. May be that there were two reasons to it. The first, as was already mentioned, the cutting of the tunnels started already in some early period and it was found to assure the flow of spring during the dry seasons and in years of drought (the longest tunnel in the Jerusalem area is of 225 meters). The second reason is the constant irrigation of the terraces that are below the spring guaranteed the supply of vegetables and even grain for domestic use even in years of droughts. In addition to digging the tunnels, the investment of labor in the springs included also the building of collecting pools and canals. According to Ron (op. cit.) archaeological findings show that the larger projects were built during the Second Temple Period of Jerusalem (538 B.C.E.–73 A.D.) and the Roman-Byzantine Period (From 73 A.D. to ca. 638 A.D.).

Some urban supply projects were built to serve the population that would have been forced to find refuge in the fortified cities during a long period of siege. The structure of these projects show that the planners had the basic knowledge about the role of groundwater table and thus there is no need to reach to the spring, which emerged at the foot of the mound on which the city was built. They knew that it is enough to go down up to the level of the spring in order to reach the groundwater table and haul the water from the layer that supply the water to the spring. And indeed in the water plants of Hazor and Megido the tunnels are not leading towards the direction of the springs that flow outside the town, but to the water level that feed them. At Hazor the gallery leading to the groundwater table was built with a system of very wide stairs that goes down in a moderate slope from the center of the town to a tunnel were the water level is found.

Another project, which demonstrates that the planners understood the principal of groundwater table, is that of Tel Beer-Sheva.[8] The ancient city was circled with a thick wall built of bricks placed on a base of stone-rocks. In the eastern part of the fortified city was excavated a deep shaft with spiraled stairs, that was leading to an underground water reservoir that was fed by a diversion canal from the riverbed. A deep well that its depth reached up to 70 meters was found outside the walls near the main gate. An additional interesting trait that is connected to water is the canal that drains the main street and passes beneath the main gate to the outside of the wall. The canal reaches close to the deep well and may have been an artificial recharge project like the one described in Arad.

During the Iron Age, the Judaean fortress of Arad was built on the top of the hill and had a system of reservoirs, a roofed canal and a tunnel quarried into the chalky rocks leading to the outside of the wall. The water supply of the fortress was based on collecting rain water from the its area and also the filling of the reservoirs by draft animals carrying the water jars or water skins

---

[8] Z. Herzog, "Water supply at Tel Beersheba in the 1st Millennium B.C.E.", in: *Proceedings of the 11th International Conference on the History of Water Management and Hydraulic Engineering in the Mediterranean Region Israel 7–12 May 2001* Schriften der Deutschen Wasserhistorischen Gesellschaft (DWHG) C. Ohlig, Y. Peleg, and T. Tsuk (eds.), pp. 15–22 (2002).

from the well of the Early Bronze Age.[9] This well became productive with the improvement of the climate during the Iron Age.

In the Negev Desert many water reservoirs from the Iron Age, are found. They are located in most of the cases on the impenetrable layer on the slope of the hill on top of which the tower or a fortress is built. The canal for collecting the water surrounds the hill along its slop and leads the water into an uncovered reservoir. There are also water-reservoirs, which are not connected to towers, but were built and were dug probably for supplying drinking water for agricultural farms. Terraces in the riverbeds near the farms were built with the purpose to stop the floods and saturate the soil in the riverbeds, which afterwards were ploughed and planted.

A water supply system from this period but of a complete different character is the Siloam tunnel of Jerusalem, known also as King Hezekiah's tunnel (described in Kings, 2, 20:20 and Chronicles, 2, 32:3, 4). According to the scriptures, King Hezekiah constructed it in order to improve the ability of Jerusalem to withstand a siege by Sennacherib of the Assyrians in 701 B.C. It connected the Gihon spring situated outside the wall of the city of Jerusalem, to the Siloam reservoir which was within the wall. The length of the tunnel is 533 meter and the difference in height between the source and outlet is 2.1 meters. The direction planof the tunnel makes an extensive detour following a karstic channel. This explains also the fact that the diggers did meet even though they did not work in a straight line and not even on the same level. The tunnel was dated by carbon-14 from plants preserved in the waterproof layers of plaster lining the tunnel and was found to be from 700–800 B.C.E. i.e the time of King Hezekiah.[10]

The most outstanding monuments of water engineering of the Iron Age are found in eastern Anatolia and Caucasia. We have already mentioned the water supply to Tushpa (now Van) by erecting barrages in the foothills surrounding the plain of the lake and acquaducts over distances of more than 40 km which include building bridges over smaller seasonal river beds. Fresh water was needed to flush the salts from the soil around the lake of Van. Remains of similar systems are found in many places which were at various perods under Urartian rule.[11]

During the Iron Age started the spread in the Middle East of the Persian method called the "qanat"[12] system (Fig. 14). This involves the development of water supply based on groundwater with the help of tunnels and shafts, or chain of wells (kharez in the Turkmen and fuqara in the Arab provinces). The origin of this system was probably in Urartu where irrigation project since the 9th century B.C.E. were based on springs which were developed with the help

[9] T. Tsuk, *Ancient Water Systems in Eretz Israel: From the Neolithic Period to the the end of the Iron Age*. Ph. D. Thesis, Tel Aviv University, Tel Aviv (2000 in Hebrew with English summary).

[10] A. Frumkin, A. Shimron, and J. Rosenbaum. "Radiometric dating of the Siloam Tunnel Jerusalem" *Nature* 425:169–171 (2003).

[11] Urartu see Chapter XIII.

[12] H.E. Wulff, "The Qanats of Iran" *Scientific American* pp. 94–105 (April 1968).

of tunnels and shafts. In Iran this method which assured water supply also during periods of droughts in the arid areas enabled the transition from the society of nomad warriors scattered among secluded oases, to an agricultural society which is ruled by people that previously were heads of tribes and later on turned to be land owners that rule on extensive areas of agricultural lands.

The qanat is a tunnel, which acts as a subsurface drain, dug horizontally from one shaft to the other. The shafts placed along its total length are utilized for servicing and ventilation purposes. The plan, depth, and gradient of the tunnel are determined from the findings of a preliminary survey, which involves digging a few exploration shafts in the alluvial fan. The length is decided according to quantity of the water needed by the qanat owner. It will end once this quantit is reached, and digging will restart once theis a decline in the flow.

The survey before the digging starts decides the general direction of the tunnel, which is marked on the surface. Digging starts at the outlet end, thereby guaranteeing outflow of the water so that the diggers can work in a drained tunnel. At the same time, another group of diggers starts to drive a vertical

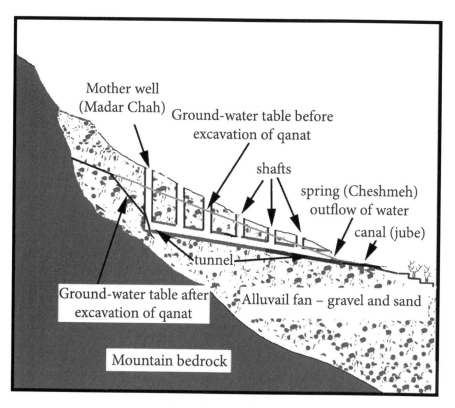

Fig. 14. Section of qanat

shaft to meet the horizontal tunnel. Since it takes twice as long to dig the same length of horizontal tunnel as it does the vertical shafts, the horizontal distance from one shaft to the other is twice the depth of the shaft. Thus the tunnel diggers and the shaft diggers work simultaneously to meet each other at more or less the same time. The shafts are used to remove the excavated sand and gravel without having to crawl the long distance to the outlet. They also keep air flowing through the tunnel. If the tunnel becomes very deep and there is not enough ventilation, a big wind catcher made of two wings of sails is placed on top of the shaft. The special mixture of gravel, sand and fine silt of the alluvial layers usually keeps the walls of the tunnel and shafts standing, especially if the dimensions are small – just large enough for a *muqanni* to crawl inside. It is usually an oval-shaped hole 90 × 40 cm. If the material is too fine and collapsible, oval-shaped rings made of baked clay are used for protection, inserted as the digging progresses. The elliptic shape of the rings enables the tunneling to continue without having to reduce the diameter. The last shaft "*madar chah*" (mother-well) is the deepest. This is where excavations restart, either if the water of the gallery is depleted, as happens in a series of dry years, or if the owner needs more water. If the gallery and the "*madar chah*" reach bedrock, the excavation may bifurcate to form a few "ranching"galleries draining into one "*cheshmeh*" (eye = spring). The length of the tunnel-qanat may reach a few tens of kilometers, and the depth of the "*madar chah*" more than 100 meters.

Water Harvesting (Fig. 15)

The most astonishing project is still under survey and partial excavation by Armenian archaeologists on the slopes of Mount Aragatz. The Aragatz (4090 m) is the highest mountain in Armenia and is situated to the north of the Araxes Valley facing Mount Ararat (Urartu, 5165 m) to the south, covered by snow throughout the year, at least for the time being. The southern and western slopes show a pattern of remains of horizontal walls in many of the gorges and valleys running down the flanks which once were dams. The barrages were linked with overflow channels, sometimes cut into the rock, more often built with large boulders, which gradually lowered the water flow to the valley near Armavir, first an important Urartian city and later a capital of Armenia in the Hellenistic period. As the winter precipitation is mostly snow which begins to smelt during spring and continues throughout summer, the supply of water for the parched plain around Armavir was ensured during the growing season.

The Hellenistic – Roman – Byzantine Period (ca. 300 B.C.E.–650 A.D.) enjoyed a relatively cold and rainy climate. During this period started the establishment of Nabatean settlements in the desert areas on the southern part of the Fertile Crescent. The agriculture system surrounding these settlements was based on "Water Harvesting" method. This method was based on catching the surface water flowing over the bare, rocky slopes of the hills. It also included the diverting floods from main riverbeds, rather than damming the all river flow.

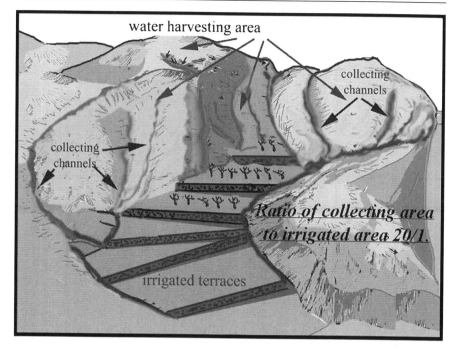

Fig. 15. Water harvesting method

The people who planned and built these systems were aware that flash floods in the desert are difficult to control because of their unpredictable volume and sudden harshness. Experience has taught them that desert flash floods are especially severe because there is no soil and tree cover on the hillsides to mitigate their impact. In addition, intense cyclonic rainstorms bring a tremendous downpour to a limited area within a rather short time and the floods resulting from the downpour of these storms are usually extremely intense and destructive. Thus the method was based on collecting the waters from small drainage basins that were extended on the mountain slopes that surrounded a valley where agricultural terraces were built across his channel. The expanse of the basin from where the waters were collected into canals that were excavated in a moderate slope on the downgrades of the mountains extended between ten to hundred hectares. The canals led the waters to the creek or riverbed, which was terraced, and collected the mud brought by the surface flow to become agricultural soil. As mentioned the sub-parceling of the slopes and dividing it according to the area of terraces to be irrigated, upgraded the efficiency of the irrigation, because as long as the parcels land were smaller, so was higher the ratio of the flow to that of the rain falling on the parcel. The ratio between the area of the collecting basin on the inclinations and this of the cultivated terraces is approximately 20:1, meaning, that every hectares of the farms land needed about 20 hectares of slopes in order to collect from there the water for irrigation.

The team that investigated this method[13] did not bring in account climate changes, and assumed that the yearly rate of precipitations was the same average as of the present, namely around 100 mm. Therefore for an average of 15% flow, each hectare is producing an average of 150 cubic meters, therefore 20 hectres will produce about 3000 cubic meters of water, having to irrigate one hectare, which is equivalent to about 300 mm of rain. If we add to this the natural 100 mm of rain than the total is about 400 mm/y. Discussing this issue with the local farmers, Issar was informed that this quantity is sufficient enough to support the plantations and not letting them to degenerate, but it does not ensure good crops. Once climate change is taken into account and that the quantity of rain – was 30% (and during good years even −50%) more then the current quantity then the average was enough to ensure an optimal crop. This, jointly with the fact that there were certainly also less drought years, might explain the fact that the earlier inhabitants of this arid area did not abandon even the smallest river basins from building in them terraces and cultivate them with agricultural cultivations.

One of the enigmas of the Nabatean agriculture is the small piles of stones that covered the slopes of the hills around the farms. The local Bedouins call them "Tuleliyat-el-Anab" which means: hillocks of grapes. There are those who claim that they were built to support the vines, as long as the stones caused to thicken the dew, as a supplement to the irrigation. Prof. Evenari and collab (op. cit.) were in the opinion that this name that the Bedouins granted to these stone piles have no connection what so ever with the real aim of the piles. In their opinion, their aim was to increase the crop of water harvested from the slopes, by way that the stones removal smoothed the surface. They showed by way of tests that such a treatment of stone collecting and turn them to hillocks brings to a situation of 25% increase in the water flow rather then from a slope where the stones remain scattered over it. In view of climate change paradigm it is suggested to reinvestigate the possibility that this piles were indeed for the use of the vines, while the additional flow was actually an additional benefit. One can see such piles of stones used for vines on the island of Thera/Santorini.

The water for urban domestic purposes, drinking or cleaning, the Nabateans collected water from the roofs of the buildings and the streets of their towns into big cisterns. In the rural areas they dug caverns in the chalky into which the water was diverted by a canal from the riverbed. These water reservoirs were used for provide drink water to the herds, and helped also by supplying drinking water to the cities.

In the cities Nizana, Ovdat and Rehoboth, wells were excavated in the chalk rock and reached the ground water level. Yet the groundwater in the chalks is slightly brackish. At the foot of the hill on which the city of Ovdat is situated, a well 60 m deep was excavated. The saline water (900 mg/Cl/l.), which was found in this well supplied the Roman bath which was equipped with all the accessories.

---

[13] M. Evenari, L. Shannan, and N. Tadmor, *The Negev: the Challenge of a Desert*, Harvard University Press, Cambridge (1971).

All these projects were abandoned during the Moslem period during the increasing process of aridity that started in the eighth, century.

Another type of water supply project based on springs, characterizing the Hellenistic-Roman-Byzantine period is that of the aqueducts. It is connected with the development of the "polis" cities and the need of collecting and transferring water over big distances from a big spring or a few springs in order to guarantee all the needs of the inhabitants of the polis. This included public and private baths and flowing sewage canals. In these projects, a great deal of knowledge was invested, resourcefulness and engineering daring.[14]

One of the shortcomings of these projects was the fact that the Mediterranean area is tectonically active. In most cases the aqueducts were repaired after an earthquake and continued to operate. A more influential shortcoming was the principal of the flow of the water by gravitation. Thus as long as the level height of the spring was above the collecting point at the intake of the aqueduct, even if there was a drop in the quantity of the water, the aqueduct still fulfilled its task. Yet once an extreme fall of the groundwater level as a result of an extreme climatic change took place, it left the intake part of the aqueduct high and dry. This decided the fate of all the aqueducts and the urban systems that were dependent upon them.

Up till now, the main blame for the desertion of the Roman-Byzantine waterworks was put on the Moslem Conqueror who destroyed the system that maintained these projects. The present knowledge on the radical climate changes after the Arab conquest brings to the conclusion that Valley, see Araxes Valley, see also the brunt of the blame has to be shifted from the human agency to the climatic cause. The aridization reached its peak in the 9th century C.E. and brought to the need of the development of water elevating systems, driven by humans or animals which replaced the aqueducts until mechanized pumps arrived.

The dry climate continued with small variations during most of the Arab – Ottoman Period (650–1920 C.E.) during which the socio-economic situation of the Middle East deteriorated as it depended mainly on groundwater. The introduction of mechanized pumps, during the 19th century enabled the introduction of modern methods of irrigation and farming which ushered a gradual improvement in the socio economic level of the population.

---

[14] D. Amit, Y. Hirshfeld, and Y. Petrich (eds.), *The ancient aqueducts in Palestine.* Yad Ben-Zvi Publishing, Jerusalem (in Hebrew) (1989).

Appendix II

Isotopic Tools in Palaeo-Hydrological Investigations

II a. Oxygen and Hydrogen Isotopes and Their Role in Hydrological Investigations

The atoms of the isotope of oxygen 18 (^{18}O) and of hydrogen (H), i.e. deuterium (D or ^2H) are heavier than of the atoms of oxygen 16 (^{16}O) and hydrogen (^1H) building the molecules of water in the atmospheres.

Although these heavier isotopes constitute a rather small proportion of the total quantity of $^1H_2^{16}O$ atoms of the total mass of water, they nevertheless affect its thermodynamic behavior and can thus serve as environmental "finger-prints". Thus, when water evaporates, a higher proportion of $^1H_2^{16}O$ atoms will leave. This causes the water vapor to be lighter, isotopically speaking, than the mass, which remains as liquid water. Consequently, the vapors, and later the precipitation, always contain less heavy isotopes than the ocean water. From the start, however, the composition of the vapor will depend on the temperature of the water from which evaporation started, as well as its isotopic composition. Generally speaking, water evaporating from warm ocean water will be heavier that that of cold regions. When condensation takes place, the heavier water atoms composed of the heavier isotopes will condense first and leave as precipitation. Thus, when a rainstorm from the sea comes over the land, the rains in the coastal areas are the heaviest and become lighter as the clouds travel inland. This is referred to as the "continental effect". The same happens with precipitation caused by adiabatic cooling when the clouds ascend a mountain barrier. The higher they climb, the lighter the precipitation. This is referred to as the "altitude effect". Also, when cooling is due to meteorological reasons, the lower the temperatures, the higher the rates of condensation, and thus the lighter the rain becomes, isotopically speaking. The opposite trend is seen in the isotopic composition of the water bodies, which undergo evaporation and are not replenished. The more evaporation the water undergoes, the heavier is its isotopic composition. A worldwide survey of isotopic composition of precipitation, carried out by the International Atomic Energy Agency in Vienna, has shown that on a global scale the $^{18}O/D$ composition of the precipitation, correlated with the Standard Mean Ocean Water (SMOW), falls along a certain line called the "global meteoric line". On the other hand, the isotopic character of the precipitation of the Mediterranean region falls on another line, denoting heavier composition.

This is due to the special marine and atmospheric regime of the Mediterranean Sea.[1]

II b. Isotopic Dating Methods

The carbon radiometric age determination method, ^{14}C or radiocarbon is based on the fact that this element, produced in the upper atmosphere by the bombardment of cosmic rays, decays exponentially, its halflife being about 5570 years. By modern methods of enrichment and by using a nuclear accelerator, dating range approaches 75,000 years. The constraints of the ^{14}C method are a function of errors arising from accuracy limitations of the laboratory methods, the range of which may span over decades or a few centuries, depending on the method of sampling and measurement, the time scale of the dating, and sample contamination. This range of error is given when the dates' data are presented. Secondly, the amount of radioactive carbon in the atmosphere was not constant during the last 12,000 years. In order to overcome this limitation, calibration curves were produced in which the measured radiocarbon ages were corrected according to absolute ages obtained from counting the rings in tree specimens.

The U-Th isotope age determination was based on measurement of the decay of uranium ^{238}U into ^{234}U into thorium ^{230}Th. The dating range is from a few years to 350,000 years. The accuracy of dating of speleothems depends on the determination of the initial isotopic concentrations and the assurance that the thorium in the sample is only the product of the decay and was not enriched from external sources like organic debris. Accuracy range for the period discusse in this book is in the order of magnitude of a few decades.

Additional reading:

R.S. Bradley, *Palaeoclimatology. Reconstructing Climates of the Quaternary*, International Geophysics series, 2d ed., vol. 64, Academic Press, San Diego, pp. 50–73, 76–80 (1999).

U. Olsson, "Radiocarbon dating" in: *Handbook of Holocene Palaeoecology and Palaeohydrology*, B.E. Berglund (ed.) John Wiley & Sons, Chichester. pp. 275–296 (1986).

G.W. Pearson, "How to Cope with Calibration" *Antiquity* 61: 98–103 (1987).

E.H. Willis, "Radiocarbon Dating" in: *Science in Archaeology*. D. Brothwell and Eric Higgs (eds.). Thames & Hudson, London, pp. 46–58 (1969).

---

[1] W. Dansgaard, "Stable Isotopes in Precipitation" *Tellus* 16:436–468 (1964).

J.R. Gat, *Stable Isotope Hydrology, Deuterium and Oxygen-18 in the Water Cycle: A Monograph*, IAEA, Vienna (1981).

Appendix III

Pastoral Nomadism

The arrival of the pastoral nomads was an important turning point in the history of our region. It added an alternative life-style to the well established way of life of the sedentary agriculturists and herder and the future town and city dwellers. The basic difficulty of tracing their origin and development is to find tangible evidence as material remains of nomadic people, including pastoral nomads, are scant and hard to find. Only in a few desert areas that have not changed over lengthy periods of time and that were not ploughed by farmers or eroded away, are we able to find remains of their camps sites and tombs.

"Seen from the outside, the nomads seem a handful of human dust blowing in the wind; from the inside, all these desert societies, so simple at first glance, reveal their complicated organizations, hierarchies, customs, and astonishing legal structures"[1] The histories of the Old World are all intimately interwoven with these nomadic peoples who can therefore be regarded as the yeast of many cultural developments.

The majority of ancient historiographers regarded any non-sedentary people as being outside the pale of civilization. Information about them is always provided by sources which a basically hostile to them and their descriptions of their way of life always negative. Already by around 2000 B.C.E. we know a Sumerian poem about the marriage of a goddess with the god Amurru, an allegory of the life of the nomads and what to expect:

Uzu nu-izi-ga al-ku-e	He eats uncooked meat,
Ud-ti-la-na e nu-tuku-a	throughout his whole life he does not posses a home,
Tam-ma us-a-na kin u-tum-mu-dam.	And even his dead companion he does not bury.

Herodotus tells that the Scythians "do not cultivate the soil, but are nomads" (IV, 2), "they do not sow at all and do not work the earth" (IV, 19), etc. Also Strabon, Arrianus, and, in particular Aristotle (Politics 1.84-8 = 1256a) describe

---

[1] F. Braudel, The Mediterranean and the Mediterranean World in the Age of Philip II. 1949, quoted from the English translation Fontana Press, Glasgow: 176 (1966).

similar attitudes emphasizing the fact of not working the soil as the clearest sign of barbarian primitiveness.

As we have no self-expression of these people, we are largely ignorant about their conceptions, religions, oral traditions or any other form of a spiritual character. Only when they occasionally succeeded in infiltrating an urban literary society and adopt their means of expression, we learn a little about their past life out there in the desert. The obvious pride which the 'Amorite' rulers of the second millennium B.C.E. took in their 'forefathers living in tents', such as Hammurabi of Babylon, the nostalgic Biblical references about the purity of the nomadic Qenites (descendants of Cain) and Rehabites, and the glorificationof the way of life in the 'Jahiliyyah' in early Islamic literature,[2] all point to the fact that the way of life of pastoral nomads was not only a socio-economic reality born out of necessity but also a deeply rooted ideology and a source of pride. There is a consensus among all travelers in the Near East that the most typical characteristic of the nomads is their pride in their life-style and their disdain for the peasants or any other sedentary peoples. Even after generations of settled life, the descendants of nomads are still aware of their superiority over any other class of people.

Many episodes in the past of the Near Eastern civilizations can often only be understood as the result of the interaction, peaceful or antagonistic, between the nomads and the sedentary natives, with many examples of nomadic leaders as movers of ancient and medieval history. To trace the origin and development of pastoral nomadism is not only to search for an idiosyncratic form of economy but touches upon much larger issues, such as the birth of civilization in general, the formation of states and complex societies which all, in one way or other, were influenced in many respects by the presence or the absence of the nomads. Frequently, not the nomads themselves but their real or more often imagined descendants within these societies were responsible for these issues, e.g. the various Turkic and Mongol dynasties in the Middle East, the Moghuls of India or several dynasties of China.

The world view of pastoral nomads is centered on their livestock and their riding animals. The idea of "Having Herds" is the one and most important pole around which turns everything in the life of a pastoralist. To increase the herds as status symbols irrespective of their economic effectiveness has not only been observed in Africa's "cattle complex" and is a much discussed, and disputed, topic in anthropologic literature. Knowledge about specialized breeding, however, is necessary and was transmitted in the form of oral traditions from generation to generation until most recently.

We can assume that the roots of religious notions specific to pastoralists can be found already in the Chalcolithic period when the 'Secondary Products' Revolution' set in. Small 'altars' made of basalt in shape of goats' heads with horns and topped with flat bowl-like depressions were found near the entrance of many houses in the Golan[3] and most probably used for some kind of ritual

---

[2] W. Caskel. Die Bedeutung der Beduinen fuer die Geschichte der Araber. Koeln-Opladen (1953).

[3] C. Epstein, The Chalcolithic Culture of the Golan. *Bulletin of the Anglo-Israel Archeological Society*, pp. 41–45 (1982/83).

resembling the small domestic St. Mary shrines in the Mediterranean countries. Other glimpses into this world are the many figurines of domestic animals, in particular donkeys and cattle, found throughout the Bronze Ages, often to the exclusion of any other representations, human or otherwise. The biblical stories about the hospitability of the Patriarchs culminated in the slaughtering of a kid, the highest sacrifice for a pastoralist. Sacrificing animals in the temples of the Bronze or Iron Age Levant is another expression of this desire to please the gods by offering the most valuable possession of man.

Another striking trait of pastoral nomadic traditions is the periodic appearance of tribal leaders, called 'charismatic leaders' by Max Weber. He claims that peasants normally follow their daily or annual routine without the need to make fateful decisions. If those are found to be necessary, they were decided upon by a council of the elders, a so-called 'primitive democracy' found among most agricultural societies. In contrast, for nomadic societies the decision where to lead their flocks after a certain territory was exhausted, prolonged drought years, or the threat of tribal wars, is of utmost importance: A wrong move and the entire livelihood of his people can be annihilated.[4] Lucky, or inspired and, therefore, successful leaders often rallying a large band of disparate followers became the hall-mark of nomadic societies which seldom were egalitarian. Contrary to a generally accepted view, authoritative rule backed by 'signs from heaven' is the most conspicuous social aspect of nomadic societies.

The leaders of nomadic peoples often became the eponymous ancestors of a clan or tribe and ancestor worship constituting an important feature of their religious concepts. The evidence of funerary monuments erected to these ancestors dot the fringes of the desert of the Old World. The tumuli, the dolmens, sacred places defined by stone circles (such as Rugm el-Hiri in the Golan) are best understood as the remains attributable to this phenomenon even long after their descendants have become sedentary. Ancestor worship is the key to tribal identity and ideology of tribal states. It was seen as an obstacle in the establishment of supra-tribal, national, or imperial states and the struggle to suppress these beliefs in the interest of a sedentary 'civilized' society is amply documented in Biblical and Islamic literature.[5]

The study of nomad-agriculturist interaction shows a consistent and predictable pattern which dominated the history of the countries of the Old World.[6] The to the outsider's apparently low level of the nomads' material

---

[4] T. Asad, *The Kakabish Arabs: Power, Authority, and Consent in a Nomadic Tribe*, Praeger, New York (1970).
R. Cribb, *Greener Pastures: Mobility, Migration, and the Pastoral Mode of Subsistence*. Production pastorale et société 14:11–46, (1984).

[5] T.J. Lewis, "Ancestor Worship" in *Anchor Bible Dictionary*. Vol 1:240–242. Doubleday, New York (1992).

[6] M.A. Khazanov, *Nomads and the Outside World*. Cambridge University Press, Cambridge (1984).
M. Baistrocchi, M. Passato e presente dei pastori nomadi del Sahel e del Sahara: Lo studio del passato come contributo alla comprensione del patrimonio culturale delle genti sahariane moderne. in *Atti del Convegno Euro-Africano sul Sahara e Sahel: Dall'indipendenza alla siccità*. Istituto di Studi Africani, Firenze (1986).

culture and their dependence on the volatile environment, their mobility and inherent predatory and belligerent character created a 'field of tension' acting as the trigger for many cultural development. The repeated infiltrations and invasions, frequently followed by conquest and subsequent occupation of territories settled by agriculturists.[7] as well as the counter active efforts of the sedentary opponents created the necessary preconditions for the early Near Eastern civilizations. The fact that the majority of these great civilizations grew in fertile river valleys surrounded by grass-lands or deserts with their unavoidable and undesirable inhabitants seems to be significant in this context.

Whenever the local sedentary population was large enough and possessed a sufficiently strong cultural resilience, they withstood the devastating influence of a normally bloody conquest and eventually overcame the invaders by integrating them into a new social framework. The first and still energetic new rulers of real or pretended nomadic ancestry contributed considerably to the expansion of their realm and the spread of the civilization they now represent. The best known historical examples were the Akkadians in the third, the Amorites in the second, and the Aramaeans in the first millennium B.C.E., followed by the Arabs in the first and Turks and Mongols in the second millennium ce. It stands to reason that in earlier periods, documented only in the mute archaeological record, the situation was not very much different.

In cases where the social, economic, and above all the environmental stress proved too much for the sedentary population, it had to join the invaders and in some cases even adopt their way of life. Instances of nomadization of previously sedentary people are rare but documented in the archaeological and historical record of the Near East, e.g. the Intermediate Bronze Age of the Levant (ca. 2400–1900 B.C.E.), or more clearly in medieval Iran and Anatolia.

Lastly the question of ethnicity is of some interest. Whereas nomadic peoples of central and northern Asia were and some still are multi-lingual (whether Finno-Ugric, Indo-European, or Ural-Altaic), the pastoralists of the Syro-Arabian Desert were always Semitic speaking tribes belonging to the Afro-Asiatic language family.[8] There is no evidence of Sumerian, Hurrian, or Indo-European speakers who turned to pastoral nomadism as their major form of subsistence economy. The same applies to Africa where, with the exception of the Tibbu and the Fulani in western Africa speaking now presumably only recently adopted Sudanic languages, or some Bantu speakers in eastern Africa, the pastoral nomads are predominantly Hamitic or Kushitic speakers and belong to the same Afro-Asiatic language family as the Semites of Asia. [9]

The multi-ethnicity of central Asian nomadic population has been described in detail by several ancient and medieval Chinese and Near Eastern obersvers. In nearly all case, the formation of "nomadic empires" depended on the

---

[7] D.J. Bates, The Role of State in Peasant-Nomad Mutualism. *Anthropological Quarterly* 44/3: 109–131, (1971).

[8] See below; Zohar 1992b.

[9] T.A. Sebeok, T.A. ed. *Current Trends in Linguistics. Vol. 6: Linguistics in South-West Asia and North Africa*, Mouton, The Hague and Paris: 237–661 (1970).

charisma and the fortunes of a particular leader who could gather around him a motley of warriors of various ethnic or genetic origin as long as the booty was coming in. They would come from eastern Siberia, could be Tungus, Manchu, Koreans, Tibetans, Indians, and everybody from the great steppes just riding by. If the coehesion proved stable, the language of military command of a group, or horde, became the language of the new "people". Until the middle of the first millennium they belonged to the Indo-European language family (see below Appendix IV), such as Scythians, Sarmatians, Avars, etc. and rom then on, the Ural-Altic languages, e.g. Turkic, Ungarian, Mongol, and others became dominant. By adopting the name and the heroic myths of the founder and his dynasty constantly repeated by the bards, oral traditions were created and became hallowed history. But with few exceptions all these new nations, their history and traditions disappeared as quickly as they were born.

The question of the earliest appearance of the way of life of pastoral nomads is under debate in a variety of disciplines, such as anthropology, sociology, archaeology, and history with no solution in sight. Various theories existed: A traditional view based on the outlook of the classical authors, formulated already by Aristotle (Politics 1.8.4-8 = 1256a), which assumes that man was first hunter-gatherer, then pastoralist, and lastly agriculturist. A second, more modern conception, and still regarded as main stream, was that pastoral nomadism came into being together with agriculture and developed parallel with the Neolithic Revolution.[10] The third and last line of thought sees pastoral nomadism a late phenomenon which, for some 'purists' such Khazanov (1984 above) appeared only in historical periods not before the beginning of the first millennium B.C.E.

It seems safe to assume that hunting-gathering as a distinctive food production declined as the domestication of plants and animals, sheep and goat, gradually established itself during the early Neolithic period. The late Neolithic and Chalcolithic societies were basically agriculturists with varying degrees of sedentary way of life due to the need to move around their small herds in order to avoid overgrazing and creating fodder reserves, a life-style very close to the transhumance found in many Mediterranean countries until most recently. In all these cases pastoral economy was an opportunistic alternative to and extension of farming.

Ongoing research in the Sahara suggests that the domestication of cattle was a local process and that a primitive form of pastoralism was among the earliest food production strategy of epi-Palaeolithic groups without the agricultural component).[11] There are some possible cases of cattle taming or perhaps

---

[10] Mellaart et al. (1975).

[11] R. Kuper, R. Untersuchungen zur Besiedlungsgeschichte der östlichen Sahara – Vorbericht über die Expedition 1980. *Beiträge zur allgemeinen und vergleichenden Archäologie* 3:215–275 (1981).
A. Muzzolini, *La préhistoire du bœuf dans le nord de l'Afrique durant l'Holocène*. Paris, Ethnozootechnie, (1983).
K.M. Banks *Climates, Cultures and Cattle: The Holocene Archaeology of the eastern Sahara*. Dallas, Southern Methodist University Press (1984).
A.B. Smith. Origin and Spread of Pastoralism in Africa. *Annual Review of Anthropology* 21:125–141 (1992).

domestication in other parts of the world, such as Anatolia or the Indian sub-continent, but the evidence is far from clear. The latest re-investigation of the cattle bones at Catal-Hoyuk concluded that they were bones of wild aur-oxen. In any case, pastoral nomadism seems to have had a longer documented history in northern Africa than somewhere else, including the Near East. The population living during the 8th to the 5th millennium B.C.E. in northern Africa were apparently the first to whom the appellation "pastoral nomads" can be given.[12] The first step was a slow adaption of people to certain animals leading to a symbiotic way of life which initially might have resembled, for instance, the half-domesticated reindeer herding of the Nenetz and some Ural-Altaic Lapp-speaking groups in the northern Eurasian steppe. With time, this superficial interaction became more and more intimate, man protecting the animals from predators and scouting for new pastures with the animals gradually lost their fear and aggressiveness and eventually became dependent on their human manipulators. To this day, the cattle herders of east and north-east Africa, who are still animistic, live of the blood and the milk of their animals but hardly ever kill them. This 'gentle' attitude changed, of course, with the coming of alien and more brutish religions, such as Islam or Christianity.

Before the 6th millennium B.C.E. bovine cattle was the only animal herded, after that caprovids, i.e. sheep and goat, originally domesticated on the outer flank of the northern Fertile Crescent, are found in increasing numbers not oly in south-west Asia but also northern Africa. The change clearly indicates that far-reaching relations with the Near East had been established and that the climate has changed for the worse, meaning drier. The nick-point in the development of husbandry was undoubtedly A. Sherrat's "Secondary Products Revolution" in the middle of the 5th millennium B.C.E. which was instrumental in establishing pastoralism as an independent branch of food production[13] and rendering it not only more profitable but essential for survival. Pastoral transhumance opened up new territories for exploitation, particularly in the Mediterranean countries and the Near East.

Besides the aspect of food production, the Secondary Products Revolution had another impact on the early herders. Developed pastoral nomadism is possible only with the help of animals carrying not only the produce, the homes and all the other possessions but also the people. The life-style of nomadic groups is shaped more by their pack and riding animals than the animals directly exploited for subsistence. The difference in the way of life of

D.G. Bradley, D.E. MacHugh, P. Cunningham and R.T. Loftus, Mitochondrial diversity and the origins of African and European cattle. *Proceedings of the National Academy of Sciences of the United States of America* v. 93(10) 5131–5135 (1996).

O. Hancotte, D.G. Bradley, J.W. Ochieng, Y. Verjee, F.W. Hill, and J.E.O. Rege, African Pastoralism: Genetic Imprints of Origin and Migrations *Science* 296, no. 5566:336–339 (2002).

R. Kuper, and S. Kroepelin, Climate Controlled Holocene Occupation in the Sahara: Motor of Africa's Evolution. *Science* 313:803–807 (2006).

[12] G. Aumassip, Le néolithique en Algérie: Etat de la question. *L'Anthropologie* 91:585–622 (1987) Kuper and Kroeplin op. cit. (2006).

[13] A. Sherrat. Plough and Pastoralism: Aspects of Secondary Products Revolution. In: *Patterns in the Past: Studies in Memory of David Clarke*, pp. 261–305, ed. I. Hodder, G. Isaac, and N. Hammond, The University Press, Cambridge (1981).

the central Asiatic nomads whose life depends on the horse, and those of the Near East and northern Africa, who depend firstly on the donkey, and later on the camel, is conspicuous.

With the exception of the early cattle (and perhaps the cat), Africa has produced only one domestic animal, the donkey (*E. asinus*). The fact that of all known equid races, the domestic ass interbreeds successfully only with one wild ass, the *E. africanus*, points unequivocally to its African origin[14] It was bred by the Egyptians around 4000 B.C.E. and is attested at Maadi.[15] In the Levant, its earliest occurrence is attested in the Early Bronze Age I, i.e. the middle of the fourth millennium B.C.E.[16] Together with the use of dairy products and wool, the moment had arrived when pastoral nomadism became an independent life-style distinct from sedentary agricultural subsistence.

As we have already mentioned, the Mediterranean countries are invaded every winter by Atlantic depressions bringing rains. These rains create extensive pastures on its eastern and southern coast lands allowing the herders to move out into the bush and grass lands which are uninhabitable for most of the year due to the lack of water or food for man and beast. The dry summers force the herders back to the vicinity of water resources of the fertile lands in order to ensure the survival of their flocks. These seasonal and predetermined movements result in horizontal transhumance which was most clearly seen until recently in northern Africa.

In the Levant, the pattern of horizontal transhumance is complicated by the other form of Mediterranean pastoralism, vertical transhumance, i.e. moving the flocks up and down the mountains. Vertical transhumance, like in Anatolia, the Balkans, Sardinia, the Alpine countries, and Iberia, is found here only in some areas of the Lebanese Biq'ah.[17] In all other areas of the Levant and along the outer flank of the Fertile Crescent, horizontal and vertical movements occur contemporarily in a complex pattern of migrations. Whereas Mediterranean vertical transhumants are always village based, horizontal pastoralists in general may remain mobile throughout the year, they are true nomads whether in central Asia or the Middle East.

Whatever the case, the pastoralists always enter into an exchange of products and services with the sedentary population. Agriculturalists can sustain a more or less stable subsistence economy even under adverse climatic conditions whereas specialized pastoral nomads require inputs from external sources. No pastoral society can survive for long without access to agricultural produce and manufactured goods which they must acquire from the farmers or craftsmen

[14] F.E. Zeuner. *A History of Domesticated Animals*. Hutchinson, London, pp. 374–383 (1963).

[15] I. Caneva, M. Frangipane and A. Palmieri. Predynastic Egypt: New Data from Maadi. *The African Archaeology Review* 5:105–114 (1987).

[16] E. Ovadia, The Domestication of the Ass and Pack Transport by Animals: A Case of Technological Change. in: *Pastoralism in the Levant – Archaeological Materials in Anthropological Perspectives*. pp. 19–28. ed. O. Bar-Yosef and A. Khazanov. Monographs in World Archaeology No. 10, Prehistory Press, Madison, Wisc. (1992).

[17] L. Marfoe, The Integrative Transformation: Pattern of Sociopolitical Organisation in Southern Syria. *Bulletin of the American School of Oriental Research* 234:1–42 (1979).

in exchange for their dairy produce, wool, meat, etc. Therefore, they often engage in agriculture themselves, or, if they can, they raid.

In the Near East, they often enter a symbiotic way of life with the sedentary population: First the grazing of animals on the stubble of harvested fields where the aminals' droppings help to fertilize the field for the coming winter. Pastoralists who have neither the possibility to graze their animals on harvested fields, nor to store provender, such as chaff or straw, are not able to survive the barren late summer months. This led in many cases to the ownership of fields and houses by the pastoralists with the result that most Near Eastern rural communities became mixed farming and pastoral groups. By intermarriage the two groups soon merge into a new social entity which still maintains the distinction between sedentary farming and nomadic pastoral segments. The latter become specialists in husbandry, herding, and breeding, also for those animals owned by the farmers. This arrangement ensured the survival of many individuals as the desert fringe has only a limited carrying capacity of animals and, therefore, herders. The natural population increase is easily channeled into the agricultural segment of the same community which can always need working hands.[18]

This symbiotic arrangement can occur also in central Asia or in Africa, but is relatively rare as ethnic and religious differences seem to run deeper. The potential for violent clashes between the two opposed ways of life is, therefore, much higher than in the Near or Middle East. A modern example is played out right now in front of our eyes: It began with the onslaught of camel-herding nomads, the Janjawid, against the farmers in the Darfur region of the Sudan. In the wake of the steadily continuing desiccation of the Sahel since the mid-20th century's climate change, the despaired nomads attacked the farmers' fields and villages and created an escalating human disaster of catastrophic dimension. Both groups are only recently interrelated but hardly distinguishable, both are Muslims. Yet, the nomads claim to be "Arabs" with ties to and support of the usual corrupt Islamist government, the others are contemptuously called "Africans", defenseless and fair game without rights with the world idly looking on. Darfur is not only an illustration of typical nomad-farmer relationship in the past, but a foreboding of dire things to come in other parts of the world and, worst of all, in an ever drier climate without an end in sight.

---

[18] G. Flatz, Laktase-Phantypen in sesshaften und nomadischen Bevölkerungsgruppen. *Homo* 35: 173–178 (1984).

Appendix IV

Middle Eastern Languages

Language is what makes us human. The beginning and development of speech is a subject to controversies and speculations in many fields. Without it there would be no abstract thoughts or complex teachings to the next generation. Besides being uniquely human, language is the one of the most characteristic constituents of every culture and ethnicity, and yet, without written records this most precious of all cultural possessions is irretrievably lost. To this day, only a small fraction of all languages were ever recorded and whenever an unwritten language becomes extinct, the entire tradition and knowledge of a people is lost for ever. As only a few of the early societies achieved the level of literary languages, these few remains are doubly precious for our understanding of the past.

Due to a quirk of nature, the oldest evidence of writing is found in southern Mesopotamia where the abundance of alluvial mud served not only for architecture, pottery, sculpture, and all other cultural expressions of its inhabitants, but also the first lumps of clay carrying impressions made with sticks or reeds. Besides some doodling, they represent a variety of dots, perhaps numbers, or strokes for simple drawings of animals or humans and abstract signs. All ancient writing systems started as mnemonic devices, not necessarily representing spoken languages. The idea of representing ideas associated with sounds probably arose with the beginning of urban civilizations and their need of keeping and simplifying economic records during the 4th millennium B.C.E. or slightly before that. The difficulty of rendering rounded lines on clay led during the 3rd millennium B.C.E. to the method of impressing small sections with the help of a reed cut like a pen. The result were triangular impressions which look like small wedges, hence cuneiform (from Latin *cuneus* = wedge).

Probably due to the Mesopotamian example which "put writing in the air", similar steps were taken in Egypt during the pre-Dynastic period (4th mill.) where petroglyphs (from Greek *petro* = stone, *glyph* = incision, hence writing) and ivory labels from pre-Dynastic tombs at Abydos show the beginning of keeping records. Abstract representations of tangible object were used to represent other words with a similar sound, and with added phonetic complements and determinatives, Egyptian writing evolved into a complex system which in its monumental hieroglyphic (from Greek *hiero* = sacred) version never gave up its pictorial character.

Sumerian, probably spoken since the early 4th millennium in southern Mesopotamia, is the oldest known language of the Near East. It had no known relatives or linguistic affinities with any other language. Most of its words seem to have been mono- or two-syllabic, with words of three syllables very rare. This structure facilitated the use of certain signs for representing not only its primary meaning but other words having a close enough similar sound but different meanings. There must have been some phonetic distinction, as for instance, the word/syllable DU has over twenty different meanings and vice versa, i.e. certain sounds/words pronounced similarly are represented in up to twenty signs. Whether Sumerian phonology used finer distinctions of the vowels or consonants than those we are normally acquainted with today, such as series of umlauts, nasalization, glottalization, aspiration, palatilization, or click-sounds, etc. is unknown. Many linguists suspect that it was a tone language but proof is lacking. Our sources, in most cases word lists of Sumerian alongside with Semitic Akkadian translations with its rather simple phonetics, is a very insufficient: The only vowels in early Semitic were /a/, /i/, or /u/ and a basic consonantal skeleton which clearly gives no indication of a realistic pronunciation of Sumerian or any other language written in Akkadian cuneiform.

Its structure is called agglutinative, each words was composed of distinct segments arranged in a fixed order. Nouns did not indicate gender but belonged into two classes, gods and humans as opposed to all other beings or things, with suffixes indicating spatial and temporal relationships. The adjective follows the noun to be qualified. The verbal system consisted of an unchangeable root which was inflected with few prefixes and more numerous suffixes indicating aspects of time (present, past, future) and action (active, passive, and ergative). The sentence structure showed a variety of positioning of words but the basic rule for the most simple phrase was Subject – Object –Verb. Taken together, all these characteristics squarely put Sumerian into a general framework of other Asian linguistic families such as Sino-Tibetan and Dravidian, Finno-Ugric and Ural-Altaic, and probably a host of others in the past we have no knowledge about (the best up-to-date and concise representation of Sumerian including bibliography is the German "Sumerische Sprache" in *Wikipedia free encyclopedia*, the English version rather mediocre).

The Elamites were the eastern neighbours of the Sumerians in south-eastern Mesopotamia (Susa, now Khuzistan) and south-western Iran (Persia). The oldest inscriptions on clay tablets known as Proto-Elamite were found not only in these two areas but at many other 3rd millennium B.C.E. sites in the Iranian highlands. The writing system is as yet indecipherable and appears to have been a local invention, contemporary but probably unrelated with Sumerian cuneiform – as Th. Jacobson once said: The idea of writing was in the air. Therefore, the question whether the language employed on these ancient tablets was the ancestor of the later Middle Elamite of the late 2nd millennium B.C.E. or the neo-Elamite of the Achaemenide period cannot be answered. The excavations at Susa proved that the Akkadian language and cuneiform with a local flavor were used from the end of the 3rd to the middle of the 2nd millennium B.C.E., when eventually cuneiform was adopted to write a local language called Elamite.

Like Sumerian, this Elamite of the second and first millennium B.C.E. was an agglutinative language. Verbs were declined by adding modal or temporal suffixes, nouns were in two gender categories, animate and inanimate followed by postpositions and by their adjectives. The normal sentence order also is Subject – Object – Verb. But here the similarities with Sumerian end: It was an isolated language without any known relatives. Some linguists suggest relations with a hypothetical primeval super-language family called Nostratic which should include practically all known languages of the Old World, a notion which is of no great help. A better case, however, can be made for some lexical links with Dravidian of southern India, such as Tamil, Malayalam, Telugu, Kannada, and others. Brahui, the language of a nomadic people on the border between Iran and Pakistan, is regarded by linguists as belonging to the Dravidian linguistic family.[1] It is tempting to think about an ancient Elamo-Dravidian continuum from southern Mesopotamia to the Indus valley and beyond but the final verdict needs more evidence.[2]

Hurrian rose to importance during the 2nd millennium B.C.E. throughout the central and northern part of the Fertile Crescent. It was written in Syro-Akkadian cuneiform and religious, ceremonial or legal texts were found from Hattusas, the capital of the Hittites in central Anatolia to Alalakh/Tell Achana and Ugarit/Ras Shamra in Syria and Nuzi in northern Mesopotamia. Here, e.g. at sites such as Sapinuwa in the Habur River Basin, was the densest Hurrian speaking population. In other parts of the Near East they seemingly represented an ethnic minority but with tremendous impact on their cultural environment. Their spiritual culture deeply influenced the Hittites and all the peoples of the Levant, where traces of their cults and laws are preserved in several cuneiform archives and became enshrined in the Hebrew Bible. Many scholars attribute also material cultural traits, such as bronze metallurgy and the introduction of the horse and perfection of the chariot warfare of the Middle Bronze Age (ca. 2000 B.C.E.) to the Hurrians. Whether they were directly or indirectly responsible is still an open question.

It is also a typical agglutinative language in which all grammatical and syntactical relations are expressed by suffixes attached to verbs and nouns in a fixed order. Certain features, such as an ergative also found in Sumerian and some modern Caucasian languages, and even Basque, led some linguist to believe in possible links with these languages. So far, Hurrian is an isolated language, except Urartian spoken in eastern Anatolia during the 1st half of the 1st millennium B.C.E. As the Hurrians in the Syro-Mesopotamian lowlands began to adopt Aramaic as their spoken language during the first centuries of the 1st millennium B.C.E., related tribes in the isolation of the Anatolian high-

---

[1] Chicago Dravidian Etymological Dictionary,
http://dsal.uchicago.edu/dictionaries/burrow/index.html.

[2] E. Reiner, The Elamite Language. in: *Altkleinasiatische Sprachen, Handbuch der Orientalistik*, 1. Abt. 2. Bd., Brill, Leiden pp 54–115, 1969.
M. Khacikjan, M. *The Elamite Language*. Documenta Asiana IV. Consiglio Nazionale delle Ricerche per gli Studi Micenei ed Egeo-Anatoloci, Roma (1998).

lands around Lake Van kept their own language, which developed into a dialect related to Hurrian but not its direct descendant. Urartian royal inscriptions are found from central Anatolia to Lake Sevan in Armenia and Lake Urmia in Iran.[3]

Hattian was an isolated language in central Anatolia. It is known from the temple archive of Hattusas, the capital of the Hittite empire since the 16th century B.C.E. The texts at hand are mostly dated to the period just prior to the burning of the temple (which helped considerably preserving the clay tablets) and mention names, titles, and many loanwords in Hittite for religious and ritual purpose. Most important, however, are a few bilingual prayers and other ritual texts addressing the local gods in their native language with their approximate Hittite translation (see below). From these transpires that Hattian was an agglutinative language structurally similar to other Asiatic languages already discussed but with a significant difference: the widespread use of prefixes together with suffixes. Verbal stems were declined only by prefixes, whereas to nouns prefixes as well as suffixes can be attached. Nouns seem to lack gender or class distinctions and were preceded by adjectives. The free poetic nature of the restricted corpus of texts does not allow us to determine the sentence structure of the spoken language. There exists the possibility of linking Hattian with the archaeological finds from the Early Bronze Age tombs of Alaca Hoyuk as well as the Trialeti Plateau in Georgia and the Maikop Culture of northern Caucasia. Separated by four millennia, any suggestions that some of the Northwest Groups of Caucasian languages (Adygean and Abkhazian) could be related to Hattian should be viewed with utmost skepticism.[4]

There are several other ethnic groups mentioned in cuneiform texts but we do not have identifiable material written in their languages. The most prominent among these people were the Kassites who ruled Babylon from the beginning of the 15th to the 12th c. B.C.E. As they had adopted Babylonian for their hitherto unwritten language we are ignorant about their language except a few exotic sounding royal names.

Except Sumerian, all these isolated ancient Near Eastern languages mentioned are known in various degrees of imperfection. Besides the small number of actual primary texts, the principal obstacle is the lack of comparisons with related or descendant known languages despite the fact that most once were probably members of larger but now extinct linguistic families. In contrast, any newly discovered ancient document in a language belonging to the larger and still existing linguistic families, such as the Indo-European or the Afro-Asiatic languages, can be read and understood with relative ease. There is little doubt that the wide and apparently fast initial expansion of these two linguistic groups into the Middle East, one from the north-east and the other from the south-west, is a relatively recent phenomenon and most likely connected

[3] E.A. Speiser, *Introduction to Hurrian*. Reprinted from the Annual of the American Schools of Oriental Research, vol. XX, (1940–1941).

J. Friedrich. Churritisch in: *Altkleinasiatische Sprachen, Handbuch der Orientalistik*, 1. Abt. 2. Bd., Brill, Leiden, pp. 1–30, Urartian 31–53 (1969).

[4] A. Kammenhuber, Hattisch. In: *Altkleinasiatische Sprachen, Handbuch der Orientalistik*, 1. Abt. 2. Bd., Brill, Leiden, pp. 428–546 (1969).

with the spread of pastoral nomadism in various ways. For example, the basic vocabulary of both language families has more terms about the animal world and husbandry with its related terminology than about agricultural pursuits and tools.

Indo-European languages are now found on all continents and spoken by half the world's population. The original home of the earliest speakers of Proto-Indo-European is not yet clear, and probably will never be, but all the evidence points to the wide swath of the open grass land corridor ranging from the steppes north of the Black Sea to central Asia.[5] The century old classification into sub-groups began with an arbitrary example of the word for "hundred", *satem* in Avestan/Sanskrit or *centum* in Latin, into a supposedly western Centum group (such as Italic, Keltic, and Germanic) and an eastern Satem group (such as Indo-Iranian, Slavic, and Baltic). Yet, the most eastern of all Indo-European languages was Tocharian, a Centum language used in Buddhist texts of the 6th and 7th century in central Asia. The centum/satem system could as well be a diachronic difference, with the Centum languages representing an older or more conservative stage in comparison with the "advanced" Satem languages. For example, the change from a Proto-Indo-European /q/-sound, retained in Latin (e.g. "five" *quinque* in Latin, *cuig* in Irish), but becoming a /p/-sound in some others (e.g. *pende* in Greek, *pump* in Welsh), not necessarily following other sound changes. The English saying "Watch your p's and qu's!" supposedly originating in the divisions of Gaelic and British Keltic but has its roots in a much older branching of the Indo-European languages.

The traditional complex genealogical system resembling a neatly trimmed tree increasingly looks more like an entangled bush with its roots deep in the world of Eurasian hunter-gathers of the Palaeolithic[6] reaching out under the umbrella of Nostratic to other linguistic families to the Caucasian and even the Afro-Asiatic languages. The morphology is quite challenging, as everybody knows trying to learn ancient Greek or Latin where, in contrast to most other language families, there are more exceptions to the rules than rules. Verbs can be inflected by changing the root-vowel (e.g. English sing, sang, sung) and by adding prefixes, affixes, and suffixes indicating persons, numbers, moods, and aspects of time in a confusing manner. The nouns belong to one of three genders (masculine, feminine, and neuter) and had originally up to eight cases that defined the function of the word in a sentence. These cases mostly consisted of suffixed short remains of postpositions and were further defined by additions of prepositions allowing an absolutely free word order in a sentence. Composite words of considerable length were also quite common. Modern languages, mostly those of trading nations like the Dutch and the English, have shed nearly all of these features and are evolving back in the direction of other Asiatic languages (see the modern internet's "Globish" devoid of grammar and complex words).

---

[5] J.P. Mallory *In Search of the Indo-Europeans*, Thames and Hudson, London (1989).

[6] J. Adams, Did Indo-European Languages spread before farming? *Current Anthropology* (2006) in press.

The oldest evidence of Indo-European languages we find in names of the (Satem) Indo-Iranian branch appearing in cuneiform archives in northern Mesopotamia of the 17th century B.C.E. Nearly contemporary texts of an apparently different (perhaps Centum) Indo-European language conventionally called Hittite were found in central Anatolia and dated to the middle and the end of the 2nd mill. B.C.E. The bulk of these texts were written in a Syrian version of Babylonian cuneiform and delt with rituals and matters concerning the royal household including some significant historical letters. A native "hieroglyphic" writing system was used for more mundane matters and became officially recognized in the royal seals of last years of the empire. Related to Hittite were other Indo-European languages: Luwian in the west and south of Anatolia, the probable initiators of the "hieroglyphic" system, and Palaic in its north-west. The lack of epigraphic material makes it hazardous to follow these languages as they probably developed into the only slightly better known Anatolian languages of the 1st millennium B.C.E., such as Lydian, Lykian, and neo-Luwian written in "hieroglyphic" and spoken by the ruling elite in the neo-Hittite kingdoms of northern Syria. An important discovery were the bilingual inscriptions of Azitawadda, king of the Danunites (Adana) at Karatepe, parallel texts written in "hieroglyphic" Luwian and alphabetic Phoenician. The population of the region probably continued to speak Aramaic or a closely related dialect of North-West Semitic.

Herodot's story that the Indo-European speaking Phrygians and Armenians both migrated from the Balkan Peninsula during the first half of the 1st millennium B.C.E. is substanciated by linguistic research and some archeological evidence. They, therefore, represent an intrusive element but the question was wherefrom? Armenian was once believed to be of Iranian origin as it had so many Persian features which by now appear to have been borrowed during the centuries of Persian rule over Caucasia and northern Iran.

Persian and Median had appeared in the Iranian highlands at the beginning of the first millennium B.C.E. Persian, first written in a cuneiform alphabet, was the language of the elite in southern Iran but when it became the language of the common people is quite unknown. The oldest religious text in existence were in Avestan, a developed north-eastern dialect written with Aramaic letters. Equally unknown is the origin of the Kurdish language now spoken in western Iran, northern Iraq, eastern Anatolia, and Caucasia and belongs to the same Indo-Iranian branch like Persian. Local traditions claim a descent from Median which is possible but as Median is unknown, a link cannot be proven. The principal dialect of Kurdish is Kurmanji but as it did not become a literary language, the various dialects are often unintelligible to each other and linguistic research is seriously hampered by the modern states. For instance, speaking the language in public was considered a criminal offense in Turkey. There are many loanwords from the neighbouring languages including traces of Urartian, such as ergativity, and several personal and geographic names of unknown origin. As the speakers of Indo-European remained predominantly in the northern and eastern highlands, their languages left no remarkable impact on the linguistic character of the lowlands and the inner part of the Fertile Crescent which remained Semitic.

Greek became the speech of the intellectual elite during the Hellenistic and Roman-Byzantine period. Despite its cultural and spiritual impact on the attitude of mind of the rulers, particularly after Christianization, it left only a few loan words in the spoken Aramaic of the time. Literary monuments written in Near Eastern Greek (Koine) are the Septuagint, a translation of the Hebrew Bible, and the New Testament.

Afro-Asiatic (or Hamito-Semitic in older literature) was, and still is, the most important Near Eastern language family among which the Semitic language group is currently the most widely spread linguistic entity in the Near East. On the Asiatic side it includes modern Arabic, Hebrew, and Aramaic, with Ethiopic with its daughter languages Amharic and Tigrinya in the highlands of north-eastern Africa. The ancestors of these languages can be studied against a historical background of considerable depth, having the longest history of any recorded language family. Old Akkadian and Eblaitic date to the third millennium B.C.E., Ugaritic and Canaanite to the second. Hebrew, Phoenician and Aramaic are documented from the first millennium B.C.E. onward, Arabic and Ethiopic from the first millennium C.E. Slightly more than sixty distinct forms of Semitic speech are known, ranging from such important literary languages as Akkadian, Hebrew, Arabic, and Ethiopic to obscure unwritten tribal languages spoken in southern Arabia, Ethiopia, and the Horn of Africa.

The epi-center of Afro-Asiatic was in northern Africa where we still find its closest relatives, including the Berber languages, ancient Egyptian, Cushitic in the eastern Sudan and Ethiopia, Chadic with Haussa in northern Nigeria, and Omotic in southern Ethiopia, but no other related languages in Asia or Europe. Most Semitic-speaking peoples seem to have had a nomadic past and their appearance in southwest Asia is debatable. The only historical situation in which adequate archaeological evidence exists to support the infiltration of nomadic tribes from northern Africa between the end of the Ghassulian and the *floruit* of the Early Bronze Age was the end of the Transitional Period, or the later part of the E.B. I period in Palestine, as suggested already in 1965 by I.M. Diakonoff. Meticulous studies of linguistic changes led him to conclude that the Libyo-Berber and Semitic languages are so closely related that they must have been a single entity until as late as the fourth millennium B.C.E. This finding seems to fit all the archaeological evidence unearthed since then and is followed by the majority of linguists [7]

Based on the speech of these first migrants from north-eastern Africa in the wake of the aridization of that part of the world, Semitic has since developed its characteristic properties in the area of the inner flank of the Fertile Crescent and the inhabitable parts of the Arabian peninsula. One of these is the principle of triliteralism in the structure of words, i.e. ideally each root should consist of a skeleton of three consonants with inserted vowels and limited additional prefixed or suffixed consonants, such as /'/ (i.e. a glottal stop), /m/, /n/ or /t/. In

[7] Ch. Ehret, *Reconstructing Proto-Afroasiatic (Proto-Afrasian): Vowels, Consonants, and Vocabulary.* Berkely, Los Angeles, University of California Press (1995).

C. Ehret, S.O.Y. Keita and P. Newman, The Origins of Afroasiatic. *Science* 306:1680–1683 (2004).

E. Lipinsky. *Semitic Languages: Outline of a Comparative Grammar.* Uitgeverij Peeter, Leuven (1997).

the majority of cases this was achieved but there is, naturally, a great number of basic words which do not confirm to this rather artificial rule. Originally, like in Berber, Egyptian, and Cushitic, the root word consisted of two consonants indicating the basic concept, with the third indicating an extension or intensification of this basic word's concept.[8]

The elasticity of the word structure allowed the Semitic languages to achieve a high level of semantic differentiation within a limited space: as no stringing together is necessary, words, and therefore sentences, are short and succinct. This quality can be lost in translations, or as consequence of close contact with alien languages. The best known examples of the latter are the Greek influence on Aramaic in the first centuries ce, or Yiddish and English on modern spoken Hebrew. With few exceptions, the conservative character of normative classical Arabic due to the "divine" Koran appears to be quite effective in repelling similar trends.

As we have already seen, the early contact with Mesopotamia ensured that the Semitic language family has the longest recorded history of all living language groups. From the hypothetical Proto-Semitic pastoralist immigrants of the late 4th to early 3rd millennium B.C.E. no records have survived. The transition to Old Semitic is known from the colophons of some Sumerian texts, e.g. at Tell Abu Salabikh[9] mentioning typical Semitic of some of the scribes from ca. 2700 B.C.E. on. Old Semitic was probably one language with dialectical differences and is known from Old Akkadian texts from central and southern Mesopotamia,[10] and Eblaitic from northern Syria.[11] Two presumed languages or dialects diverging during the later half of 3rd millenium could have been Old Canaanite and Old South Arabian but the lack of epigraphic evidence only allows speculations. In Mesopotamia two distinct dialects evolved which were to become Assyrian in the north, and Babylonian in the south, both known as Eastern Semitic. The Old Assyrian correspondence from Kultepe in central Anatolia throws a bright light on the socio-economic and political circumstances from the end of the 3rd millennium B.C.E. until the mid 18th c. B.C.E. not only on the trading city state of Assur and the Anatolian kingdoms but the entire central part of the Fertile Crescent.

The absence of texts obscures the evolvement of the language of the so-called Amorites, pastoral semi-nomads living on the fringe of the Fertile Crescent and the assumed ancestor of the North-West Semitic languages of the 2nd and 1st millennia B.C.E. Their way of life and their turbulent politics are known from the 20th to 18th c. B.C.E. archive of Mari. These texts concern

---

[8] L. Herzfeld, *Einblicke in das Sprachliche der semitischen Urzeit betreffend die Entstehungsweise der meisten hebräischen Wortstämme*, Hahn'sche Buchhandlung, Hannover (1883).

[9] R.D. Biggs, *Inscriptions From Tell Abu Salabikh*, Oriental Institute Publications XCIV, University of Chicago Press, Chicago (1974).

[10] I.J. Gelb, *Sequential Reconstruction of Proto-Akkadian*. Assyriologcal Studies No. 18. The University of Chicago Press, Chicago.

[11] G. Pettinato, *Ebla: Un impero inciso nell'argila*. A. Mondadori, Milan (1979) Several English translations are available.

L. Cagni ed. *La Lingua di Ebla*. Atti del Convegno Internazionale, Napoli, 21–23 aprile 1980, Napoli (1981).

the administration, rituals, and many political aspects of the kingdom were written in Akkadian, but the names of many of the actors are clearly distinct from Akkadian and have more resemblance to later Canaanite names known from the Levant and some Biblical Hebrew names. Several names of the Hyksos kings of the 15th and 17th dynasty ruling the Delta of the Nile in the 18th and 16th century B.C.E. are clearly North-Western Semitic, such as Yaqub-har or Anat-har. Apart from a few grafitti on potsherds or on rocks in a linear alphabetic script derived from Egyptian known as Proto-Sinaitic, most of our knowledge about Canaanite is derived from the archives of Ugarit and its port Ras Ibn-Hani in northern Syria. Ugaritic is written in a cuneiform alphabet but, unlike Akkadian, without indication of vowels. It foreshadows the arrangement of the single letters in an alphabetic sequence still found in Phoenician, Hebrew, and Aramaic script.

Hebrew is a descendant of the Northwest Semitic Canaanite dialect spoken in the southern Levant during second millennium B.C.E. The scarcity of epigraphic material (most famous is the so-called Calendar of Gezer) obscures the transition to Biblical Hebrew, known by the Siloam inscription from Jerusalem and the Stela of Mesha', and a few others, all dated from the end of the 9th B.C.E. onwards. The oldest extant Biblical texts date to the Hellenistic period, most prominent among them are the so-called Dead Sea Scrolls from Qumran on the northwestern shore of the Dead Sea. Under the influence of Aramaic vernacular and Greek, the language developed during the first centuries of the common era towards a less compact structure known as Mishnaic Hebrew, the language of the Palestinian Talmud.

The revival of Hebrew as a vernacular language in the State of Israel since the early 20th c. has no parallels anywhere.

The apparently youngest member of the North-West Semitic language group is Aramaic. Its origin is, as usual for pastoral nomads, unknown but its emergence in the same geographical area, the outer flank of the Fertile Crescent, suggests a direct descend from some eastern dialect of Amorite and distinct from the other North-West Semitic languages of the Levant. During the 1st millennium B.C.E. it began to supplant all the other languages, most notably Akkadian, Hurrian, Luwian (or neo-Hittite, see above) and was adopted as the common language of administration and commerce first by the Assyrian and, from the 6th c. B.C.E., followed by the Persian empire. The intervening neo-Babylonian empire during the 6th c. B.C.E. was founded by an Aramaean tribe, the Chaldeans or Chasdaeans who lived at the Persian Gulf since the 9th c. B.C.E. The Aramaean alphabet spread eastward and became the ancestor of the Iranian, Indian and central Asian alphabets, such as Uigur, Mongol, and Manchu, such inspiring even the Hangul alphabet of Korea.

Aramaic, known as Syriac, continued to be spoken throughout the Fertile Crescent during the Hellenistic and Roman-Byzantine period in the west and the Parthian and Sassanian period in the east developing slightly differing dialects. Western as well as Eastern Syriac produced voluminous literatures by Christian and Jews, studied and sometimes even spoken to this day.

We have no knowledge about the fate of the Old South Arabian, or South Semitic language groups during the 3rd or 2nd millennia B.C.E. Only when settling and founding trading towns and networks from the 9th century B.C.E. on, the need for a written language was felt. Inspired by the Phoenicio-Hebrew alphabet, a local Old South Arabian alphabet was designed and preserved only in a lapidary style for royal propaganda. Its oldest inscriptions are difficult to date but most scholars seem to agree that by the end of the 8th c. the kingdoms of the Minaeans, Qatabanians, Sabaeans, Himyarites, and the Atramitae (Hadramaut), all occupying and often succeeding each other in the southern part of the Arabian peninsula. The descendants of traders crossing the Red Sea and settling in the northern Ethiopian highlands created a Semitic speaking and writing island with its center at Yeha and later Axum. The modern Ethiopic alphabet is a direct and only descendant of the Old Arabian alphabet.

The nomads and farmers in the oasis of the rest of Arabia spoke a variety of South Semitic dialects and sometimes used characters derived from Old South Arabian for rock-inscriptions, grafitti on sherd or other items. Whether Thamudic, Dedanite, Lihyanite, and Safaitic ranging from the Nejran to the southern Syrian Desert were actually spoken dialects, or became stylized magic tribal signs burned on the hides of their camels, is unclear.

When the dialect spoken at Mekka during the 5th or 6th c. needed to be written, a reduced Nabatean version of the Aramaic alphabet was employed. The oldest Arabic documents were written in a style called Kufic (after the town of Kufa in southern Mesopotamia) and date to the 9th c. As not only the vowels are absent but also the all-important diacritic points distinguishing between the consonants, a horizontal line punctuated by six differently shaped little hooks or knots represent twenty-four basic sounds of a Semitic language. Early Arab writing is, therefore, more a mnemonic device to assist the memorization of an oral text than writing. Classical Arabic first developed under influence of Syriac after the conquest of Mesopotamia (Luxenberg, C. *Die Syrisch-Aramäische Lesart des Koran – Ein Beitrag zur Entschlüsselung der Koransprache.* Berlin 2004) and then under Persian scholarship. The developed Arabic script dates to the 10th c. and was used to represent a variety of Muslim non-Semitic languages, such as Persian, Urdu, and some Turkic languages. As the Arabic script is unsuitable for the vowel-rich Turkic languages, most now use Latin characters.

Appendix V

Egyptian Chronology

Egyptian chronology for the New Kingdom hinges on the date of accession of Thutmoses III: 1504, 1490 or 1479 B.C.E. These three dates differ by as much as a quarter of a century, known as the high, middle and low chronology respectively. The differences result from the imperfect knowledge of the length of the reign of some of the kings, in particular Thutmoses III and Raamses II. For each chronology convincing arguments can be brought forward, but with an increasing uncertainty factor going back in time to as much as several decades. As Egyptian chronology is the cornerstone for calculations of the time charts of other Near Eastern and Mediterranean countries, it has far-reaching implications and remains a vexing problem for the archaeology of our area, allowing each student to manipulate her or his dating of events according to need – or taste. For instance, calculating these cross-relationships, several scholars "lost" the 10th c. B.C.E. in Palestine which fuelled the already heated discussion about the historicity of some Biblical figures such as David and Solomon by abolishing their established "time allotment".

Additional reading:

P.P. Betancourt: "The Chronology of the Aegean Late Bronze Age: Unanswered Questions" pp. 291–295.

P. Warren: "Aegean Late Bronze Age 1–2 Absolute Chronology: Some new Contributions" pp. 323–331; both in: *Sardinian and Aegean Chronology*, ed. M.S. Balmuth and R.H. Tykot, Studies in Sardinian Archeology V, (Oxbow Books, Oxford, 1998).

Index

Printed by Publishers' Graphics LLC